On the origin of Mind

Life's need to re-represent itself

Part II

Martin Wurzinger

10th anniversary print edition

Original digital edition

Website
http://www.otoom.net

Contact
http://www.otoom.net/contact.htm

Printed by CreateSpace

Available from Amazon.com and other retail outlets

ISBN 978-0-9944053-1-9

Cover:
The image on the front cover is a Mandelbrot set, especially configured to represent the fractal nature of complex dynamic systems.
The large image on the back cover is part of that Mandelbrot set, magnified almost 1 million times.

Contents

13. A change of pace

We have come to an important waypoint in our journey. The unpacking of the semantic structures which a philosophical tract contains revealed its nature in a very specific format. It is the format we, as thinking organisms, use in order to extract the meaning we have come to expect.

That last sentence is loaded. When we had reminded ourselves to use care with the choice of words, then such advice is even more significant now; it applies not only to the articulation of words but also to the selection of meaning we attach to what we read. Technically speaking, both cases are a form of articulation, with the first a translation into muscular movements to create speech and the second a creation of further abstracts to build thought structures. To speak about a 'format' then implies a pre-configured arrangement which is being accessed by ourselves and others and allows us to partake of its information value. Without this generality any message would be incomprehensible. The 'extraction of the meaning' can only occur if not only the format but also its symbols, that is the detail we have learned to respond to, become available to our minds. And when we 'expect' something, its representation must in some way or another already be in our possession, otherwise foreknowledge would be impossible.

The semantic content of the philosophical texts had lost much of its esoteric nature because human society has achieved a degree of interdependence far beyond the possibilities of the past, certainly as far as we in the West are concerned. As to their cognitive-functional aspect, the descriptions throughout the book's chapters enabled the minds of readers to create their own observational tools. These tools, for instance the label 'PAT-chain' and its attached concept, make it possible to place the word on a page within a meaningful environment of affinitive abstracts, where the abstracts themselves are imbued with the associative potential they need in order to become a productive element of the mind's processing space. The use of such tools is part of the mind's dynamic; so is their content, and together they sensate a set of perceptions which have allowed you, the reader, to see the work of a Descartes, Hegel, or Popper in a very specific light.

Before we embark on the next stage it is helpful therefore to define the significance of our present position. Just as the first chapter was dedicated to prepare the newcomer for what was to come, so do the present paragraphs accentuate the goals reached and the perspectives still in the waiting. Without Part I behind us this would be confusing at best and useless at worst.

Descriptors such as 'PAT-chain', 'illusion tree', or 'functional association' are some of the planks alluded to in chapter 1 when I introduced the first two, 'thought structure' and 'thought pattern'. Apart from their self-referencing nature (which is one major reason for the diligence we need) they evoke a particular representation of what it is we want to communicate about. Just about any field of knowledge uses certain words and phrases for its purpose. Although they can also serve the games of power and identity, they have come into being because in their abstract form they make use of the functional domain in which they have been placed. Consider for example an explanation about woodwork, picture painting, and nuclear physics. The functional domain of the word 'chisel', in the cognitive-functional sense we are concerned with here, is first and foremost a first-level abstract denoting a very particular artifact, used for a very particular purpose. Depending on the demographic complexity of its user, it may also be the entry point to further abstractions illustrating the relationship between human and object, the role of crafts in society, and so on. Now take 'paintbrush': it certainly is a first-level abstract too, but for the artist it also represents an extension of their inner selves and, in conjunction with the visual forms that flow from it, becomes a symbol for the creative power of the mind, its master. Recalling our past encounters with contextual associations, denominative and resonant thought structures, horizontal and vertical PATs and such, it is easy to see how a cognitive space emerges from the application of such terms when harnessed to the concept of 'painting'. Not only do they represent, they interpret the sub-set of our conceptual domain we own and bring to bear upon the subject. So, although we do not live amongst the landscape of a Dalí or a Rembrandt, we can discourse about its finer points because we have the intellectual tools to do so, applied as they are to their own functional domain. Now take 'the atom': its concept underwent modifications over the centuries because the experiences available to the people using it changed as well. The labels and descriptors were derived from whatever perceptions were open at the time and on each occasion the development progressed from the first-level simplicities to the extremities of the conceptual space that was capable of being utilised. From grains of sand to the immutable particles of the universe, from spheres suspended in a force field to abstracted differentials circumscribing space-time variants, their nomenclature reflected the uses the content was meant for. It is an important point: just as an idea prompts the creation of symbols for its articulation, the functionalities of the symbols in turn determine the ultimate usefulness of the idea. Effectively then it is not the real-life counterpart of this or that notion which provides us with its benefits or otherwise, but it is the inherent potential of the abstracts assembled as a TS complex which moves us humans.

What or where is the functional domain of this book? It is a layer, a conceptual band stretched between the borders of neuroscience on one side and those of cognitive science on its other. Neuroscience deals with the dynamics of neurons, their internal workings and their relationships with each other as they transmit signals of varying nature across their world. Its elements are proteins, enzymes, voltage potentials. On the other side lie the dynamics of cognition as we have come to understand them by now, described by texts about memory, about idea associations, emotions. Further afield would be the realms of psychology with its emphasis on behaviour and further still social science, political science, and so on. Given the sheer amount of work in these areas combined with the increase of interdisciplinary foci one might argue with my exact placement of the groups in relation to each other. Nevertheless, this book deals with the hitherto grey area where our conceptual grasp of neurons becomes insufficient to ascertain their combined complexities, while at the same time our understanding of cognitive dynamics is too coarse for a technically useful deconstruction to show their functional detail. This is the region we are dealing with here; our terminology is designed to cultivate its space, the functionalities of our descriptors arranged to explore its extent. I will use the acronym of the book's title, Otoom, to designate its existence. Thus Otoom resides above the neurological and below the cognitive. Although borrowing from its neighbours every now and then, I hope that the specificity of its realm will remain protected and linguistic precision be maintained. True to its representational nature, it also is subject and object together—and so, self-referential.

As we move amongst Otoom's scope the intellectual tools are available; more importantly, it is their functionalities which make the difference, because it is through them that we gain the kind of understanding our minds are made for. In other words, a label such as 'PAT-chain' denotes an event, an instance of a particular dynamic. The conceptual domain of that label allows us to imply a presence, but in order to construct the entire notion of what is occurring there and then we also need to place such an instance within the framework of the overall dynamic. Ordinarily, to do that we must add the scale of its occurrence and relate it to its environs, that is to say, its statistical measure. Yet this is one purpose for which our model is ill-equipped. If a PAT-chain, say, is an event to be mentioned, and the ramifications of its existence are significant enough to be listed, the actual number of such instances become only relevant once we leave Otoom and concern ourselves with the effects appearing in behaviour, or in the wider social context. Therefore it is not a question of whether such behaviour can be found at one specific time or another; rather, the relevance lies in the fact of its existence at all. The traces such an event would leave in terms of the underlying detail would be explored by the neuroscientists who have different functionalities at their disposal, just as psychologists conduct their experiments in terms of their conceptualisations.

To forego statistical evidence in favour of logical/functional correctness has already been a concern for Popper, as we have seen in chapter 12. In that sense we can paraphrase the analogy of white and black swans: suppose we know the exact nature of genes which enable the growth of black feathers, but have not encountered any swans wearing them. Under conventional auspices we would have to be quite careful how we proceed—after all, as far as everyone knows, "black swans don't exist". Under the current circumstances however such presumption is unwarranted. If we know the functionality of the genes we also know what possible birds there can be, and that is enough; whether there exists a black swan in some place on some planet we call Earth is neither here nor there. Note that we are not engaged in the creation of either fantasies or pure abstracts: our prediction of possible black swans resides on the firm ground of our understanding of genes and how they interact with the rest of the organism (the actual existence of black swans ceases to be trivial once we get to the cognitive-functional domain of biology). It is only the specific instance of one particular set of event chains that can be debated, not the possibility of the set as such. While this approach might still be frowned upon by some people, I would argue that in the era of nano technology, gene research, quantum physics, but especially complexity theory, these considerations have their rightful place. As they do in Otoom.

As much as we need to preserve the integrity of Otoom's conceptual space, it should not be forgotten that such a space is nevertheless embedded within the wider context of reality, which of course includes the electro-chemical dynamics of neurons as well as the behavioural phenomena of cognitive science and so on. Thus there are connections at various levels of abstractions without which the entire complex would become ungrounded and we would find ourselves once again in the realm of spiritual or secular ideology. It is precisely to articulate these connections in a productive manner that we ought to appreciate the purpose for each domain's functional integrity. The precision enables us to traverse the entire landscape, just as it is important to understand the difference between 'water' in the desert and 'water' in a rain forest.

Keeping this in mind, not only is it possible to look beyond Otoom but also necessary to relate to regions which constitute the daily grist of individuals, groups, and societies. At the smaller end of detail we will look at some of the research which is steadily uncovering the processes of neurons. At the higher end we are going to observe the events that shape our single lives and the existences of their conglomerates. In line with our approach the focus of our attention will not be this or that person but this or that representative of particular circumstances, in terms of the functional configuration we were able to ascertain. Therefore I will not mention the names of people unless they serve to identify the source of some comment or action through which they, after all, would want to be known. On the higher level it would be similarly unproductive to deal with the categories of nation and race; they will be referred to as demographics, for this is what they are in terms of our perspective (again, the exception being the source of some act through which the performers wanted to be known as a race or nation). One brief example should illustrate the usefulness of this approach: consider the daily characteristics of a chemical engineer, working for a successful company in an urban setting. His colleague somewhere else on the globe will more likely have more in common with him than with a fellow citizen tilling the land in some rural outpost—and this applies no matter what place names we chose to include. At the current stage in our history the demographic becomes more important than the designation of traditional officialdom. While on the level of social interactions the use of demographic identifiers serves well in the context of globalisation, it serves us equally well here because it allows the specific nature of cognitive detail to be articulated without having to disentangle it from the usual personal or cultural semantics. On this note the forty countries hinted at in chapter 1 are given here as a geographical pool and for general reference only: Algeria, Australia, Austria, Barbados, Belgium, China, Djibouti, Egypt, France, Gambia, Germany, Greece, Guyana, Hong Kong, India, Indonesia, Israel, Italy, Liberia, Luxembourg, Malaysia, Morocco, Netherlands, Oman, Philippines, Senegal, Singapore, Spain, Sri Lanka, Sudan, Surinam, Switzerland, Taiwan, Thailand, Trinidad & Tobago, Tunisia, Turkey, United Kingdom, USA, Yugoslavia. These countries have been visited between the years 1970 and 1996 and are listed under the names current at the time.

As such the critical analysis of human-driven events opens them up for us, regardless of their placement in time and space and unencumbered by traditional interests concerned with the preservation of biological identity. The functional possibility to do such a thing ought to stand alongside its ethical counterpart; I do feel it is time to make that move. Sir Joseph Rotblat, nuclear physicist and 1995 Nobel Laureate for Peace, expressed himself along those lines at the World Conference on Science held in Budapest in 1999, supporting a type of Hippocratic oath for scientists at the time of graduation [1]. It should not stop there however.

If there is a system which features an event space, and if there are types of events that can be grouped in terms of their abstractive complexity (recall the effectiveness chart introduced in chapter 10), then we find not only an interdependence between the members of a single group but also amongst the groups themselves. That means cause-and-effect relationships spanning across differing levels of abstraction and so a justifiable attention crossing the borders of our domains. Therefore, just as the globalisation of politics and trade made national borders transparent, so does our use of the underlying functional detail enable us to step through the demarcations imposed by behavioural culture.

Under these circumstances it should be pointed out that the observations, descriptions, and analyses presented throughout the following pages are not affiliated with any group, organisation, political entity and/or religion; the observations etc are made on the basis of the latter's performance within Otoom's domain. Similarly, the interpretations of any statement or act produced by anyone belong to the conceptual space of Otoom and not the entity having made it. Whether it is possible to ascribe to someone an understanding in line with Otoom's perspective is not for me to judge nor is it my concern.

Given the above mentioned interdependence within systems and my own identity as a member of human society, the question could be asked what, if anything, prevents my own personal characteristics from intruding into the conceptual space of my subjects. As an answer I would refer the reader to Patrick Süskind's novel Perfume, the story of Jean-Baptiste Grenouille who, in 18th century France, distinguished himself by an extraordinary sense of smell [2]. Such ability was heightened by the virtual absence of his own personal odour, thereby allowing him to perceive his surroundings that much more clearly. Just as body odour defines a person, so do one's personal references to and from the outside determine the person's presence. The more those are tuned down, the less noise there will be interfering with the perception of the inner as well as the outer world. Put another way, the cultivation of detachment is a worthwhile exercise.

These then are the preparations to be made ahead of the second half of our undertaking. They are more detailed compared to those in the first chapter, but what follows is quite different from the open and accessible texts of philosophers. We are going through regions where the customary laws of perception and response stand aside; where the object and subject are more entwined than ever; and where the medium itself, the very air the traveller breathes, has turned into a fog impenetrable to ordinary eyes. In order to see, to perceive, and to understand, we need the conceptual tools we already acquired plus others yet to come within our reach. They are not artificial things, but are part and parcel of that reality we inhabit. What this means we are going to find out.

The road is waiting.

Part II:
The Unstated

14. From: Chaos (?)

Neurons - complexity - post-bio

Imagine that every neuron in the human brain is the size of a chicken egg, and that all these neurons are arranged in the form of a sphere. How large would such a sphere turn out to be?

The number of neurons in our brain is estimated to be approximately one hundred thousand million [1]. In North America this is usually referred to as a hundred billion, but to avoid confusion and to make references to this and other figures easier let us use the descriptors adopted under the Système International d'Unités, the SI units (see Appendix). A thousand million (10^9) is thus 1Giga (1G) and so the above number can be expressed as 100G (10^{11}). If we take the average volume of an egg to be almost 40cm^3 then our 100G of eggs would fit into a sphere of 4,000,000m^3; its radius would be just under 100m. As it happens, there are several bridges across the river girding the city centre of Brisbane, Australia, and under one of them the distance between the two river banks is almost exactly 200m. If our egg neuron-filled sphere was made of glass and would straddle these banks as if some alien object had descended upon the city, then the scene would look something like shown in Fig. 14-1.

Fig. 14-1: Illustration of a glass-like sphere assumed to have a volume equivalent to as many eggs as there are neurons in the human brain, straddling Victoria Bridge in Brisbane, Australia.

But it doesn't end there. The neurons act as individual units yet are connected to each other via cell extensions, summarily called *neurites*. Such a configuration, known as the *neuron doctrine*, is significant for the brain's dynamic because it hints at the sheer complexity of the system [2]. For the neuronal units to function as information processors they need to talk to each other, and the points of electro-chemical interchange, the *synapses*, add to the system's scale. There are between one and ten thousand synapses per neuron, which is to say that each neuron is capable of communicating with up to ten thousand others through vesicles which in themselves contain about five thousand molecules of *neurotransmitters*, the chemical symbols employed inside the synapses [3], although certain areas of the brain feature as many as 250,000 connections per single neuron [4]. For a neurotransmitter to be created and used there has to be a particular state within the neuron cell. Such states are defined by the existence of *proteins*, other molecules that also come in many varieties. Just how many can be gleaned from the fact that a neuron manufactures several hundred thousand of them, and they all contribute to the definition of a current cell dynamic [5].

The discrepancy between the amount of neurotransmitters available and the number of protein molecules responsible for their manufacture suggests a greater latency on behalf of the neuron compared to the possible articulation inside a synapse. While right now we are situated firmly amongst the physicality of neuroscience we can nevertheless recall our discussions in the context of the e-chart, pointing to the relationship between a product and its realised and unrealised potential as well as the phenomenon of homologous TSs (see chapter 10 for the former, chapter 9 for the latter). The greater the complexity in a system, the lower the latency of its members, and vice versa—where it should be remembered that 'complexity' in this case referred to the system's functional dynamic, not its physical counterparts. Still, the functional dynamic relies on the multitude of molecules for its unfolding, and from the figures it can be seen that for all the operations inside a cell there is a relatively smaller number of combinations available in the form of neurotransmitters. To put this another way, if the process inside a neuron can be defined at a particular level of variety, when it comes to the articulation of the process within the synaptic vesicle the level is lower. Therefore, in terms of the articulation of a given message its specificity can relate to several protein states, which means that its unique symbolic value is relatable to a diverse set of cell conditions—the principle of homology. This does not necessarily mean a diminution of potential however; within the super cluster of 100G neurons sub- and sub-sub-etc clusters are being formed for the purpose of conducting a certain set of processes and no other. More of this later.

Without going into the mathematical details of these configurations it becomes evident that the overall complexity of such a system is huge. Much of the work the proteins do remains to be discovered, as is the case with the molecules involved in the transmission of signals between the neurons themselves. Yet enough is known to speculate about the intrinsic nature of the system, its functional structure and its hierarchical arrangement, for us to make certain observations. We will go through them in this and the following chapter.

To begin with, the authors of the already quoted book *Neuroscience* make the following remark in their text on neurotransmitters [6]:

> The signaling network within a single neuron resembles in some ways the neural networks of the brain itself. It receives a variety of inputs, in the form of transmitters bombarding it at different times and places. These inputs cause an increased drive through some signal pathways and a decreased drive through others, and the information is combined to yield a particular output that is more than a simple summation of the inputs.

Already we can see the general hierarchy of the entire dynamic. Just as we were able to draw comparisons between thought structures and their equivalents in wider society, there is the implication that in principle the stepped interaction of information, its symbols and its processes is reflected in the most basic operational units. It should not surprise us: a behaviour occurring in wider society is in the end generated by the mental processes of the people conducting it; each of those people has a mind where TS complexes go through their paces; and each TS complex derives its existence from the myriads of neuronal interactions which in turn represent the almost infinite confluence of protein molecules. We are dealing with complex behaviour, our next topic.

Much can be said about complex systems and the ways in which they can be described and modeled, but for our purposes we concentrate on the underlying theme (one short but useful introduction to the main issues is Keith Clayton's *Basic Concepts in Nonlinear Dynamics and Chaos* [7]). We all are familiar with the flow of water as it leaves the tap, and the way in which it changes as we continue to open the valve. From a relatively smooth flow the current becomes progressively more uneven and as the speed and pressure of the water increase disturbances are introduced which turn an almost cylinder-like column of liquid into a spluttering, uncontrollable spray. What happens here is easily understood if we consider the state of each water molecule as being defined by the forces that act upon it, one time unit after another. When the speed is slow there is little pressure and the resultant forces influencing the molecules are kept low. This means as each molecule enters the valve region it will be pushed this way and that, but so gently that the effect from one—small—force vector may cut across the domain of another one but not much else. The overall result is a lining-up of molecules according to the shape of the pipe in which they are travelling plus the effect of the inherent attraction to each other, less so due to the pressure exerted. Although the force vectors do represent a certain energy, the time it takes for this energy to be transmitted to another molecule (and for that affected molecule to transmit *its* energy to yet another one) is less than the time taken by the particle to travel through that region. By the time the energy field representing all the force vectors would have made itself felt across the volume of liquid inside the valve, that parcel of water has left the area and the energy will have dissipated. Now take a much faster flow rate: immediately the force vectors are larger, they influence a greater volume of water (ie many more molecules which themselves are affected by their own vectors), and there comes a stage when the time period necessary for the volume of water to traverse the space has become too short for the entire energy to have dissipated as before. The result is an energy input which keeps adding itself to the already existing state, causing ever more disturbance and so creating the conditions for even further effects from the vectors still to come.

The field of fluid dynamics offers a number of mathematical equations which describe the situation with great detail (and difficulty one might add), but the essential points have been covered by the above. For us the salient features are these:

there is a dynamic system;
there is a form of input;
elements exist on which the input acts;
the input acts within a limited time frame;
as an element is acted upon, its now updated form becomes the recipient of further input.

Since in our case the 'elements' are not molecules of water but live entities we'll leave tap currents aside and consider a different analogy, that is a crowd of moving people. Somewhat similar to our water pipe, the people have to move through a gate and down a lane (let's say they've been watching a football game and are now leaving the stadium). If there is no untoward urgency and they can choose their own pace they will move through the gate in an orderly fashion. In other words, the pressure exerted by those who are trying to get to the gate is not enough to upset the ones who are already there and are now in the process of stepping through it. On the other hand, if there should be some panic that drives the people then each individual will exert a certain force towards the one in front, and this force will add itself to that other person and their own force. Because we are dealing with a living system there is an added factor: not only will each element (ie, each person) be subjected to a force, there will also be a response. In pure physical terms (that is, leaving aside such things as anger, shouting, and so on) any response will translate into further pressure as the individual is trying to create more space for themselves and repel the push exerted by the others—all adding to the already existing energy field. Therefore we are left with force vectors pointing towards the exit and other force vectors which have been introduced as a response and are acting in any kind of direction. While some of this energy ends up being targeted away from the exit the overall effect however is one of compounding the total pressure since each vector is in itself the trigger for a response which then moves in the opposite direction, in other words, back to where the trigger came from in the first place. And, as in the previous water example, we also have a time period defined by the number of units it takes for any individual to clear the gates. As the pressure increases however and the erstwhile directional movement of the people gives way to a more incoherent one, that time period begins to stretch as less and less of the crowd manages to negotiate the exit. Again we have a development already seen with the pipe: while it becomes harder for each person to actually go through the gate (they have to overcome the multitude of force vectors acting against them), the time needed to do so and *not* be influenced adversely along the way increases as well—two opposing curves on the graph which sooner or later cross, from which point onwards it becomes impossible to move in any direction whatsoever. For anyone who has ever wondered how it can be that people end up in a great heap around an exit under conditions of panic with resultant loss of life, here is the answer. As can be seen, it does not matter that the exit has been open all the time; all it takes is for it to be narrow enough to cause a sufficient build-up of pressure.

Therefore, when it comes to living systems we can add two more points to our list from above:

the acted-upon element creates an autonomous response in terms of that element's situatedness within the system;

the response becomes an input in its own right, adding complexity to the system overall.

We label the response 'autonomous' to illustrate its direct independence from the outside; that is to say, while the element derives sustenance and information from the outside, it generates a response based on its own dynamic and not that of its surroundings. Regarding the ongoing state of complexity in the system, we have already mentioned the inverse relationship between latency and complexity. It stands to reason that as more and more people respond to the pressure building up around them, their contribution to the entire crowd adds to the overall confusion, which is another way of describing an increasingly unmanageable complexity. If they did not respond, the potential of the crowd remains undefined—after all, nobody knows what they would have done otherwise.

Our two analogies seem to suggest a range of behaviour which starts at zero and climbs the complexity scale until it reaches an extreme value, thereby defining some kind of band bounded by its two end values. This is misleading. In terms of the state of the system it does not matter what the behaviour at each given complexity level might entail. Its elements act according to the parameters they encounter, and it is only our human interpretation which bestows a particular judgment upon a given state (such as trampled bodies at a stadium exit). While this may seem a callous approach, it should be remembered that even at the functional scale of human beings the participants in the stadium accident do not wilfully engage in turning themselves into injured or dead bodies; they end up this way because they are part of a system that abides by laws which under specific circumstances form such and such dynamics. If one can disengage from the human/personal aspect at that level, it is even more necessary at the smaller scales of mental processes and neuronal interactions.

A system which provides an input to its elements such that they in turn articulate the input once again but now under slightly different configurations is an iterative system. Let us be clear what we mean. In non-living systems one example of an iteration would be

$$x_{n+1} = x^2 + c$$

where x is some variable and c a constant parameter (by sheer coincidence, the above equation describes the well-known Mandelbrot set provided x is a complex number). We can see its iterative character: every value of x consists of the square of its previous value plus c, and in turn becomes the basis for the next operation. No matter what elements there are in the equation and no matter what their mutual relationships, its current variable value (x_{n+1} in this case) is well defined in terms of the expression on the right. But suppose we let x stand for a living unit, where 'living' indicates a dynamic, pattern-seeking, self-regulating agent (see chapter 12 for this definition of Life). Now something like the c becomes the input and enriches the unit, turning it into some x_{n+1} but much less well defined than before. Let's say the unit is a neuron, packing a few hundred thousand proteins in its internal chemical factory as well as being connected to a few thousand other neurons of like character and communicating with them via a protocol containing five thousand discrete entities. Whether these connections are axons or dendrites, that is sending or receiving information, is not the point right now—in so far as they articulate a modification to the present state their contribution can justifiably be called an input. I will not even attempt to construct an equation that comes close to describing such a situation, but to some degree the hopelessness is caused by the perspective induced by formulae such as the one above. When dealing with complex systems the usual approach is to take the variable and observe its life cycle under the conditions of constant, iterated alterations; as such the variable is held to be passive.

Turning this around however, that is to say considering the variable as a living, *active* agent which takes or rejects input and/or processes it this way or that, opens further possibilities. Recall chapter 12 and our discussion about pattern recognition. There we looked at a series of dots in a newspaper and recognised them as a picture because the resolution was of a sufficient degree to enable an interpretation by our mind. Or, we observed a meadow from a certain distance away and also knew what it meant because our mind had been trained to respond to a pattern of earthy colours. In other words, the *pattern* contained within the input, posited against the pattern recognition facility our mind provides, permitted the correct interpretation. Yet the 'pattern recognition facility' is nothing else but a set of neurons which are capable of responding positively to the input because they themselves have been organised in a configuration that resembles the pattern they are meant to see. We can express the situation in terms of an affinity between the input and the recipient, where the input is represented by a particular pattern such that its symbolic value is meaningful for the functional dynamic expressed by the recipient. Thus it is less a matter of finding the intrinsic, technical detail of the dynamic (although to know it would be an achievement in itself), but rather the issue becomes one of identifying the efficacy of such an affinity relationship. To take this further, whereas in the original view our complex system would be defined by the manner in which it settles into one or the other state and so trending towards this or that attractor, for an active agent the interest lies in the degree to which an affinity relationship can be realised. And it is an interest too: live systems actively seek affinities and in the process re-organise themselves to suit the contingencies of the moment—all throughout Part I we dealt with just those effects.

With or without that shift in perspective the similarity between brain events and chaotic systems in general is certainly attractive (in fact, affinitive). Frederick Abraham in his article *Chaos, Bifurcations, & Self-Organization* sees the functional descriptors employed to articulate mental dynamics as useful metaphors to lead us into chaos theory, suggesting as they do the existence of attractors, bifurcations, and vectors, all in the service of their neuronal mechanisms [8]. If one considers an ongoing mental process as a vector field (reasonable, given the electro-chemical energy used to affect signal propagation amongst neurons) then the types of bifurcations identified can make sense in that domain. A bifurcation occurs when the control parameter in the equation defining the dynamic (such as the c in our formula above but now not so constant anymore) changes slightly, but in doing so affects a considerable change in the system's behaviour. As such bifurcations and attractors are intimately linked, the former indicating the nature of the latter, and the latter influencing the nature of the former. The attractor can change its type (a subtle bifurcation), it can either appear or disappear (a catastrophic one), or experience a sudden shift in magnitude (plosive) [9]. Substitute 'attractor' with 'set of neurons' and the allure of such a view becomes evident once we regard the set as a functional complex rather than merely a bunch of cells. Of course, as soon as we bring in the underlying physical details of these sets the immensity of the definition begins to take shape. Even an approximate description becomes hopeless and the text tends to flatten into more or less useful generalities.

To demonstrate the magnitude of the problem, Heinz von Foerster gives us an example of a trivial and non-trivial machine, operating on a simple task—or so it seems [10]. A trivial machine functions according to pre-determined rules which are followed consistently. A non-trivial machine changes those rules according to some other rules, a rule-in-a-rule behaviour. I will present a shortened version of Foerster's illustration by supposing that the task is to take four letters (for example A, B, C, D) and associate them somehow with letters from another set containing the same number (E, F, G, H). If the machine is given an A it could pair it with an E, if given a B with an F, and so on (eg, pairing each letter following the same sequence). If we do not know this rule however and are presented with the result only we need to be able to analyse the sequence in order to deduce what kind of rule the machine had used. To do this we need to identify all possible rules, which means that we need to be aware of 256 different combinations (4^4)—this applies to a trivial machine. In case of a non-trivial machine however our letters from the first set can not only be combined with any of the four other letters, the sequence itself can change at every step by choosing any one of the four letters after the other. Suddenly we are faced with $4^{4\times4}$ = 4,294,967,296 possible combinations!

Especially in our present context it is important to realise that a simple analysis based on several samples of letter pairs is not enough. To observe, say, the combination A-F and assume a linking of the first letter of the first set with the second letter of the second set would only be correct if we know that there are only four letters in each set and that each set contains the elements listed above. If we came across something like B-M for instance we would have to ascertain a different case and therefore a different rule set. Without the knowledge of what possible rules are applicable it could not be done and therefore we are stuck with the methodical declaration of the entire range of alternatives.

Note also the particular state of *our* minds as we discuss these very details: our internal knowledge base contains the ideas and concepts mentioned in these pages and so we are in a position to consciously evaluate what is before us. *We* choose to discuss complexity, *we* create an example with pairs of letters, and it is *us* who calculate the value of $4^{4\times4}$. Without knowing maths, without having an interest in how the mind works, in fact without being able to read, these considerations would be neither here nor there. While it is true therefore that even a single nerve cell with its hundreds of thousands of proteins and its army of messengers represents the above problem on a hugely different scale, the challenge exists only for us who make the decision to bring to bear our mental powers upon the issue. We are the conscious observer and as such can avail ourselves of the conceptual overview which allows us to muse about the underlying issues. Like Chris Smith from the Department of Vision Sciences at Aston University in Birmingham for instance, who tries to place the brain's dynamic into some kind of framework to tease out a meaningful value in a format we can understand [11]. From the biologist's perspective, he describes one attempt at a complexity definition in the following terms [12]:

> The best known definition of complexity is the KCS (Kolmogorov-Chaitin-Solomonoff) definition which places 'complexity' somewhere between order and randomness; that is, complexity increases as P_{min} (the shortest algorithm which can generate a digit sequence, S) increases to equal the length of the sequence to be computed; when the algorithm reaches this incompressibility limit the sequence is defined as random. The KCS definition brings out the distinction between "highly ordered" and "highly complex" structures. In highly ordered structures $P_{min} << S$. The implication that the more complex a structure is, the closer it is to being random, is difficult for biologists to accept. The biologist, moreover, needs to know what the sequence of digits specifies. Papentin points out that for the purposes of comparing biological complexities the KCS algorithm need only generate a description of the entity in some agreed language, L, rather than generate the entity itself. But what are the biological entities which should be compared?

He then goes on to mention a few organic compounds and their molecular structures, pointing out the difficulty to express the characteristics productively through linguistic/chemical descriptors. As an alternative he offers 'organisation', which is interesting because it shows a move from a contextual articulation to a functional one. He suggests "we recognise molecules, organelles, cells, cell layers and cell columns, cortical areas and maps" [13].

If all we wanted to do was to find some means through which we could articulate the degree of complexity to ourselves, then the reference to Papentin's advice would suffice: choose a convenient language, formulate a particular complexity situation, and there is an expression useful for further comparisons (assuming for the moment we could actually agree on a language that satisfies all the interested parties). But we have seen in the four-letter problem how a ridiculously small functional space can quickly turn into almost unmanageable proportions—and we haven't even considered a third or fourth layer of rules yet. Of course, merely addressing complexity is not enough; since we want to explain what is going on we need to identify the rules and conditions under which the event materialises, and its complexity description only delineates the search space for us but not much else.

To give an example, there is a protein by the name of kinesin which enables material to be transported inside the axons towards the synapses [14]. It looks like two wires twisted together with the ends acting like 'arms' and 'feet' respectively. The 'arms' carry the material enclosed in vesicles and the 'feet' shuffle along tubular structures until they get to the synaptic terminal (see Fig. 14-2).

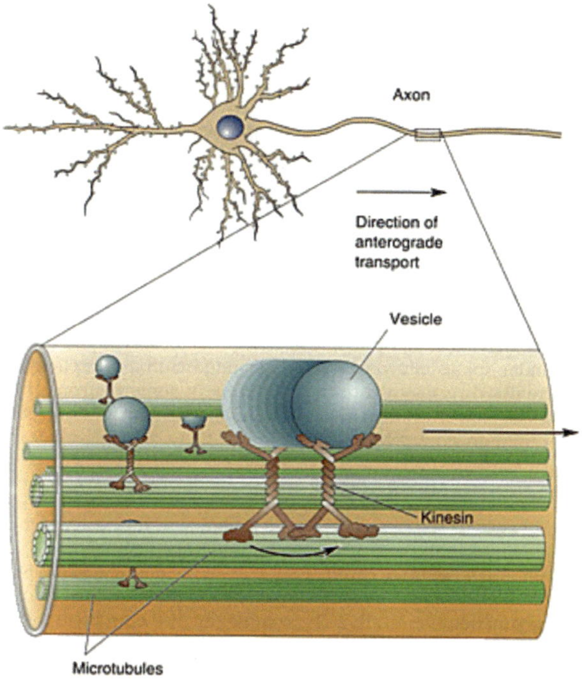

Fig. 14-2: Illustration of the protein kinesin as it 'walks' along the microtubules inside a nerve's axon towards the synapse, and transports material in a vesicle. (Adapted from Neuroscience - Exploring the Brain)

The protein is known and so is its activity, but what exactly drives it? Even assuming we could map the entire set of proteins contained in a cell (and remember, there are a few hundred thousand of them), these chemicals must be seen in the context of the pro- and reactive requirements imposed by the surrounding system, and that includes the neighbours to which it is connected (similar nerve cells with their own sea of chemicals), the event history down at the synapse (defined by the five thousand odd neurotransmitters), and last but not least the not insignificant activity inside the cell's nucleus *plus* all the other cellular bodies that surround the neurons and which are just beginning to be focused on by researchers.

A team from the universities of Edinburgh and Antwerp set about modeling the firing patterns surrounding Purkinje cells, which are a type of neuron found in the cerebellar cortex, a layer of tissue in what is commonly called the 'small brain' [15]. Because of the scale required the researchers used a Cray supercomputer running specialised modeling software. Summing up the logistics they wrote,

> The final parallel model required 128 processors. 16 processors simulated one Purkinje cell each. The 244,000 granule cells were spread between the remaining 112. Total run time for 2.0 seconds of simulated time was 2.5 hours, of which 1 hour was spent constructing the distributed model in processor memory. [16]

Despite the huge computing power (by today's standards), their final comments are telling:

> The most interesting, and perhaps distressing, scientific conclusion is that there are neuronal systems, viz the cerebellum, for which radically scaled-down simulation models cannot achieve realistic behaviour. In this case we believe that a simulation at 10% of real scale was necessary because realistic Purkinje

> cell firing requires an appropriate large number of relatively weak synaptic inputs which is only achievable with very large numbers of afferent units. There is no reason to suppose this characteristic is limited to Purkinje cells.

And these difficulties arise in relation to the rat's brain, a significantly lower level of complexity compared to the human equivalent.

The problem lies in the type of conceptualisation we apply to what we observe. For the biologist, or the computational scientist who tries to articulate the dynamics within a space delineated by various types of units which rely on their internal electro-chemical processes, the descriptive format rapidly assumes the unwieldy dimensions we have already encountered. How can it do otherwise: each unit, if presented in terms of its composition, needs a commensurate re-description to account for its movement through various states. To identify these states properly we need to know the composition of their participants, in other words the total structure of the rules at each and every point in time. Fig. 14-3 shows how that state space has been visualised through the software when run on the 128-processor Cray [17].

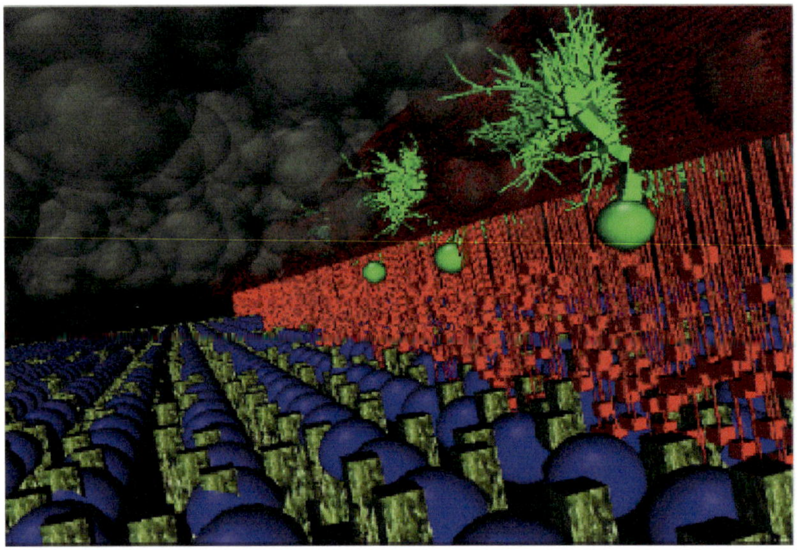

Fig. 14-3: A largely functional visualisation of one slice of cells from the cerebellum of a rat brain, run on a 128-processor Cray supercomputer. The view shows 1% of the total. (Source: A large scale model of the cerebellar cortex using PGENESIS)

Although the appearance of the cells in the picture has been designed to allow for easy differentiation and identification on the screen (after all, the main purpose of the exercise was to portray their functionalities and not draw biological specimens), it nevertheless serves to underline the kind of environment we encounter when descending to the micro-scale of cellular processes. It is here that the fundamental communication occurs, the generation and shifting of data that we, through what we call consciousness, sensate on a much higher, coarser, and far more distant level. Down there however consciousness has nothing to say, and none of our mental imaging has any purchase whatsoever. It is

...a claustrophobic world of being,
a space that is, and yet is not—
of air-less breath, of darkness seething,
dimensionless, a life-long vault.

What would a Wittgenstein have made of it if, instead of mantra-like feeling himself into the deeper and deeper meaning of words, he had exercised his desire for conscious depth by thinking himself into *this* world—surely he would have experienced more than "slight giddiness"! And yet, it is the ultimate home of the Self.

Not everyone despairs at the confrontation with complexity. For Ranulph Glanville there are benefits to be had, in fact he suggests that without a minimal scale of multifacetedness there would be no productive variety let alone creativity [18]. Using a calculation similar to our four-letter problem earlier he asserts the difficulty of deriving any real meaning from an array of inputs because of the rapid escalation of possibilities one has to consider—the task becomes transcomputable, it is no longer realistically computable [19]. He then mentions the Law of Requisite Variety developed by Ashby in 1956. The law states that for a system to control another it must have at least as much variety as the one it is to control; later the condition has been refined as being of exactly the same variety. Glanville quotes an analogy from the classroom to illustrate the relationship:

> Let us assume that all human brains have about as many states as each other. The teacher finds him/herself faced with a class of students who have the number of states his brain has, exponentially raised to the power of the number of students in the class. It doesn't need many students with much brain power for this to become transcomputable.
>
> To control the students, the teacher must obviously reduce their variety to his: ie to the variety of one brain.
>
> In old classrooms this was done by means of devices to destroy individuality: uniform, single sex, one age, serried rows of desks facing the front, no talking in class, only the chosen student answering, etc.

The comparison cited here comes from 1979, a time when the field of education was just one of many which experienced a renewed probing and questioning of its traditional tenets. For us it would be quite possible to delve into the cognitive-structural characteristics of such sentiments under the mantle of Otoom, but there is no time; whether the desire to remove authoritarian principles in this case overrode any factual co-relatives connecting to Ashby's law must remain a rhetorical question for now. Glanville does see that analogy as being "beautiful" however, despite a certain incongruence, at least from our point of view. The problem is not trivial, so let us consider it more closely.

Since a simile to the four-letter calculation has been used we can assume that in terms of complexity and its use the focus has been on *content*. That is to say, a collection of numbered units (letters, or lights in Glanville's text) is posited against the total number of possible variations these units can be used for. The result, an immense figure, is meant to demonstrate the sheer extent to which a functional space which contains such a system would have to go in order to accommodate the dynamics. "The question," writes Glanville, "is whether this is necessary":

> Does the teacher need to control the students: can the students teach themselves: what is the role of the teacher (surely it is not to know: that is the student's job. The teacher has only to get them to know: but that is the topic of another paper). [20]

By the type of juxtapositions used in that paragraph we can glean a hint of social theorising at work, but this distracts from the main point. Were the scenario about non-living items, the question of how many students it takes to overtake the teacher's brain in the complexity stakes could be valid. Then the whole system would be comparable to an air conditioning control having to deal with hundreds of individual rooms in an hotel for instance. In living systems however the problem is not only one of pairing one control against so many inputs, but also the affinity between the proffered inputs coming from the autonomous units on the outside and the identifying and responding processes residing within the recipient. To return to the teacher in the classroom, in terms of controlling complexity it is not a matter of how many students the teacher has to confront; rather, whether the incoming complexity of the data offered by the students will or will not overwhelm the teacher. If only content was the issue, the potential for unmanageable escalation is addressed by what has been referred to as "old classrooms" where "no talking in class" etc seems to have destroyed the individuality of students. But "only the chosen student answering" relates to content, ie the number of possible interjections. To view that as a complexity problem is to reduce the conceptualisation of the entire scenario to first-level abstractions, a move that has been amply described in Part I as an example of a simplistic perspective. Living systems, on the other hand, go further: they produce an input which actively *relates* to the environment, that is to say represents a configured response across the existing detail on both sides—our already mentioned affinity relationship. As such the teacher does not deal with so many students but with their qualitative offerings. While this may be enhanced through the interplay of many, the outcome is not defined in terms of its numerical base however.

The question of how it is possible for a system containing a large number of members to maintain control over another like system expands therefore to include the *state* of the system as well as that of its counterpart. Yet in its technical sense, the word 'state' implies not some ambiguous 'being' but a precise delineation of each and every sub-unit and its sub-units and so on until the entire set is covered. Just because it is impossible (transcomputable) to describe the whole domain through a point-by-point articulation of all its states does not mean those states do not exist nor that they are ineffective. On the contrary, they are effective within their specific sub-domains in relation to the specific sub-domain of their counterpart, and so on and so on until the entire functional space has been traversed. There is no overall 'control' to apportion types of behaviour at whatever appropriate location—the system itself interacts with the other through its multiplicity of contacts, where a 'contact' denotes the *effective* approach-and-response dynamic manifested at the given level of scale. What is meant by 'level of scale' refers to the eventual conceptualisation by an observing mind, but is meaningless to the system in question; what 'effective' means depends on the affinity relationship established between the two contacts. Glanville looks in the right direction when he says,

> In current cybernetic understanding, control exists between the systems within a control system. The act of control is neither action nor reaction: it is interaction.
>
> Unmanageability is a result of hoping to control in circumstances in which the demands of the Law of Requisite Variety cannot be fulfilled.
>
> Therefore, unmanageability exists at the same site where control exists, between the (nominally) controlling system and the (nominally) controlled system: unmanageability lies in interaction. [21]

It is us who look at something under the microscope and declare it to be a molecular interaction. The molecules do not possess the means by which to explain their behaviour, but they act in accordance with their properties because their individual characteristics won't allow them to do otherwise. Our idea of 'control' is one of centralised perception, the hallmark of a mind, but does not apply to molecular, even cellular, mechanisms. Thus the 'unmanageability' alluded to here remains a problem only for the observer but not for the active participants in the process. In terms of the system's overall efficacy it could be said, 'all for one, and one for all', since it defines itself through the multitudinous interaction amongst its members, and these follow their inherent natures.

The transposition of the meaning of control in terms of our personal history as humans towards something which we want to describe is nothing new of course. In a contribution to the *Harvard Mental Health Letter* David Kahn and Allan Hobson talk about the 'homunculus', the 'little person', who sits in our heads and issues orders to the rest of our self [22]. As they rightly observe, having posited one such master the need arises for another who directs the first, and so on ad infinitum. Instead, they write, one should see the mind as an emergent system, where the transition from one state to another is triggered by external influences. One particularly pertinent example they provide is the amoeba Dictyostelium discoideum. It exists as immobile single cells until the food supply runs out, at which point the cells emit an enzyme which causes them to clump together through chemical attraction. Now, as a multi-cellular organism, it swims towards some new food [23].

The theme of Kahn & Hobson's article is 'dreaming', and they point to the similarities between the amoeba behaviour and others as analogies circumscribing the fundamental issue: a complex system, reacting to outside input and producing a multitude of responses which nevertheless form a functional whole. In the case of the brain it is the neurons which come together as temporary clusters, where each cluster is representative of a combinatorial sensation we associate with 'thought' (or with dreams, as in the article). But we are getting ahead of ourselves.

Perceiving a complex system is not enough. Stocking it with affinity relationships is insufficient too. We can always ask, "affinity with *what*?"

The *what* remains relative to every step of the functionality scale. In Part I we associated one thought structure with another and defined the relationship along certain lines within the context of that TS's environment. The neuroscientist measures the response of one neuronal cluster against the output of some other set, being mindful of the biological significance of both. And somewhere in between those descriptions there exists the emergent system of the mind. The *what* comprises the compositional detail at the respective level, although for the TSs its resolution is too coarse to be helpful, and with regards to neurons the biochemical matrix is meaningless for our consciousness (assuming we are indeed able to articulate its entirety). So let us step up to the functional domain of Otoom, but at a time when thoughts are not sensated and complexes have not yet formed; a time when the system is already functional but occupies a featureless dynamic space.

There is a universe composed of thousands and thousands of millions of singular, interconnected, expressionless, entities. The internal mechanism of every one of these provides for an awareness of the beyond, the creation of an effect, and the articulation of a response. The many thousands of connections enabled by each entity allow the results of its processes to be communicated to the multitude. That billion-fold multitude carries the energy supplied through its support, but in the absence of any disturbance the field, defined by the electro-chemical interchange across the units, remains suspended in an unstable balance.

Then—via conduits from the outside a perturbation affects the space around a locale. The even distribution of metabolic energy vectors falters to make room for the dislocation of the former matrix. Via the activating channels which link the temporal centre of activity the resonance spreads to neighbouring regions. It may dissipate in some, but in others the motion folds back on to another incoming wave, an equally articulate and rhythmic progression of a field disturbance. Its amplitude and frequency, the physical summation of a dynamic, becomes enhanced or diminished by the traces left by its predecessor. The combination of either effect, distinct in its historical characteristics yet related through their common fundamentals, becomes a further localised field.

The stream of perturbations from several such conduits continues to unseat the balance amongst the entities. The disturbances pair with those with which they are able to resonate, and accentuate their disparity towards others with unlike energy distributions. The field has turned from a monotonous whole to a diverse landscape of effects. Now that the inertia determined by the initial balance has been overcome, the diverse oscillations coalesce into entities of their own emerging centres of representative states. The communicative potential of each centre defines its relation with every other, and the outward-spreading field vectors together with the backwash of their resonance acquire the distinctive appearance of their source; thus the channels turn into flows of communication.

Just as the formation of distinct domains contributes to a further enhancement of localised specificity, so do the movements of information distinguish themselves according to the processes that sent them. Growing and disbanding affinities, and the assimilating and differentiating representations of such events, now characterise the previously amorphous system. The gradients of similarity and difference in themselves accentuate the creation of conglomerates, whose position within the range of those articulations place them in particular regions of functionality. As such a region manifests its source-related nature, and the information it is able to acquire as well as dissipate reflects its very particular characteristics. From now on, impressions from the outside will be differentiated.

And still the process of emergence continues. Bands of similarities broaden, but as they do so form interstices with their own accompanying distribution of field effects. Islands within islands form, and flows within flows connect them. Those localised perturbations, initially seeded by the outward wave of imbalances, have grown into self-defining domains of representational clusters. As affinity-seeking as their incoherent origins, they build cascading separations of reflecting, resonating structures—and all in their own right.

While the pervasive fabric of information flows keeps communication throughout the system going, a certain region becomes the receptor of the distributed fields from the newly-grown centres. Although similar in terms of responsiveness and as interconnected as the rest, there is one feature separating it fundamentally from all the others: it is able to sensate to its host, and its host is able to sensate to it. From this point onwards the domain-specific field effects across the functional space of this universe have found one resonance that, in relation to the system itself, can be active. The representational quality of a cluster has become translatable into a sensated expression: consciousness.

For the resonance to be re-directed it needs definition in relation to the shifting domains it attempts to interact with. In the beginning this definition lacks accent, and so the impressions are responded to with ambiguity and dissonance. The general existence of consciousness increasingly becomes an act of deliberate choice between one articulation or another, and between one targeted domain or another. From the erstwhile meaningless yet significant perturbations the field effects have matured into an accessible array of experiences, addressed in terms of their pronounced interpretation by the host. The sensate representation creates a 'light', 'dark', 'red', 'blue' and 'green'; in liege with accompanying movements of limbs which contribute their own sense feedback it defines a 'near', 'far', 'big' and 'small'; supported by sight and sound it learns about 'stillness' and 'movement'; and eventually, at an even more evolved domain structure, there comes the recognition of 'I' and 'you'.

The process of cluster formation and its derivatives together with the parallel developments of articulation and refinement leads to a tiered representation of content. Based on repeated implementations of affinity relationships between field effects, the corresponding structures align in terms of specific resonance between one state and the next. Raw data from importing conduits do not matter as much anymore and are relegated to different channels due to their dissonance. The connectivities between higher-level representations relate to their own functional distributions and no other, and whether they become sensate or not, their informative qualities elevate the perceptional power of their host to an extraordinary degree: abstractions are born. - -

The words used in the foregoing description of the emerging mind have been deliberatively kept at a high degree of generality. I wanted to show that it is possible to articulate such a development without relying on definite labels which have been gleaned from the field of neuroscience, the traditional home of these building blocks. At the same time, although not sufficiently accurate to serve as a hands-on manual for building a mind, they are precise enough to identify the actual features of the process.

A general expression is general because it encapsulates the underlying principle without becoming fixated on a specific context. In that form the original circumstances can be left behind and its functionalities are able to be transposed into a different environment, thereby extending the usefulness of the model.

Leaving the wet-ware in our skulls for the moment we turn to software running on a computer, and implement the functionalities identified above as far as is feasible. A computer model would therefore have to contain the following features:

a set of nodes which are capable of holding unique values that can act as representations of incoming data;

a system that enables the nodes to be connected to a given number of other nodes, with traversable paths that form a definite structure;

a process that allows an input to be transmitted to a certain number of nodes, such that the transmission is unique, traceable, and formal;

a process which permeates the set of nodes such that the state of the system at any given moment is a function of this process, distributed across the set in a formal manner;

a facility for an output which derives from the system's state and is characteristic of it.

Above all, we need to ensure that

the program abides by the dynamics known to be representative of emergent systems; and

there are no more rules governing the internal processes as are absolutely necessary, in fact the ideal case would be one single rule only.

The last two points are vital: if life processes are to be simulated there can be no patterns that had been articulated with a special purpose in mind. Emergent systems do not evolve to suit any particular hindsight, and the inclusion of any form of behaviour which is designed to address a limited environmental format is bound to fall over sooner or later. Biological minds do not configure themselves under specific auspices—they may trend towards this or that environmental bias but retain their underlying functionality regardless of any given situation. The only constraint is the complexity of the system itself, but that is a matter of scale and not functionality. There is also a deeper reason: for a mind to emerge, the representative resolution of any input from the outside must be sufficiently fine to allow for its abstractive decomposition at a level which we can identify as intelligence. It can only do so if the underlying processes are so basic that any functional eventuality can be addressed. The principle can already be found in Ockham's Razor which states that entities should not be multiplied without necessity; it is a problem facing the developer of any robotic device which is meant to have any semblance of independent behaviour; and it can be seen in every scientific endeavour that seeks to uncover the underlying principles governing nature.

We are going to analyse the Otoom computer model (from now on referred to as OtoomCM) and observe its behaviour. It will serve as a reference to the bio-version of the mind and demonstrate that its principle functionalities are not restricted to the type of neurons we have in our brains but can occur anywhere provided the medium is capable of supporting its dynamics. We will also use the program to confirm and/or emphasise certain aspects of mental dynamics because it gives us the means to conduct repeatable experiments under formal conditions. For all the technical details of the program see the Appendix.

One of the simplest ways to hold a set of data in a configured, stable, and traversable structure is to embed them within a matrix. The structure is configured because it consists of a defined number of rows and columns, with all the elements related to each other through a strict geometric layout. It is stable because once defined, the parameters do not change under the influence of processes dealing with the elements. And the formal arrangement of rows and columns makes the matrix traversable in any direction where every point inside becomes accessible from any other point. This makes it also attractive from a topographical perspective, adding to its efficiency.

Due to its stable format the points of a matrix can be used as another structure, and since in our case the advantages relate to the particular uses which we are going to discuss very soon, the structure of the points in the matrix constitutes yet another matrix.

From a software engineering point of view there is another benefit to be had: unlike other designs, indeed every other anatomy found in nature, any point on a matrix can be accessed instantly through reference to an index if so desired. Other formats need to be sequentially traversed, which means that no location can be arrived at before having negotiated a path along certain edges, an effort that costs energy and time.

A system that deals with information must have a facility for input, for processing the data, and output. The matrix for OtoomCM reserves two sides for input nodes and the two opposite sides for output nodes so that the rest, the inner matrix which is committed to the actual information processing, is surrounded by a rim of nodes serving entry and exit. The basic configuration then looks something like in Fig. 14-4, with input nodes red, output nodes blue, and the central processors shown in green.

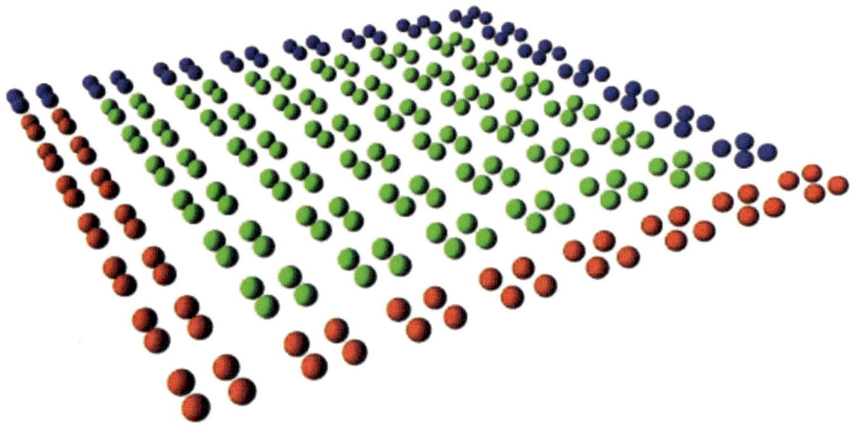

Fig. 14-4: View of a 8×8×2×2 OtoomCM matrix, with in- and output regions and the inner matrix.

The illustration features 8 node rows and 8 node columns, with each node consisting of more nodes which form another matrix, in this case of 2 rows and 2 columns. We will refer to such a configuration as a 8×8×2×2 matrix.

As indicated before, the idea is to simulate the bio-version as much as possible while retaining the engineering advantages of the artificial design. Thus the nodes can be likened to neurons, and the members of each node to the masses of proteins a neuron cell contains. Under present conditions we won't be able to approach the huge complexity of the brain, so hundreds of thousands of members are out, as are millions of nodes for that matter. Nevertheless, we will see that this layout, as simple as it is, can demonstrate the principle dynamics we would like to expect.

When studying neuro-transmitters, certain facts become clear: for a molecule to act as a transmitter, it must be made and stored in the pre-synaptic neuron; it must be able to move from the pre-synaptic terminal to the other side, the post-synaptic receptor; and the response generated by that movement must be distinct, that is under experimental circumstances it must produce the same effect [24]. It is also obvious that the molecular compounds necessary for the building of neuro-transmitters are generated by the chemical factory inside the neuron cell such that their existence is consistent. If it is not, if there is a change in their composition then the effects can be observed for example in diseases based on chemical imbalances in the brain or after the intake of mind-altering substances.

In the computer model this becomes: transmission must be possible between the nodes of the matrix; there has to be a discrete unit that moves in a distinct direction from one node to another; the response generated by this movement must again be distinct, which means that a particular datum will generate the same response under the same circumstances no matter where in the field the action occurs.

It is also known that in the bio-version the growth and deployment of synaptic connections depends on the physical as well as the functional environment within the organism; dendrites grow according to a plan specific to the species [25]. Not only that—contrary to more traditional beliefs it has been found that the brain continues to generate nerve cells even at the adult stage, as Colin Blakemore, Professor of Physiology at Oxford, has outlined when discussing some of the issues surrounding Huntington's disease, a genetic disorder which results in the degeneration of neurons [26].

Considering the phenomenon of transmission in conjunction with the abundance of proteins as the molecular basis for the building of neuro-transmitters and the difference in the degree of variety between the last two, we can infer certain characteristics which are fundamental to the existence of phase states across the neuronal network. If a neuro-transmitter is an agent which modifies the state of another neuron such that a significant result eventuates, then (a) the previous state had to be defined in order for the difference to be ascertainable, and (b) that definition is part of a cause-and-effect relationship between the proteins making up the cell and the nature of the neuro-transmitter. If this was not so, an interdependent system such as the nervous system could not have a life cycle that consists of the manufacture of agents which trigger dynamics in the recipients which in turn produce more agents which continue with the same consistency. There is also another implication: since the number of proteins in nerve cells is substantially higher than the number of agents used for communication between the cells it follows that there is a process which produces a relatively low number of results (ie, the neurotransmitters) but itself possesses an inherent potential which, although residing at the same level of functionality (ie, the particular molecular compositions characteristic of neurons in general), nevertheless allows the multitude of proteins to participate in the formation of such agents under a considerable variety of conditions. It is the conditions which form the functional environment of neuronal elements, it is the resultant agents which in their combination are affinitive with a similar multitude in yet another cell—and so the cycle continues.

Thus we have a further reason to require a minimal set of rules, if not a single one. The complexity of multiple affinity relationships between two sets of participants, one numbering in the thousands and the other in the hundreds of thousands, would make the existence of several rules impossible at different levels of scale. Recall the example of letter combinations earlier in the chapter. If there is a rule that governs a set of, say, four hundred letters, what are the chances the same rule can be productively applied to a set of four letters—unless it is designed to operate at both scales *and* is capable of satisfying the dynamic at either. The problem becomes anything but trivial when trying to imbue robots with some degree of independence: the greater the variety of situations the machine is expected to master, the more complex the rules become unless they can be decomposed into a set of instructions sufficiently fundamental to encompass the generality of expected conditions. The problem assumes frustrating proportions when approached from the top down; evolution on the other hand does not need to bother—when systems develop from the bottom up every manifested next step is a solution otherwise it could not have eventuated in the first place.

In the OtoomCM the proteins in the neuron are the nodes in the matrix that in turn represents a node of the entire system, which of course is also a matrix. We call the first type of nodes 'elements' to distinguish them from the components of the wider network. These elements are integers, ie values assigned to each index of the element matrix according to a formal pattern. As such they are discrete, they are traceable, and their modification can be performed under defined conditions—all characteristics which have been found significant in the bio-version. Furthermore, they allow us to devise a form of transmission which is distinctly representative of the in- and output and can be made synchronous to the complexity relationship between proteins in general and their neuro-transmitters. Here is how it works:

We take a particular algorithm (let's call it AL) which can be applied to two numbers such that the first acts as a reference value to the other. When processed by AL the second number should move towards the first in indirect proportion to the numerical distance between them. So for example, if the first number is '90' and the second '28' then one application of AL will get us '51', a value that's now closer to the reference. Since the initial distance to the reference value was '62' (= 100%) and the new one is '39', then AL shifted the value of the second number (the resident value) 37% towards the reference. If we apply AL a second time (now between '90' and '51'), then the result will be '74', which represents a shift of 59% across a distance of 39. Repeating the same procedure once more we get '87', a shift of 81% across a distance of 16. As can be seen, the smaller the distance, the larger the proportional shift towards the reference value.

If that was the whole story, we would end up with a straight-forward modification that applied equally across the entire element matrix and so the entire system. While we cannot identify the exact details of neuronal dynamics because of the sheer complexity involved, we do know that for different thoughts to exist the set of proteins available must offer a certain degree of unpredictability, otherwise there would not be any differentiation between neuronal complexes and the whole brain would think in tandem. Clearly, AL needs the capacity to be inconsistent at times, and yet follow a formal process. The only way this can be achieved is through chaotic behaviour, where a precise procedure may induce a type of outcome which ranges from predictable values to completely unpredictable ones. AL does that, and what this means can be seen in the following excerpts from graphs that chart the movement of resident values when presented with certain reference values. Shown are the three types of behaviour we have come to know from Chaos theory, where our reference value can be termed an attractor and, depending upon the result it produces, can therefore be labeled a stable, a periodic, or a strange attractor.

If a resident value, under the process of AL, moves more or less towards the reference value in a steady fashion, the reference value acts as a stable attractor in relation to the resident value (such as in the top section of Fig. 14-5). If the application of AL results in an oscillation of the resident value and so does not converge, the reference value can be called a periodic attractor (centre section). Should there be neither a steady convergence nor an oscillation we are dealing with a strange attractor.

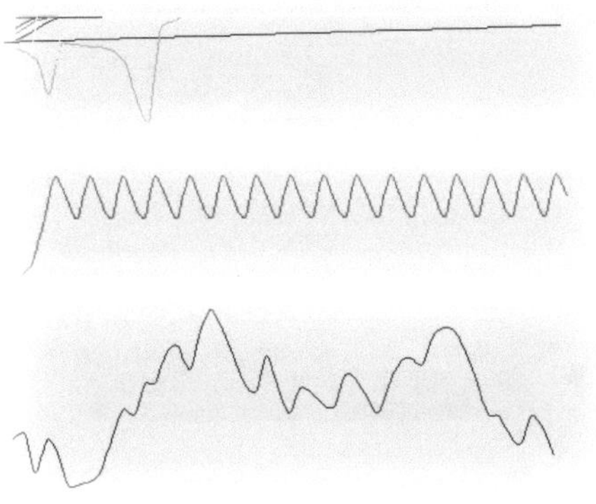

Fig. 14-5: Examples of the effects by an adjustment algorithm (AL) on a series of numbers which, from top to bottom, achieves stable, periodic, or chaotic sequences.

Amongst all the possible number combinations, AL triggers one of those three types. Since the number of possible combinations will always be greater than the elements contained in a node, the situation is comparable to the proteins and neuro-transmitters in nerve cells. It is not necessary to actually implement a huge number range; the modification by AL of any given set of values in an element matrix generates a variety of integers far in excess of the original set once the system continues to go through its cycles.

In the OtoomCM the nodes are connected to each other similar to the neurons in the bio-version. Every node is connected to a given set of other nodes, where the size of the set is kept constant across the entire matrix and each member of the set is unique, that is no connection is duplicated within the same set. However, a node can and usually does share connections with other nodes. As each node consists of elements which represent integers within a certain range, those integers can be made to interact with their respective counterpart in another matrix through AL. The process moves systematically from one node to another, following the connections up to a certain depth and adjusting the elements at every node in terms of reference and resident value. We therefore have a relatively small amount of numbers which are 'active' (ie, they act as reference values), but have a much larger pool of potential numbers (those which *could* have been generated by AL if presented with a different reference and/or resident value). As such our functional synchronicity with the biological configuration is maintained.

Fig. 14-6 illustrates some of the paths taken. In technical terms it constitutes a depth-first traversal of the nodes, and at every node each element of one particular matrix index acts as an attractor to the element at the same index position in the target node.

The picture assumes each node in the inner matrix is connected to two other nodes, and therefore there would be no theoretical limit to the distance one could travel along the connection paths from one node to the other. In a very real sense the traversal would be endless, and it may well be that in the brain one could indeed follow the neuronal interchanges in such a fashion. Since every node has multiple neighbours, sooner or later every node would have been encountered, albeit after a very long time. But is this approach necessary, and above all, is this a realistic conceptualisation of the biological processes?

Recall the adjustment procedure represented by AL we described just before. There are three possible outcomes, a stable convergence the most likely. If periodic and chaotic outcomes are relatively exceptional they are so because the interdependency of nodes, making AL accessible across the entire network, has the very real potential of turning these exceptions back to a more normal scenario. That is to say, letting AL do its work will most likely produce a convergence within a certain node domain, and once a convergence has occurred that particular area is now removed from the immediate influence of any periodic or strange attractors (naturally, exceptions can recur at a later stage).

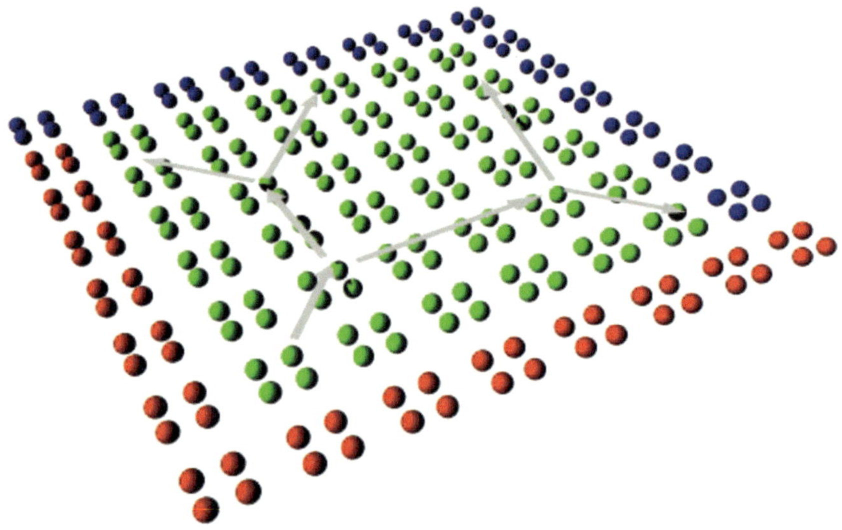

Fig. 14-6: An example of a layout where each node is connected to two other nodes, followed to a depth of 3 (only one instance of the first path is shown).

Since the system defines itself by the creation or destruction of one of three states, and these instances are a function of the interaction between reference values and their resident values positioned in a certain way, it does not matter whether we come across this or that state via an endless traversal or by stepping through the nodes along the edges of the matrix. Two points need to be made however: firstly, the positional relationships between reference and resident values must be the same in both cases, and secondly, we need to follow the connections to a certain degree in order to allow the effects of AL to take place in terms of the connectivity plan of the system. The former condition is met by the formal configuration of the matrix and the connections applicable to each of its nodes; the latter is addressed through the connection depth parameter which tells the system how far a connection ought to be followed before continuing with the next node in the matrix. For instance, in Fig. 14-6 the connection depth happens to be '3', and so once the nodes have been covered as indicated by the arrows the second connection is taken on and followed similarly, after which node #2 in the inner matrix is accessed with its connections explored.

For these reasons we would not have to pursue the infinite paths in the biological brain either; for all practical purposes, whatever neuronal states are encountered after a certain traversal of connections, they would also have been observed moving through the network without worrying about the actual dendritic attachments across its landscape. The salient phrase of course is "certain traversal". Without some idea of those states it would be rather difficult to identify the point beyond which a depth-first traversal becomes superfluous. For our purposes right now however it is enough to recognise that in principle an endless depth-first approach is unnecessary.

In OtoomCM we have the benefit of a user-defined setup which allows us to analyse the possibilities various connections and their depths make possible. So far the program's output are circles drawn on the screen, one for each output node, where the size and colour is a function of the affinity values found amongst the inner matrix nodes the output node is connected to (affinity values are integers which result from comparing the elements in one node with the corresponding elements in the other and so give us a picture of the relative similarity between nodes). While the circles are thus strictly defined, the underlying numbers are merely the result of the adjustment processes that had gone on before; therefore they have no inherent meaning as far as the output is concerned.

This is an important point: in terms of the system's dynamic, an output is of a particular nature because of the processes which are part of the system. It has been *derived* from those processes, but is not directly *representative* of them. The only reason we, the observers, are able to imbue an output with a certain meaning is because we follow its pattern development through from some beginning to some end and then assign a link between one interpretation and another to some pattern we find suitable. In strictly technical terms, an output is nothing more and nothing less. The same applies to the biological version: because there is a neuronal complex which consists of certain patterns amongst its proteins we are able to think 'rose' for example, but not one of the proteins identifies with 'rose' or anything else for that matter—all they do is follow their affinity relationships and it is our consciousness operating at a higher level of interpretation which makes 'sense' out of their patterns.

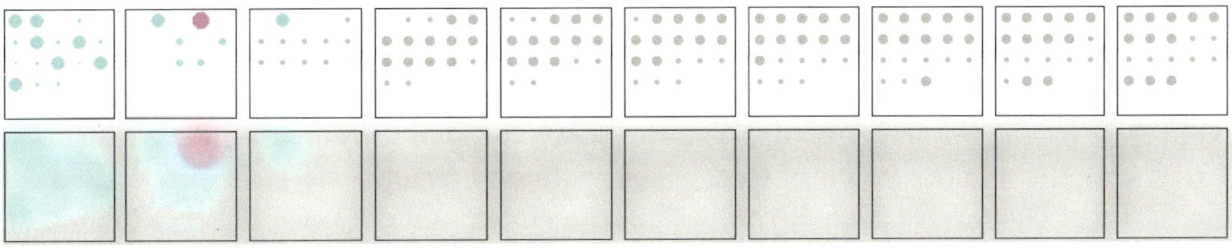

Fig. 14-7: Detailed (above) and distanced (below) output from 10 cycles using a 10×10×10×10 matrix, connectivity 5%, element number range 1 to 50, traversal through to a connection depth of 2 with one repetition, and input being '60' throughout.

In Fig. 14-7 some of OtoomCM's outputs are presented. They are taken from a series of 10 cycles with a consistent input, and the connection depth was 2 which was repeated once per cycle. This means that after the distribution of the input value from the nodes in the input region to their inner matrix connections, each node in the inner matrix was accessed once with each of its connections followed through to a depth of 2 levels. The first row shows the outputs as they appeared on the screen, in the second the circles have been blurred—rather like a picture in a newspaper whose parts can either be seen as the dots they really are or from a distance. Our eyes, indeed our interpretative faculties need the distance to make sense of the picture.

At first the input produced some perturbations amongst the element matrices but since the value did not change and was applied consistently throughout the cycles the system had settled into a stable state by cycle #4. The detailed display shows minor variances in the circle sizes at some nodes but hardly noticeable in the distanced view. Still, they indicate their cyclic movement rather like a wave through the system. Note the smaller circles: it is a group comprising seven in number and it wanders from the bottom towards the top, right to left, where the larger ones can be seen as a 'filler' that stretches beyond the boundaries of the square.

Compare the above with a similar run but now having a connection depth of 5 and this traversal being repeated 6 times at every node of the inner matrix in Fig. 14-8.

Since the input value did not change we again find the system settling into a stable state, but now it occurs from cycle #5 onwards. In addition the shapes at cycles #1 to #4 are more consistent than in the previous example and their outlines are better preserved as well.

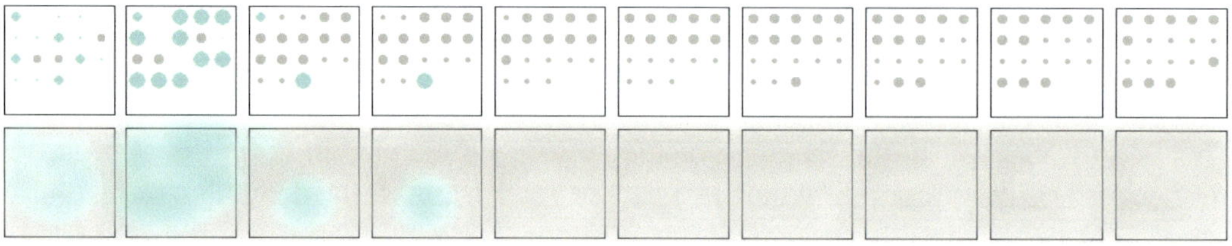

Fig. 14-8: Detailed (above) and distanced (below) output from 10 cycles using a 10×10×10×10 matrix, connectivity 5%, element number range 1 to 50, traversal through to a connection depth of 5 with six repetitions, and input being '60' throughout.

Clearly, pursuing the adjustment algorithm to a greater depth (and so reaching more nodes) and doing it repeatedly (thereby consolidating the changes AL is capable of producing), makes for a more pronounced output. We also find the group of smaller circles snaking their way from the bottom to the top, right to left, although it measures now eight instead of seven. Note that the wave-like progress of the light-grey pattern only occurs at the relatively stable and settled phases; in contrast there is no such movement where disturbances have triggered a more pronounced output. This seems to be a general phenomenon which can also be observed at other configuration parameters: regions of distinct formations tend to hover at their locale, but more regular patterns march through the network.

Up to now the number range of the element values at every node had been limited to 50. While the input could have been anything, the adjustment process was only able to draw from a relatively narrow contingent, particularly since a node contained only 10 elements anyway (if any statement in this section seems obscure, the reader is advised to refer to the Appendix where the technical details are explained fully). Because the connectivities amongst the nodes are shared, that is each node is participating in more than one connection path (exceptions are the in- and output nodes which only feature one connection each to the inner matrix), when AL goes through the system not only are input values propagated via the traversals, but so are other values which may or may not have been affected from the outside. If those other values come from a small pool they will act upon their targets in a more direct fashion (recall that a closer distance makes for larger changes, and therefore a narrower range causes faster convergence than a larger one). A wider range on the other hand provides for a bigger variance of outcomes, thus a higher possibility of distinctive states which in turn make their endurance more probable.

Fig. 14-9 shows such a scenario, where the matrix, connection depth and repetition have been kept the same, but the number range has been extended to 300. In addition there is one more input besides the basic value of '60', namely '10' which occurs halfway through at cycle #5.

Here the input frequency is 2, so the value '60' is presented to the system at cycle #1, #3, #7, and #9. The appearance of dark-green at cycle #4 and its presence at #5, #6, #7 and again at #10 suggests the formation of a domain which is able to maintain its character despite other influences, notably the exceptional input at #5. While '10' causes some modification at #6 there is a trend back at #7 (it should be mentioned that given the different circle sizes and colours available to an output node there are between approximately 100 and 400 variants possible, depending on how many nodes have to fit into the square).

Fig. 14-9: A similar configuration to Fig. 14-8, except the number range is 1 to 300 and in addition to the basic input of '60' at every odd cycle there is '10' at cycle #5 (emphasised). Red arrows indicate cycles with basic input.

An increase in connection depth already provides us with a better definition of the output. However, 100 nodes with 100 elements per node is really quite tiny. The next illustration demonstrates a significant change in subtlety once we enlarge the matrix to a dimension of 26×25×10×10. There are now 20 cycles, input is '60' at a frequency of 4, with a '10' at cycle #5 again, and the connection depth has increased to 10 repeated 5 times.

Fig. 14-10: A 20 cycle output of a 26×25×10×10 matrix, featuring a connection depth of 10 repeated 5 times, input frequency 4 with '60', and one different input of '10' at cycle #5 (emphasised). The system has been seeded 5 times with '200'. Red arrows indicate cycles with basic input.

In addition the system has been seeded with a figure of '200' before the cycles proper started, and this value had been presented 5 times. Thus the first output display reflects an input of '60' against a network which had its input connections modified by AL using a seed value vastly different from the rest.

Two things are becoming apparent: there is a basic ambience represented by a purple/blue formation, and the input at cycle #5 produces a yellow pattern which undergoes certain changes as the cycles continue. At first purple/blue consists of smaller blue spots plus a larger purple one at the bottom of the square. The latter is strongest at cycle #2 when the input had been presented and the system is now settling down. It reappears at cycle #9 when the basic input is applied again and from then on at every 4th cycle. The yellow spot appears first at cycle #5 and can therefore be ascribed to the input occurring there. It then subsides but establishes itself again at cycle #9, but this time not as a response to '10' (its original trigger) but to '60', the basic input which at first had nothing to do with yellow at all.

An input to the system does not in itself produce some output. Only when a procedure such as AL has traversed the node connections and was able to affect the elements accordingly will an output be possible. Even then the interdependency of the network ensures that whatever modification emerges will be subjected to continuous attenuation in terms of the surrounding element matrices. But we have seen that 'attenuation' does not necessarily mean a scaling down of the integers. In our case it means the deconstruction of the matrix state which had been induced by the previous input—regardless of whether this means an increase or decrease of the actual integer values. Yet it can also mean a move towards a periodic or chaotic state in line with the relationship between the resident and reference values, so that the outcome is not measurable in terms of some input and its manifestation, but to what extent the relationship actualises the potential for one of the three states—stable, periodic, or chaotic.

To see this relationship let us examine the previous set of cycles through a depiction of the node states, rather than the output itself. As mentioned, a node represents a series of elements containing integer values. It is also connected to a set of other nodes (and, as we said, the same applies to all these other nodes too). If we identify the integers at each index of one element matrix and compare them with the corresponding index of every matrix the former is connected to, we arrive at a picture of an affinity relationship between those nodes. We can then ascribe particular colour values to equal blocks of the resulting value range and this gives us an approximate idea of how one node fares in relation to its connections. Using the setup from Fig. 14-10 the result can be seen in the next illustration.

Fig. 14-11 shows four samples taken from a total of 20 cycles. Each coloured rectangle in a sample represents a node of the inner matrix, and the darker the colour the more dissimilar this node is to its connections in terms of the respective element values. Thus regions of a dark colour indicate a complex whose nodes are similarly different when measured against their connections, and lighter colours imply lesser differences. White means all the elements are the same. Comparing the node displays with the outputs in Fig. 14-10 it can be seen that cycle #1 is characterised by a relatively high dissimilarity across the entire matrix, although categories of dissimilarity can be identified. The second sample shows the matrix state at one cycle after input '10' has occurred and the yellow spot had appeared. The greater differentiation of the output is equally reflected in the affinity values of the nodes, with almost the entire network in a state of maximum dissimilarity. In the third sample a number of regions have begun to form, induced by a further basic input at cycle #9, and the last sample confirms the further establishment of stable regions (ie somewhat higher similarity) which coincides with the appearance of the distinct upper yellow pattern from cycle #16 onwards.

Fig. 14-11: The same configuration from Fig. 14-10, where the node states are shown at every 5th cycle, beginning with #1.

What do these pictures tell us? If there is a system in which its members can be influenced from the outside such that the effects are a function of the quality of such an influence, then the complexity and inherent instability of the member states have the potential of creating an outcome that either converges upon some value (a stable state), falls into an oscillation (a periodic state), or begins to fluctuate unpredictably (a chaotic state). Because the system is interdependent, that is to say each of its members has a relationship with any of the other members on the same functional level as exists between the input facility and its members, the dynamic which causes the emergence of one of the states (ie, stable, periodic, or chaotic) operates with the same significance everywhere in the network. Therefore, just as the input acts as an attractor in terms of these phase states so does each member, and hence it can be described as an autonomous part of the whole. As the emergence of some state depends on the relationship between any two members that have entered into a functional relationship, the stability of such a state is a function of the affinity relationship between the two members.

In OtoomCM the members are the nodes, and the displays we have seen illustrate the affinity relationships between them. Note however that the coloured rectangles are not representations of the processes under the adjustment algorithm; rather, they reflect the *potential* affinity between the connections, in other words the relationship that would eventuate *if* no change had occurred between the moment that particular snapshot was taken and the time AL had done its work through subsequent traversals. As the outputs show, a pattern may continue, become modified, or disappear altogether. Or, to put it differently, an affinity potential may be actualised, it may be attenuated, or it may not be realised at all.

Since actualisation and attenuation require a certain degree of stability, and since a region of stability emerges from a distinct phase state as we saw in the graphs earlier on, we can say that an affinity potential represents an abstraction in relation to a phase state of a particular node complex. If the potential emerges into a definite state within the domain, the abstraction can be identified—if not it might as well have never existed and no-one will be the wiser.

Here is the output from another 26×25×10×10 matrix running to 30 cycles, with basic and intermittent input plus seeding and once more a connection depth of 10 repeated 5 times, in Fig. 14-12.

Fig. 14-12: Output from 30 cycles on a 26×25×10×10 matrix, featuring a connection depth of 10 repeated 5 times, basic input '30' at every cycle except for the additional input of '80' at cycles #10 and #12, and '10' at cycles #25 and #27. Initial seeding with '200' repeated 5 times. Extra input at emphasised borders.

The matrix has been seeded with a value of '200' presented 5 times, so the first normal input arrives at a stage when the system already has been configured to a certain degree. There is an immediate response, namely a purple/blue pattern with a double yellow addition to its right. This formation gradually decays, although its yellow contingent remains longer than the purple/blue one. There seems to be a re-emergence at cycle #9, but more likely that is a coincidence because at cycle #10 an input of '80' begins to destroy the patterns as can be seen over the next few cycles beginning with #10, when blue starts to disintegrate and green takes over from yellow. However, from cycle #14 onwards purple/blue begins to re-assert itself, gives some way to green but becomes quite strong at #21. This continues until cycle #25 when input '10' gets rid of it and by #27 we find a return to a significant green pattern, although the current input is different from what had triggered the pattern before.

The re-appearance of distinct patterns indicates the resilient nature of certain state complexes, which may become suppressed temporarily in terms of the final output but do not disappear from the system. We also find the emergence of a particular pattern which in turn becomes replaced and then is evoked again, but this time not through the original input value but another, its affinitive counterpart. Without the stability induced by distinct states that form within defined domains, the underlying element characteristics could not sustain themselves. The larger the space within which these states can exist, the greater the possibility of any one of them surviving.

Apart from a larger scale of the entire system, stability is enhanced by a greater connection depth and repetition of the adjustment process. Thus, the more other nodes a node is connected to, the longer the path the system allows to be taken, and the more often the process can be performed, the more profound the results. The connectivity of a system may only fully present itself when its processes can be comprehensively traced, but a comparison between animal and human brains ought to hint at some basic difference. That's why Guy Elston, a neuroscientist from the University of Queensland, is led to think that those areas of the brain which handle the more complex processes of thought are also wired differently [27].

Or consider for instance the neuronal layout concerned with visual input. From the two retinas the nerves connect to an area at the back of the brain, known as the primary visual cortex. It is divided into several parts, labeled V1, V2, etc. One can map the retinal field to the corresponding terminations in the cortex, where an interesting feature becomes evident. A kind of deformation can be found there, meaning that the centre of the retina with the highest concentration of receptors and used for focusing on detail, also relates to a relatively large amount of cortex. The peripheral part of the retina has a relatively smaller area to its own [28]. We would expect a more equal arrangement if the connectivity was based only on the physical properties at the source. Greater detail prompts a need for 'deeper' processing however, and this is reflected in the larger volume of nerves acting as the recipients of the data. Not only that, experiments have shown that perceptual processing takes place in various places; there is no 'master' area conducting a grand theme of understanding. Instead, many locales are connected to many other centres in a distributed fashion, and they all receive *and* send signals [29].

The relationship between processing depth and the differentiating capacity of the entire system is further confirmed in the bio-version by studies in psychology which show that "among adults and high-IQ groups abilities are better differentiated (ie, there are multiple abilities and/or a weaker general factor) than among children and low-IQ groups", writes Lazar Stankov in the *Australian Psychologist* [30]. The purpose of his article is interesting in itself. There is an argument about the reasons behind cognitive abilities, particularly as they relate to ageing and the intelligence of an individual. One side sees processing speed as the deciding factor, the other goes for capacity, that is working memory. Stankov reviews both types and comes out in favour of capacity in both, ageing and intelligence. Capacity, ie working memory, implies distinct thought complexes; distinct complexes need space and connectivity for their creation and maintenance.

Our computer model is still tiny—the colour blotches it comes up with are hardly the stuff of high intelligence. Nevertheless, a complex system needs the dynamic homogeneity of its parts to function, and when it can be shown that affinity relationships based on one single common algorithm are able to produce distinct effects which are comparable with the principle phenomena we have learned to associate with intellect, it is time to move away from our traditional insistence that cognitive behaviour can only be performed by biological entities.

The scale of the phenomenon, replicating the same principle *functionality* regardless of content, can be observed again. The affinity complexes are being created by a single rule and are therefore sufficiently fundamental to permit the assembly of structures. It is those structures that form productive conglomerates capable of representing an input from the outside. At what level of decomposition the smallest unit inhabits the functional realm becomes therefore highly significant. Its parallel can be seen at the emergence of life itself. In a research project run by the Institute for Complex Adaptive Matter based at the Los Alamos National Laboratory in New Mexico, USA, physicists and chemists are trying to push at the boundaries of what is known about the essential difference between dead and live matter. According to them, the special properties of matter that result in complex self-organisation, the precursor to biological compounds, exist at the nano-level—not the microscopic, and not the macroscopic scale [31]. Therefore, in any description of phenomena care has to be taken to articulate the *content* at the corresponding level of abstractive sale.

The functionality pertaining to the driving force behind the assembly of affinity relationships could have a possible counterpart at the nano-level, where the configurations of protein structures are organised. A research team drawn from the Australian National University in Canberra and Griffith University in Brisbane has proven that a hitherto theorised reversal of the 2nd law of thermodynamics is in fact possible for small systems over short time scales [32]. In our large-scale world entropy is generated by any system which dissipates heat, a process that cannot be reversed. Thus the 2nd law of thermodynamics ensures the age-old search for a perpetuum mobile is futile. However, when it comes to the realm colloidal particles inhabit it is another story. Using an optical trap (a device in which a laser beam draws transparent, micro-sized particles to its centre due to a force that is a function of the particle's higher refractive index) they showed that entropy is actually being consumed, rather than being generated as 'ordinary' circumstances would have us believe. As the authors write, if the probability of such energy reversal at these scales becomes greater (which it does), the implications for nanotechnology and "indeed for how life itself functions" are important. (A curious aside: there is a Latin-German dictionary used for high school students in Austria, and in its section listing the translations of commonly used words and phrases we find an entry on *Perpetuum mobile*. After telling us it means "continuously moving" and a brief pointer to the expression's history the last sentence reads, "Also describes an instrumental piece of music played very fast with small note values" [33] [transl. by the author])

In our sample outputs the coloured splashes do not have any meaning by themselves. They have not formed in order to give us the impression of green or blue or whatever but they derive from the affinity relationships existing inside the matrix. In other words, it is not some need for this or that colour which prompts the system to come up with a particular number set across certain elements, but what one group of numbers means in relation to another group of numbers. In OtoomCM the configuration is fixed at the beginning of the test run. In the biological version however the plasticity extends beyond our present focus—the entire system is a malleable entity responding to this or that environmental factor provided the response is within the range of a sub-system's potential. Thus affinity relationships can prompt not merely a re-formation of protein complexes, but the connectivity itself is subject to the demands of the network. Experiments involving a mapped layout of areas on the cortex of a monkey representing its limbs show that the removal of a finger for example results in the reorganisation of the cortical map such that the area previously devoted to the digit is now used by the adjacent fingers [34].

If the connectivity of the neuronal network answers to the attractors that are determining its phase state, where does that leave the representational content? Not only does the hue of our blotches hardly matter to the system, on a higher level of operation the contents of thoughts are secondary as well. For whatever length of time our computer program is allowed to run, it will happily form and re-form its affinity complexes and give us the output to go with it. At its current maximal size of 552 inner matrix nodes there is not enough room for sophisticated abstractions and so there is no identity and no character (but there is a hint of an *emerging* character, as we will see in the next chapter). Still, even on a very much larger scale the principle dynamics are the same and when it comes to thought, what drives a structure to grow or shrink is ultimately based on its characteristics at the micro-level.

Hence any definitive assurance about the picture we have of our surroundings simply does not exist. For the most part, what is in our minds is only marginally a reflection of the outside, as Gerhard Roth, from the Institute for Brain Research in Bremen, Germany, explained in an interview [35]. Far more important he says, at least as far as the brain is concerned, is the usefulness of this picture in our daily affairs. It is from that angle that concepts are born, and so our view of the world does not confirm reality, but what we are able to make of it—at whatever level of competence. Translated into our context this can be expressed as: our brain does not care whether it represents some outside phenomenon faithfully to its reality or not; rather, it produces, adjusts and readjusts the phenomenon's representational format according to the already existing affinity relationships within its network. This is the reason why no child, however highly developed its neuronal circuitry may be at birth, can ever understand its environment from day one. It has to gradually build its comprehension by creating layers upon layers of representational structures, composed of step-by-step additions of instances, a process which is particularly unique to each individual. The longer this process, the more sophisticated the result. Whether the outcome resembles reality in any way or form is not something the brain cares about. It may become an issue for the *mind*, the conceptual overlay to which we ascribe the qualitative assessments of that what the brain holds amongst its 100G of neurons. And of course, what the mind is able to tell us remains once again the affinity-dependent property of neurons involved with the system's output.

Let us take another look at the scale of such a development. As one representation gets added to another and forms an affinitive whole, the reason it *can* form something more complex is due to the inherent harmony of the domain. The harmony exists because of the affinity relationships, and the relationships exist because of their underlying algorithm. We can say then that whatever outcome results from a series of phase state changes, will be a harmonious whole resting upon the agreements amongst its parts—it cannot be otherwise. Moving up to a higher scale we leave proteins behind, pass thought structures, and arrive at the societal behaviour of humans. Just as the formation of a protein complex or an idea follows the affinity plan, so does the emergence of a public sentiment, a cultural meme. But since the meme knows as little about the inherent technicalities of its nature as the neuronal complexes know about theirs, what reflection there can be is not an articulation of some algorithm, but yet another representation of the phenomenon, expressed in terms of the general affinity relationships existent in their cultural environment. We have a name for it, and it is located fairly and squarely within the expressive domain of subjective representation: teleology.

The general question of how a system can take itself through its phases is not a trivial one. It has occupied philosophers and scientists alike, and for good reason; whether it is human behaviour, the actions of bacteria, or the fundamental processes of life per se, in principle the problem is the same. For Vladimir Nabokov it was the issue of insect mimicry which prompted him to look for an overall plan and a design, as Victoria Alexander discusses in her thesis on the man probably better known for his novels than his impressive scientific work [36]. In a different context Margaret Boden from the University of Sussex writes about the concept of autopoiesis, a label given to the form of self-organisation found in life. Her very first sentence states that there is no "universally agreed definition of life" (in fact, "notoriously" so) [37]. Since characteristics such as self-organisation, autonomy, development, reproduction, and metabolism (to name some) have come to mind at various points, it is through the perspectives evoked by these descriptions that life has been interpreted. If one concentrates on self-organisation for example, what is the underlying mechanism—and if the idea suggests ever more complex arrangements, what about self-destruction? Or take metabolism—immediately we are into the field of energy consumption, distribution, and energy packets serving some (what?) purpose.

This is not a critique of the works of biologists, chemists, or system analysts. I merely want to point out the sheer situatedness of these approaches, drawn as they are from their respective areas and so structured along the format within which they have evolved over time. They cannot do anything else; neither it seems can the people inhabiting their domains.

Just as one thought differs from another, so do the wider domains of concepts. In Otoom's parlance we would express this as affinity complexes on one side and background domains on the other. The situatedness we referred to above belongs to the background, a general environment of states which has grown in response to input and the mutual feedback induced via the adjustment algorithm. In Figs. 14-7 and 14-8 we have seen examples of a background wave. The larger the physical configuration of the system, the more chance there is for the wave to develop. There seem to be some parallels with the bio-version. Walter Freeman from the University of California at Berkeley proposed the idea that the brain uses neuronal rhythms to coordinate the activities between its regions [38]. While the exact circumstances that accompany EEG patterns would have to be researched further, the occurrences of wave-like patterns in the OtoomCM could suggest a similar coordination, except that the phenomenon would not be the active kind as assumed so far, but instead would constitute the effect of a 'background radiation' which appears as congruent blocks simply because it refers to neuronal sub-systems which happen to be in sync. If this should turn out to be true, it would also diminish the significance of EEG oscillations. While they do indicate the existence of neuronal processes of one kind or another, they in themselves may not necessarily be representative of thought structures but would only reflect a general level of activity which accompanies cognitive dynamics.

Little colour patches are just the beginning. They stand for the modified values of elements that have arranged themselves into specific states of various stability. They cannot do otherwise. In OtoomCM it is AL which sees to it, in the biological brain it is the electro-chemical potential of protein packs. No wonder that the creation of like clusters is as intrinsic to the system as are its functional elements, whether they be integer values or atomic valencies.

Steven Pinker, in his book *How the Mind Works*, writes light-heartedly about the human need for categories, for creating classes no matter how ludicrous the exercise may end up to be [39]. Semir Zeki, whose thoughts on human vision we have already encountered, describes the work of the Dutch painter Jan Vermeer and quotes Proust as calling him "an artist who remains forever unknown" [40]. With Vermeer "very nearly all is implicit", says Seki. There is mystery in his paintings, and there is nothing that the human mind likes better than something mysterious. In due process this will be filled, and the material the mind uses are abstractions. With them we bring order into our personal world.

15. To: Order (?)

Abstractions - clusters - consciousness

"Close your eyes and open your mouth!"

Surely this must be one of the more intriguing and exciting things one can say to a young child. It's all there—the expectation of tasting and intimately feeling something nice, the mystery of not being able to see it, and all of it wrapped up in a scenario that features a trusted person and a gesture hinting at a surprise any moment now.

The longer one can make this list the more reason for the senses to perk up. Conversely, the less capable the mind is of harbouring any of these sensations, the milder the anticipation. And yet, nothing has actually happened—it's been all in the mind.

If there were no thought structures which held some memory remnant of chocolate, if the act of opening your mouth cannot be associated with pleasure of events past, and above all, if any of the words spoken do not trigger some translative process from sound to personal experience, the response would be nil. The ratio between evocative TSs and superfluous associations (or worse, inhibiting ones) has to be just right, otherwise the fun gets watered down. For an infant the TSs may not be enough, an adult may have gathered too many. In a newborn the cognitive faculties are rudimentary because there is nothing to associate with; the neuronal network represents a feature-less, although dynamic, space. That's where we start.

What exactly is an abstraction and why should there be any? We know that sense organs convey information, and it has become clear that the information is in direct relationship to the medium through which it is gathered. Light waves excite the nerves in the retina in a very specific way, so does the sound in the inner ear, or the chemicals washing around our tongue. Eventually however the impulses are transmitted through the synaptic interchanges, and so end up influencing protein formations inside the neurons. If the resulting representations of input are at all able to differentiate one from the other they must likewise have the capacity to group. They clearly do both, and even without going into any more details the development of characteristic formations (whatever their composition) and their potential for congruence become evident.

The presence of such formations must relate to the general throughput of data around their domain. It is still impossible to follow the transmission path of a particular thought, but it can be deduced that a more complex articulation of any given intent (regardless of the intent's actual format) would reveal itself by a commensurate spread of progression from some source to some target. Conversely, a simpler form could be managed through more direct travel. There are parts of the brain which can be categorised along these lines. For example, a region in the cortex called the supplementary motor area is involved in the voluntary movement of limbs. It is known that its synaptic connections with motor neurons are comparatively strong here [1]. Since the final activation of a muscle does not require a sophisticated thought pattern (although the lead-up to it might) and its neuronal source is correspondingly wired, it stands to reason that other areas do not communicate as directly and so contain more elaborate modes of transmission.

Even a relatively straightforward command to a muscle is not a unilateral affair. Studies on individual neurons in the brain's motor area have shown that the eventual movement is the outcome of a combination of firings, where each cell contributes to the overall effect at a different activation level, thereby producing a smooth and accurate result. The entire movement is created from a tally of the single contributions of individual cells [2]. If the movement of a simple muscle is based on functional complexes amongst neurons, how much broader would the participation be when it comes to perceptual representations?

One of the major findings in the field of visual cortex research is the existence of many additional centres apart from the main region, and without them our normal human vision would not be possible. Semir Zeki puts it this way:

> And it is this discovery, and the train of thought precipitated by it, that was instrumental, if not unique, in ushering in the view that vision is an essentially active search for essentials. [3]

What he has to add is interesting in terms of cultural inertia:

> But these new facts have come to light only in the last twenty-five years; they were not available at the time the early neurologists speculated about the functioning of the visual brain. It is therefore not surprising to find that they have played no role in any theory of art or aesthetics.

Evidently, one generation can be too short for new kinds of musings to take hold.

Neuroscientists from Darmouth University in Hanover, USA, discovered that when objects which are capable of being handled were shown to individuals a reaction of regions other than motor areas set in [4]. In addition those visual areas became active which deal with spatial orientation. Such combinations point to complex pathways that include a variety of cognitive functions and all at the same time. While it is not a direct proof of abstractions in our sense of the word, indications are that tasks, at whatever level of depth, not only avail themselves of multiple brain regions but need their participation in order to be performed at all.

Against this background, considering information and the deconstruction of its parts leads us to abstracts. We have touched upon the issue in the previous chapter regarding the complexity facing robots for instance. The instruction set of a machine needs to be flexible when having to deal with unexpected situations. But what is 'unexpected'? For a vehicle that travels on an even floor and whose decisions only ever involve left, right, front and rear an unexpected case would be 'up' or 'down'. If we were to add such cases to its arsenal, 'unexpected' could be 'half-left' or even '23.5 degrees left from the current direction'. The latter are refinements on the theme, but they do hint at the principles involved—the wider the field of possible responses, the more flexible the approach that needs to be taken. And here we are faced with two fundamental choices: either we widen the instruction set to include the largest possible range of eventualities, or we redesign the decision making process altogether. The second option is better in the long run, because it is more radical. It also carries a price.

If we were to enlarge upon the alternatives of our floor-travelling cart to include direction by single degrees we have to construct a controller which possesses a mechanism to identify directional units. We also need to give it 3-dimensional awareness so that a degree value in terms of a x-y-z axis combination is possible. Perhaps the overall rule set is now more complex than our original left/right/etc option, but its potential has grown considerably.

Let's put this another way. Next is a list of ever increasing potential by an entity when presented with a task. It is arranged in terms of its decision making capacity against the environmental requirement. Moving from the first version up to the last we can see an unfolding of complexity, a steady heightening of generality of expression, and an accumulation of options. If we were to describe the actual algorithms accompanying each stage we would find that as we go up the ladder we resort more and more to common principles; indeed this is reflected in the generality of language. Generality therefore implies optionality, provided the interpretative quality of the system possesses sufficient complexity (after all, the options must be implemented somehow). The price the system pays for this type of enhanced capacity is the enforced step-wise only recognition of situations; it can only respond to something which can be accessed through a limited widening of its perceptual domain (eg, it cannot contemplate moving to a point in space if it has not learned what 'space' is). Suppose we wanted to increase the generality even further. Not only do we need a system that can substitute one rule for another should the need arise; it would have to be able to deconstruct rules to identify their parts with a view to selectively reassemble them so that the new rule applies in a fundamentally different manner. The only way it can do that is to hold a part in the form of a representation and manipulate that image as if it were the real thing. Such an image can then be 'looked at', it can be 'attached' somewhere, it can be 'observed' how it works in its new location. The entire operation becomes representative, and it is composed of what we call abstracts.

Capability	**Task**
Go left, right, forward, reverse.	Distinguish between left, right, front, rear.
Go in any direction on a 360^0 circle.	Identify any given direction with single degree accuracy.
Go in any direction on a 360^0 circle and use the existing landscape to get there.	dentify any given direction with single degree accuracy; observe and analyse the possible paths in terms of suitability.
Follow an order to go from the current location to another somewhere in space, using the existing landscape to get there.	Interpret a command that identifies a path between the current position and another location located somewhere in space.
Follow an order to go from the current location to another somewhere in space, using the existing landscape to get there, but avoid damage through using unsuitable features.	Interpret the 3-dimensional surrounds, and analyse the available features in terms of their suitability to reach any destination within that space.
Recognise anyone present and follow their command, analyse and interpret any results from a given action.	Understand the environment and interpret commands coming from within it; follow a command if there is no damage to yourself and others.

Most probably the reader's mind is engaged in just such an exercise that very moment; maybe the internal vision is akin to having pieces of a Lego set moved this way and that, and functionally it would be quite correct. But the complete action in the reader's mind is even more multi-faceted than this: not only are pieces of items moved about, the ability to 'move' an item or a rule, to 'attach' a cube or a principle, they all need to be general enough to accommodate the differing contexts in which they are used. For them to be used they have to be learned, from simple, straightforward applications to esoteric ones. To 'go to' therefore can mean being at one paragraph now and then at another one; it can mean dealing with one person now and then approaching someone higher; but without having been experienced first as literally having had one's finger on one spot on the floor and shoving it across to another spot no extension will be possible. The same goes for the concepts of 'attach' or 'fit' and so on. They all are representations of some act which had been experienced and relayed through the synaptic transmissions until it has become embedded in a particular neuronal complex somewhere in the network.

On this note it is interesting to observe the articulation of metaphors or indeed any phrase which lends itself to multivariant applications. Every single one of them can be employed literally, but hardly any should be used that way (unless one wishes to sound childish). They represent one of the travails of growing up, since not to recognise a deeper meaning can be embarrassing at best, dangerous at worst (in fact this whole paragraph is full of them).

In technical terms we can say then that an abstract is a representation of some input which has become stable enough to remain in memory for it to become affinitive to some other complex. Without such congruence it will not be used further. For instance, an object which had been considered fixed will not be associated with 'movement', and only additional experience (which in itself may have to overcome similar thresholds) will open the pathways to the region representing the item. In chapter 10 we have come across the famous instance of Kekulé von Stradonitz, the chemist who discovered a ring-like formation of molecules because he allowed his mind to wander and engage in hitherto undesirable associations; another treatment can be found in chapter 7 when we discussed what it takes to solve a puzzle.

Since associations are caused by the affinity relationships amongst regions, and affinity relationships are a function of a realised potential in terms of phase states, we can identify the emergence of an abstract as the effect of an attractor within its domain. Therefore there is a tendency for abstractions to be formed, in fact the system is built around the purpose of abstract creation. As more complex tasks require a larger domain and so more attractors, the trend becomes stronger the wider the span of the theme in question. The wider the span, the more autonomous the complex, and the more likely it will invite affinitive connections. In OtoomCM the result is a more encompassing region of similar phase states, but the bio-version can also opt for a physical modification of its dendritic organisation—nerve cells reshaping their connections.

John Byers from the University of Idaho and Curt Walker from Dixie State College in St George, Utah, investigated the role of play in cats, rats and mice during the development of the cerebellum [5]. It is one part of the brain which is particularly important for motor skills. In all three species it was found that just as the brain reached its maximum stage of synaptic generation, the play periods become especially intense. Playing of course is one activity where the emergence of abstracts is facilitated through pretend situations, whether the mammals happen to be young kittens or children.

Once again we find a stimulation of the system to enhance the presence of physical structures which in the end serve one basic purpose: the creation and maintenance of the forms we call abstractions. As the neuronal network grows, its ability to generate them does likewise. To show how direct the relationship between structural volume and representational quality is we can observe the output of various configurations in the OtoomCM. Below are four sets, based on a 5×5×80×80, a 8×8×45×45, a 20×20×10×10, and a 25×25×3×3 matrix. Fig. 15-1 illustrates the numbers of inner matrix nodes and the elements per node in each set.

In the first set there are 6400 elements per node but only 9 nodes which are actually processing the input (input and output nodes do not process input). While the potential for articulation within a single node is high, the system does not have the capacity to differentiate the result across the network. Therefore we have barely a hint of anything. In the second set the number of elements are sharply reduced, but with 36 processors there are the beginnings of some distinct output. Now consider set #3. Although the elements are fewer still, 324 nodes are clearly capable of formulating a response. For the first four cycles such a response is even since the input did not change. With the presentation of ‘10’ at cycle #5 however the previous distribution of states had to make room for further articulation and as the outputs show, the dominance of ‘yellow’ disappears in favour of a central shape of purple and a blue spot on the top. These manifestations are largely retained when the input reverts back to the original ‘60’, although its accompanying ‘yellow’ regains some of its presence. In the fourth set the potential for articulation is even higher (529 nodes), but since each node has only a mere 9 elements to work with the actual formulation of an output is restrained. Hence we do see something definite but hardly pronounced (the detailed output consists of dots, not circles, because the differences between the element values had been minimal).

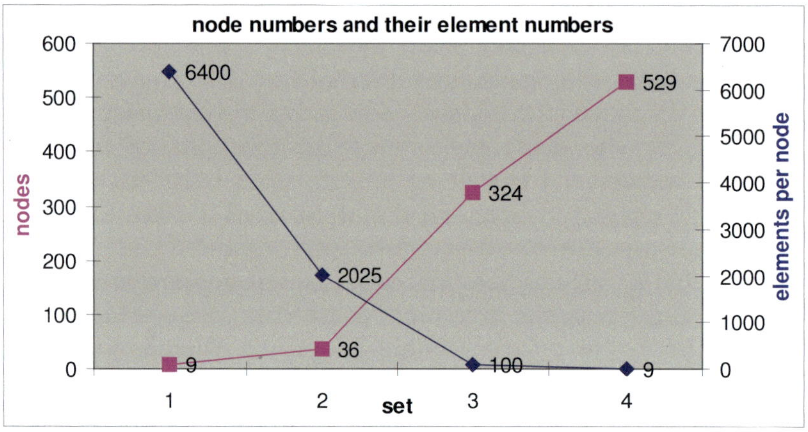

Fig. 15-1: Graph showing four sets of OtoomCM configurations with their number of inner matrix nodes and elements per node.

Fig. 15-2: Outputs from the four sets shown in Fig. 15-1, in sequence. Basic input at every cycle is ‘60’, other input ‘10’ at cycle #5, ‘110’ at #6, and ‘150’ at #7. Cycles with other input are emphasised.

In Fig. 15-2 we see the outputs from these four sets, arranged in sequence.

While the samples contain two variables at the same time (ie, node *and* element numbers), the steady changes in their sizes do indicate the tendencies of articulation and differentiation. In the previous chapter we mentioned the adjustment algorithm (AL) which goes from one node to the next in a systematic manner and modifies the values at each element. After each cycle the system is left in a particular state that is defined by the values of all the elements across the nodes. The relationship between one node and another in terms of the connection plan can be expressed as a potential, because after the processing of the values in the corresponding nodes there will be one of three possible outcomes during the next cycle: (1) the values in both nodes did not change at all, (2) with respect to the reference node the values changed towards the relative similarity ascertained before that process (ie, the similarity has become stronger), (3) the values changed away from the previous state (ie, the similarity has become weaker). What the outcome will be no-one knows until the cycle has run its course, but over a period of time a tendency will become apparent because AL does cause the number sequences to trend, as Fig. 14-5 in the last chapter showed—notwithstanding the possibility of a change in the trend itself over a particular time scale (for a detailed outline refer to the Appendix).

As the potential incorporates these possibilities, and the number of nodes and elements have a bearing upon such outcome, we can further define the potential of the system in terms of its configuration. The three variants just mentioned are a very rough guide; after all, there are only three categories to choose from. Yet inherent are the detailed developments which will articulate the output, since the output nodes draw their characteristics from the node states they are connected to. Therefore, as far as the articulation of any output is concerned, we can say that the potential of a node domain represents the inherent *latency* of the system. We have returned to the concept raised already from chapter 10 onwards.

Is it possible to define the unrealised potential of a system, in other words its latency? A particularly adjusted node domain has a finite latency due to its specific configuration. In other words, the attractors which are being accessed through AL operate under constraints developed under the influence of specific integer values. The set of possibilities under those conditions can be circumscribed by the fractal border of this or that attractor. It is the fractal border which determines the outcome at each cycle, featuring as it does a limited number of *fitness peaks*, ie the realisable results from AL's process. If there is only one outcome (ie, the convergent line in Fig. 14-5) it means there is only one peak in the domain's fitness landscape, and so the fractal border will be a circle encompassing this peak. On the other hand, if there are several or many outcomes (ie, the periodic or chaotic lines in the illustration), the fitness landscape of that domain will therefore consist of many peaks, and the fractal boundary will be a multi-directional, zigg-zagging line, its degree of 'zigginess' an indication of the latency inherent in the system represented by the line. We can say then that the degree of latency of a system is proportional to the dimensionality of its fractal boundary—the more such a line zig-zags about the greater the unrealised potential of the system it belongs to.

Note that the line itself is not a direct indication of what purposes the system will be put to—at that point the applications lie still in the future. Rather, it is the applicability we are interested in and which is expressed through the associative capacities of the constituent elements. The applicability depends on the object elements of the system and what can be done with them through its co-existing functional elements. The associative capacity of a system is thus defined by the number of meaningful relationships that can be entered into by the system's object elements with their functional counterparts.

This capacity is nothing else but the affinity relationships we described earlier. Therefore, in any given system:

the latency is defined by the dimensionality of its fractal boundary;

the fractal boundary is a function of the associative capacity;

the associative capacity is a product of the object elements and their relatable functional elements;

hence, the latency is defined by the product of the object elements and their relatable functional elements (normalised for standardising the result).

In OtoomCM the object elements are the element values at every node; the functional elements are the nodes of the inner matrix, since they are the processors of the system.

If we want to arrive at comparable expressions of latency we can use the following procedure, where L means latency, OE stands for the number of object elements and FE for the number of functional elements:

$$L = OE * FE$$

normalising:

$$OE = oe * \frac{1}{\sqrt{oe^2 + fe^2}}$$

$$FE = fe * \frac{1}{\sqrt{oe^2 + fe^2}}$$

Applying it to the four sets from before the values are shown in the graph of Fig. 15-3.

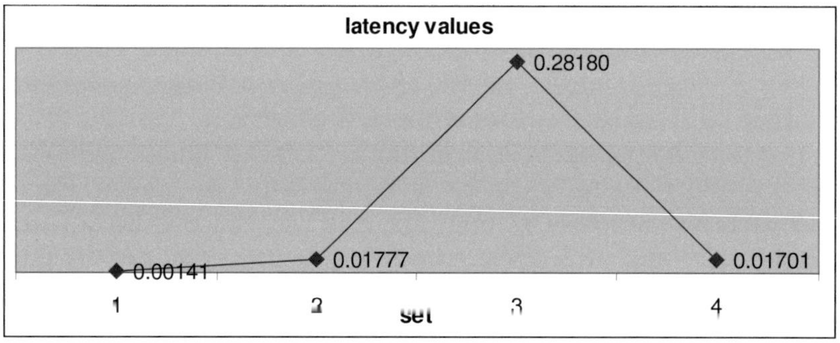

Fig. 15-3: Latency values for the four sets used in Fig. 15-1.

Set #3 with its highest latency value corresponds with the 'best' output in Fig. 15-2.

So far the latency figures related to the entire system. It may be argued that in 'real life' the system should not be regarded as a functional whole, but as a constantly changing mixture of subsystems, re-arranging themselves according to the contingencies of the moment. Indeed this is so. The object and functional elements refer to the constituents of a certain domain, which in the bio-version can be expressed as a neuronal complex which happens to experience a particular congruence in terms of firing patterns and associativity. But functionally speaking there is no difference to the fixed matrix in OtoomCM. It is not the whole network we are concerned with, but those nodes which at any given time have formed groups expressing their affinity relationships produced by AL. Therefore these domains are flexible too, and their latency parameters (number of nodes plus elements per node) are just as relevant. Furthermore, in both versions the object elements are pre-defined; neither the protein numbers alternate nor do the indices of the node matrices.

At what point a latency value begins to reveal the first onset of intelligence is hard to say. It would be easier if the question "What is intelligence?" could be answered for a start. Even at the level of human cognition the standards are not as simple as they might seem at first glance. While the *results* are more or less discernible, the underlying parts are less so. For example, to compose a symphony needs certainly more intelligence than writing a pop song (at least in my considered opinion!), but suppose one were to listen in to the composer while a bar is being constructed. Is it possible to identify the lofty standard by this or that, seemingly wayward, note? And maybe the first attempts actually *are* rubbish—yet do they not lead to definitely something, whereas the other stays at wherever it began? We are back at latency, a non-concrete (for some reason I hesitate to call it 'abstract') yet real entity which works in the subconscious background and sooner or later offers an output we in our respective cultures have come to understand.

For 'culture' is the salient word. Arguably the most fascinating science fiction stories are those which deal with an alien world; not just alien in the sense of some other sun or a technology a few generations ahead of ours. I mean truly alien, where even the most basic concepts need careful translation to make them accessible to human minds. Consider the following few snatches of text taken from a story about such beings:

> There was a Changer who had spent three life cycles planning a particular cycle climax and who had come to the moment of action.
>
> Pur waited till his colour and tone showed that consciousness had returned, then moved in a light wave-motion calculated to draw the other two back into calm, reasonable discourse.
>
> He was hardly recognisable as one of them, this ancient Oldest. He was ... not quite energy any longer. He was half matter... [6]

The words are from *The Dance of the Changer and the Three* and they tell of a planet where a colony of humans gets destroyed through the utterly reasonable actions by its inhabitants. The visitors understood nothing and so their fate was sealed.

In the end we can only resort to substitutes, designed to evoke sensations which are comfortable for us; whether they truly reflect was has been articulated is another matter. When someone plays with their dog this process of interpretation can be observed all the time. At least the canine brain is of a complexity to permit some reasonable parallels in terms of the emotive quality of the output. When the dog wags its tail it is happy and presumably so is its owner, who then usually proceeds to turn the simple emotion into contexts quite out of reach for the animal. Nevertheless, repetition (ie, the dog keeps coming back for more) can be associated with joy; inactivity (ie, sitting still and looking at its master) with expectancy; and changing focus (ie, moping about and starting to do other things) with boredom.

Parallels in the physical brain of several species, from humans to monkeys, bats, and rabbits, suggest a likewise differentiation according to the system's output is possible. In a paper Guy Elston from the University of Queensland (we met him in chapter 14) compared their cortical circuits by identifying the neuronal connectivity across various areas of the cortex, focusing on pyramidal cells [7]. These cells feature up to twenty-three times more dendritic spines in the prefrontal region than in the visual processing area, and a similar increase in the network's density can be observed from the rabbit upwards. Although it would be difficult to imagine the conceptual world of a flying fox or a gibbon for instance, the difference in cognitive processing right up to humans is obvious.

To complete this section I will present an interaction with OtoomCM at a configuration which is simple enough to carry out without having to wait for hours, yet sufficiently complex to begin to differentiate and develop habits. The setup is a 7×7×10×10 matrix, 3 active input and 3 active output nodes, connectivity 15% (meaning in this case a node is connected to 3 other nodes), element number range 1 to 200, and connection depth is 3 repeated 2 times. As usual the input was from the keyboard, and the output consisted of 3 ever-changing circles (one for every output node). The session is related by showing the input (some key press, with its character meaningless to us), the output (what appeared on the screen), and some form of anthropocentric interpretation (which is of course completely wrong...). There are various pauses between the inputs.

`	3 green circles, steady	It's just waiting, to see what I would do.
m	Green, blue, purple, yellow circles, changing sizes, moving around	It likes it! It is dancing about, really happy!
``	A few small circles, back to the 3 green ones again.	O oh, it quickly looses interest, now it just sits there
m	3 green circles, sometimes blinking yellow	It wonders - do you really want to play again?
i	First same greens, then a black circle moving about, gradually a blue one appearing, then a red one, getting bigger, more lively	"Are you sure?" it asks. After some hesitation it understands it is for real, so it is warming to it.
p	Back to the greens, a few black circles popping up here and there, then just greens	Seems to like it first, but quickly decides it's boring, so just sits there again, waiting.
m	Becoming livelier again, different colours, then immediately-	Oh yes, that's it!
p	Old greens, but interspersed with blue, yellow, red big circles	"Mmmh, ok then, let's do it this way if you want."
mpmp	A few colours, but then back to the greens, steady	Ah, it won't play along anymore, can't be tricked through a few 'm's when it's really the 'p's it's getting, can it?
m	Nothing, just 3 greens	It's angry now, won't play anymore.
m	just a blink of blue, back to the greens	Nope, doesn't want to.
mmm	Slowly, a blue, a black, then a red, blue, yellow, various sizes	Needs steady and careful nurturing, must not be disappointed now, only gradually develops trust again.

And so on and on.

Although this exercise may seem somewhat quaint, it only does so because it is obvious one deals with a machine. The trick is to forget about the computer, letting the fingers rest on the keys and intuitively pressing here or there in response to what the circles are doing. Actually, they are not really doing anything that you or I understand. Still, after a while one particular response gets patterned with one particular group of circles, to be followed with another particular key press which just happens to evoke something that is increasingly becoming familiar... so there is an instinctive reaction from the fingers and yes, there is the response like before, and so the next keystroke becomes a *reply*, which gets answered by the... *thing*?

From a random presentation of patterns there comes, through the response from the outside (this is *my* perspective) and the feedback from its output (this is *its* perspective) a development of articulations which steadily assume *meaning*. In this case the meaning does not refer to anything else except the immediate environment which consists of circles and some keystrokes by someone. But suppose it grows in differentiation, and suppose the responses grow with it, and then there is further complexity, followed by another rise in the articulation level—and we have evolution.

For OtoomCM with that setup it had been the ‘`’ which caused it to hold still and the ‘m’ that got it going. Obviously, these are integers which the keyboard transmitted to the registers of the silicon chip, there to be processed by the software I had written. At some other configuration some other value would have done something else, and the ‘story’ would have turned out differently, or maybe not. The significant step was for one specific input to trigger a specific response, which I in turn met with an equally specific further input, and the situation progressed from there. From then on a ‘`’ was not just another number; to the network it represented a meaningful value that made a very particular impact, so much so that a second input was responded to not arbitrarily but in terms of the existing format: an *identity* had come into being. This identity could only emerge because there existed a set of nodes which had entered into a certain affinity relationship that had become stable. It was an abstraction, the core entity of as many others as the system was able to accommodate. In this case 25 inner matrix nodes made for three circles with a limited set of colours, but already it was enough to transcribe the output into some kind of scenario (albeit a strained one). Take 20K nodes (like the Aplysia slug) and the thing will move when prodded; make it 300M (like an octopus) and it can open a bottle; and by the time we have 100G—well, turn on the news and you be the judge.

The expanse of the network and the size of the matrices at each node together with the pre-set number range across their elements determine the responsiveness of the system. Various charts in this and the previous chapter have shown that. We can even put a mathematical value on the performance through the latency parameter, but what we can’t do is forecast the character of the system’s behaviour. When before the ‘m’ key prompted a lively response, then it did this because under the ASCII character set that particular key is assigned the value of ‘109’, and when AL processes the integer ‘109’ in conjunction with anything from ‘1’ to ‘200’ the eventual outcome translates into some circle within a region of fractal instability. Anyone repeating the parameters exactly again will get the same results. If the system consists of many millions of nodes with thousands of elements each, not only will the repeatability depend on the same exactitude but the chance of a difference here or there will become significant—if not in the execution of the process then in the transfer of in- and output. For the biological wetware the possibility of divergence is even higher. It is unlikely that even a mouse brain will have precisely the same number of neurons than any other of its species, and in the case of humans the probability turns into a guarantee. In addition we have the hundreds of thousands of proteins which do not come in exact sets either, and so any input (which in itself of course is as diverse as you please) will meet environments which are substantially different from person to person.

The process of abstract forming begins to lay down the fundamental regions of attractor domains which become the precursors to later additions, and no matter what these additions are they will be embedded in specific, origin-dependent formats. It is the first instance of progression lock we have identified at a much higher level in Part I; the initial region of stability forms the very first container. And the attractors, coming to life during these moments of growing fractal borders, turn into the prototypes of denominative thought structures and patterns we have already encountered on so many occasions.

As so often in life, evidence of something is found in its occasional absence. Two neuroscientists, Hanna Chainay from Paris and Glyn Humphreys from Birmingham in the UK, researched the circumstances around a neurological condition called apraxia [8]. It distinguishes itself by a difficulty to perform motor movements which cannot be explained by any other causes such as intellectual impairment or muscular disorders. It comes in two not mutually exclusive versions. For example, persons may perform a sawing motion with their hands, but the gesture occurs in the wrong spatial plane (ideomotor a.); or someone may hold up a tooth brush as if it were a mirror (ideational a.). Tests with patients suggest that there are several types of information involved in bringing a certain intent to a successful conclusion. There has to be an intact verbal communication regarding the object, that is a word (eg, 'saw') must be understood in its correct meaning; one also needs the appropriate set of gestures that belong to the object; and the visual representation of a still *and* moving version must be stored in a way that can be retrieved productively. In other words, static as well as dynamic form information have to be matched with presemantic representations for the proper hand movement to be made. The article mentions an "action input lexicon" and an "action output lexicon", conceptualised faculties that deal with the arraignment of "abstract programmes for learned motor responses to objects". In terms of our nomenclature such labels should have a familiar ring to them. If a motor response is learned it implies the storage of its representational format which would be affinitive to those objects with which it has been associated during the interaction with one's environment. That there are static and moving versions, whether for objects or gestures, similarly points to the existence of specific types of abstracts which we termed object and functional (OAs and FAs). Essentially they are no different from each other; we merely chose the labels to make it easier distinguishing their role at a higher level of conceptual domain. The technical detail of their formation (ie, the importation of sensory data which become associated with existing structures) is the same.

The demonstrated need for a cooperation between semantic content for example and its visual counterpart (with reference to the same object) leads the writers to posit a "convergent route account". They cite a simulation through an "energy minimisation network" that creates a response based on some action output lexicon from separate semantic and perceptual representations. In other words, the network holds a verbal and a visual set of data driving it towards a stable state which acts as an attractor. The final state can be seen as a basin of attraction with reference to the initial states.

Chainay and Humphreys propose that the same concept be transferred to the human brain, where one part of the representation acts as a vector space which is driven by the other part towards the final state, ie the contextually correct articulation of a movement. Comparing their description with the stable, periodic and chaotic integer movements in OtoomCM the similarities become apparent. If the outcome depends on the affinity relationships between attractor domains, it stands to reason that more intact affinities produce a better result. This is born out by apraxic patients who "perform better when gesturing to a visually presented object than when gesturing to a name only", as well as the "common finding that actual object use is better than gesturing in apraxia" [9].

The attractor domains mentioned above are only approximate descriptors; they serve to outline the situation in summary fashion. But as we have seen from the examination at OtoomCM's element level the phenomenon exists at every dimension. The more flexible a representation is to be, the further we should be able to descend in scale and still find examples of fractal borders (indeed, the whole idea of fractals is their scalability). Thus before an object can be moved, there has to be a representational form of 'move' in the first place.

The need for basic building blocks in the creation of OAs and FAs can be seen in the findings by the linguist Eric Pederson from the University of Oregon. He examined monoliterate, biliterate, and nonliterate Tamil and English speakers in southern India by showing them complex line drawings and other images which either faithfully reproduced the original shapes, presented them in reverse, or not at all [10]. The subjects were asked to identify the shapes in the altered versions. It turns out that 86% of biliterates made correct identifications, whereas mono- and nonliterates were successful only 59% and/or 35% of the time. Since biliterates possess a knowledge of both the Tamil and Latin alphabets, Pederson infers that the higher success rate is due to the presence of reversed characters in English (such as 'p' and 'q', 'b' and 'd'); Tamil writing does not have such characters. Familiarising yourself with the concept of reversal through your daily diet of words would open you towards a similar idea at other tasks. It makes sense: once the FA of 'reverse' has been stored, in any further decision-making process it can be evoked by appropriate affinities. If it does not exist that option is closed. Note that for the affinity relationship to work the actual content is irrelevant, since it is the underlying functionality which matters.

Other indicators for the distinction between object and functional related domains come from the study of aphasia, the loss of language ability. In one variety, Broca's aphasia, the loss of function words prevents the speaker from articulating articles, pronouns, and conjunctions which serve to connect the parts of a sentence [11]. In another context lesions in the brain can affect either the declarative or the procedural memory in rats, as experiments have shown [12]. Declarative memory concerns the retention of object representations, whereas the other deals with movements and actions. Or take the brain's activity patterns during sleep. Several periods have been found, the most distinctive being REM and non-REM sleep. In the REM (rapid eye movement) state the brain is more active which manifests in the movement of the eyes and limbs (hence the name). Using positron emission topography (PET) one can see the active areas shifting across the cortical regions as the individual changes from one period to another and REM-type sleep gives way to non-REM sleep [13]. A similar deduction can be made from the study of visual object agnosia, in which patients are unable to recognise an object when stationary but have no such problems when the object starts moving [14]. Clearly, certain areas deal with movement, others do not.

The question of associativity between ideas has always occupied researchers into mental phenomena. To find out why one particular thought triggers another and why some other reference remains bare leads to many fascinating contemplations. While the reasons for their curiosity are understandable, a major stumbling block towards a clearer view is the very mechanism they are trying to unravel: the self-referential nature of their own mind. If I want to know why it is that 'red bus' brings up 'holiday' I might dig into the details of my past and discover some connection between bus tours and my itinerary. Yet I could always keep asking 'why' at each and every junction of my line of investigation, and ultimately it is the reason why I am asking 'why' which needs to be explored—but how can I do this if the tool I am looking for to uncover the secret is the very one I can't recognise?

An example of how this situation manages to trick people is the discussion surrounding the nature of perception and its ongoing interpretation. Take visual illusions—what makes them illusions, and exactly where does the visual input stop being a visual input and become an interpreted object, albeit a wrong (?) one? In chapter 7 we saw one drawing which is open to misinterpretation because the mind uses experience to expect circumstances to be similar (see Fig. 7-2); the 'circumstances' being affinity complexes created under particular conditions. Since they exist prior to the new experience, the current process follows these leads when constructing the latest association. Only subsequent input (whether from outside the mind or from outside the specific complex) can possibly change the associativity because up to that point no trigger to do so exists. Either the latest input negates the result (such as in Fig. 7-2, where the removal of the bar shows what the lines really are), or it supplies additional information (ie, input by further associations) which readjusts the conceptual relativities, as in Fig. 15-4.

Fig. 15-4: Opera Queensland face to face with conceptual relativity: the singing hippo?

Once again we are back at latency. While there *are* possibilities of affinity they can only be realised if and when an appropriate complex emerges; and because any articulation relies on the definition of the data it is meant to re-represent, the domain under that definition becomes the final arbiter of the output's quality.

Consider the debate about representation at different levels of sophistication. In his article about the "theory-ladenness of perception" Athanassios Raftopoulos looks at the main positions and comes to the conclusion that a more precise categorisation is necessary [15]. When the lines of a drawing get transmitted via the optic nerves to the brain we can assign such data to a basic level and label them accordingly. Once they become a coherent object consisting of a platform and four protuberances we still manage to find a name. By the time we reach the concept 'table' the conceptualisation is not that easy anymore. Although the participation of data peripheral to the vision are accepted, their definitive placement within a meaningful category that is at once descriptive of the object yet also tells us something about their functional role in turning the thing into a *table* is far less assured. This is the reason Raftopoulos decided to differentiate between sensation, perception, and observation. But even here we run into the problem of accurately bounding these categories so that their function can be settled once and for all. He correctly points to the two opposite perspectives used in such cases: a bottom-up approach when considering the fundamental representation with a view towards further processing, and the top-down alternative when looking for the influences of higher-level concepts in classifying the input according to experience. The conventional view which insists on seeing any one of those types of manifestation as some object-specific version of the thing itself leads to nowhere when it is called upon to articulate the inherent nature of them. How can it do otherwise—its own vocabulary is intimately tied to the nomenclature employed at the highest level of cognitive interpretation, the very realm it does *not* inhabit.

On a somewhat more esoteric level the uncovering of fundamental system dynamics in neural networks (biological and artificial) has left the previous symbolicism vs connectionist stance behind and concentrated on cognition as an essentially dynamic phenomenon. In this sense 'dynamic' refers to a changing quality not only in terms of the system itself but also as far as its functional characteristic is concerned. As Chris Eliasmith from Washington University explains in his paper on the dynamicist theory of cognition, the attractive feature of the new approach is its ability to model and/or conceptualise an internal state of a cognitive event through recourse to such things as state space, attractors, and the topology of its geometric parameters [16]. In the end he concedes that a true description has not been realised, although Robert Port from Indiana University in a later article on the same subject seems more confident [17]. Such optimism for the future is not entirely misplaced. Whether neuroscience can deliver a more detailed picture of neuronal dynamics and match them with the mathematical contingencies of vector space remains to be seen. Whatever the success, for a thought to be defined other than by its content, the difference across the various layers of functional scale need to be addressed.

Our object elements in the latency equation do whatever their characteristics allow them to do; form molecular reactions in the brain's neurons, or serve AL in OtoomCM. In any case, they can only then contribute to the formation of meaning somewhere down the line if some sort of distinct grouping becomes possible. As mentioned already, it is the grouping via affinity relationships that determine an ordering of representations, and it is the ordering at their respective domains which many processing steps later leads to some meaningful mental image. Just as the erroneous articulation above produced the misleading categorisation of neuronal phenomena, so did it pretend an absolute colour scheme in our visual world when there really is none. Semir Zeki writes about the perception among most scientists that our mental representation of colour is due to the perceived light frequencies of the visible spectrum, and those frequencies only [18]. To give the scientists their due, their assumption was based on a comprehensive analysis of light and its subsequent decomposition, but it also derived from the idea that our cognitive content is object-specific and nothing else. However, as Zeki points out,

> Slightly less ineffective was the work of the master of Gobelin tapestries, Chevreul, who emphasised that the colour of a patch is much influenced by the colour of the surrounding patches, a fact known to every artist. Scientists took note of it, paid lip service to it and then promptly forgot all about it.

That there is no real 'colour' in the world is not disputed these days [19], but one could go as far as to say that there is no real  in our minds either. Rather, they are sensated output, overlaid with linguistic descriptors, and referenced through constant feedback which ensures that I won't call these rectangles brown, yellow, and grey (but a colour 'blind' person might!).

If a concept has its source in visual input but is subject to other associations as well we would expect its pathway through regions that suggest additional influences. Such instances have been found. The optic nerves, on their way from the retina into the rest of the brain, pass through a section referred to as the left and right lateral geniculate nuclei (LGN), since they occur on both sides [20]. An analysis of the area's nerve strands shows that only 20% of its volume actually comes from the eyes, the rest is connected to the primary visual cortex, a major visual processing centre located at the rear of the brain. As the latter is further enmeshed with other sections it becomes obvious that the nature of retinal data undergoes a significant modification along its path.

Another difficulty that arises from the mismatch of conceptualisations concerns the nature of negation. It leads to the tendency of regarding some ‘A’ and its ‘not-A’ as both standing for the same degree of meaningfulness. While the creation of the complex ‘A’ may have been due to a lengthy and involved process, even imbuing its re-representative version with the same profundity is not justified, let alone its opposite. As we have said, the formation of affinities occurs from the most elementary level on upwards, and the experience of ‘not’ would have to arrive as soon as the mind becomes capable of input. No matter whether articulation of some sort takes place, the general sequence of [‘now this’] followed immediately by [‘now not this’ ‘now that’] would imbed itself into the earliest affinity complexes (the square brackets are meant to indicate the input per same time unit). Even the infant in the first stages of tactile exploration will have a ‘not’ in their neuronal network as soon as one sensation gets replaced by another.

To demonstrate the sheer simplicity of ‘not’ refer back to Fig. 9-1. Reading the two statements we know firstly there is a box which contains several items, one of which is represented visually only, and secondly that someone suffered a condition which again was referred to in visual terms. Suppose none of the items in question had been given a label by the reader but exist solely on the basis of the visual input; that is to say, the only way they could be re-articulated is to point at the image and say “that’s the one”. Now suppose we are told that these two things are *not* what was meant; what exactly has changed in our mental re-description of the items? Nothing—all we do is think “there is a box which does not contain [...]”. When Jung therefore talks about the “inner polarity” of the psyche (see chapter 10), the opposing forces in one’s mind, and conceptualises them in terms of archetypical phenomena he was probably closer to the point than his religious mind would have admitted. For him they reflected a spiritual truth, associated with the paraphernalia of ungrounded TSs. Technically speaking they are the result of the associative property of a lowest-level abstract, assigned to other complexes in abundance.

On the other hand, to contemplate the notion of ‘honesty’ for example and to equally understand ‘dishonesty’ is a different matter altogether. Here the two ideas are linked through a high-level articulation that makes certain parts accessible to each other. If they be seen as opposites they only do so because of multi-faceted abstracts bridging two separate groups. The mere utterance of the word ‘not’ is not an indication of their presence.

The ease with which the addition of a ‘not’ can form a suitable complex without having to rely on a substantial evaluation of what its real counterpart entails, leads to the building of personal and cultural profiles. There is hardly a nation which has furnished a truly comprehensive analysis of what it means to be its member—but they all look around them and know perfectly well what they are *not*, as Samuel Huntington has observed in his *Clash of Civilizations* when writing about the need of humans to define themselves [21]. Every individual and every group has their identity, but the advice of sages to “know thyself” lives a lonely existence.

From the very first beginnings of container formation the mind evolves associations which obey the demands of affinity relationships. As we have seen with OtoomCM the content is less important than their functional characteristics. There is no inherent law that says “combine ‘red bus’ with ‘holiday’”, but if one day I pack my bags and head for the mountains and the carriage I am in happens to be a red bus, the connection gets made. The need for affinities is met every time, and while the relationship is upheld anyway it is up to the conscious self to make sense of it all. Tor Nørretranders in his book *The User Illusion* cites an interesting experiment which illustrates that very point [22]. His introductory remarks are worth quoting:

> The lessons we learn from studies of split-brain patients is that the self or the *I* (as we call it in this book) lies like crazy to create a coherent picture of something it does not understand in the slightest. We lie our way to coherence and consistency we perceive in our behaviour.

These "lies", he goes on to emphasise, are a form of self-deceit.

In the experiment two pictures were shown, one to each eye of a patient whose corpus callosum, the bridge connecting the two brain halves, had been severed. The left eye and thus the right side of the brain saw a snow scene, whereas the right eye and left half saw a chicken claw. At the same time the person had to choose one of several picture cards that matched what he saw. When a shovel was selected with the left hand (ie, the right brain) and he was asked why, the answer was (articulated by the left hemisphere), "Oh, that's easy. The chicken claw goes with the chicken and you need a shovel to clean out the chicken shed." The right half did not see what the left half was seeing, but the left half nevertheless observed a body movement induced by the other but needing to interpret it on its own terms, and so it connected the act within the context of 'chicken'. Particularly remarkable was the readiness and certainty with which the answer was given. The connection was all so obvious to the conscious mind, but in terms of the exterior facts it was plainly wrong.

If we consider that affinities are not only enacted through singular entities but can also be the result of sets of multiple representations interacting with each other, the abstractive level such sets inhabit becomes significant. Therefore, since abstracts at various levels of complexity are only capable of accessing others at similar levels, their resultant interaction and/or computation will be at appropriate levels also. And as the above experiment indicates, once a formal system of abstract-complexity has been established it will be tailored to the degree of complexity a particular interpretation requires. The implications are profound, and in the next chapters we will have ample opportunity to observe them.

As the split-brain experiment shows, the conscious awareness of one's internal mechanism is largely a myth. I say "largely" because it does open itself to analysis once the variants of its functional scales are taken into account. This difficulty of finding the right descriptors for activities which do not operate on the same cognitive plane where words are made and undone, lies at the heart of what has come to be known as the 'grain problem'. In one way or another it has been with us for centuries, and so I think it is worthwhile to remain there for a few moments. Michael Lockwood, in his contribution *The Grain Problem* to a book on physicalism in general, uses the metaphor of a note played on a violin [23]. That comparison is somewhat ironic, given that he moves immediately to its manifestation as an electrical potential amongst "billions of neurons of the auditory cortex"—the very instance which is the source of puzzlement from the start. Once again: it is the conceptual functionality which first and foremost is the problem; rather like trying to take apart a shifting spanner using that same spanner. The irony lies in the essential usefulness of the comparison once we stay with the instrument and leave the brain with its neurons aside for a while.

Suppose then the violin is the object under focus. It basically consists of the wooden soundboard, four strings tied to a tailpiece at one end and wrapped around pegs at the other, which is held in place by the fingerboard. If the bow is moved across a string, making it vibrate, a wave is generated by the soundboard (let's not get lost in the detail of all the other vibration-generating parts). Under the auspices of the grain problem we observe the vibration and wonder how a bunch of horse hairs can make a wooden body change its shape in a periodic manner: just what is the relationship between horses and timber? At this stage the sound engineer will tap us on the shoulder and point to the bow as it moves, no, grates across the string, making it move backwards and forwards by its resistance. That movement in turn transmits to the bridge, another wooden piece keeping the strings tensed above the violin's body. The bridge affects the soundboard, and so it finally vibrates. Note I have desisted from labeling anything a 'note', a 'tune', or some such thing. We do not interpret any event, merely record its presence. Therefore it is wrong to speak of a wave-inducing movement of the bow, just as it is misleading to regard the bridge as the violin itself, as much as it is a part. Being methodical we have no difficulty restricting the bow dynamics to the bow, the bridge dynamics to the bridge, and only when it comes to the soundboard will we allow ourselves the beginnings of a certain idea that has to do with auditory experience. We can do all this because the violin is an external object and our mental faculties encompass the situation from start to finish.

Let us now move the whole scenario across to the grain problem proper and to do this let's pretend we are a member of an alien race charged with observing the behaviour of humans. There is a sound therefore (we have ascertained that); our apparatus reports a dynamic in the auditory organ of our specimen (equally confirmed); and moments later there is a broadbanded, wave-like neuronal transmission across strata of brain tissue (we can see that too).

Our scientists analyse all these observations and describe an ongoing process of data transfer from one location to another, spanning several media and consequently undergoing corresponding changes in the inherent characteristics. After work we socialise with the earthlings and as we mention our research they suddenly tell us about some major trouble they are having. They call it the 'grain problem'...

To be sure, Lockwood does enumerate a number of phenomenological entities linked to the brain's activity, and offers them as the multi-variant background to the issue. But then, like snatching defeat from the jaws of victory, he still places them as obstacles before us as if their very nature is the source of our difficulties. Neuronal firings, brain states, sodium and potassium ion switches, they all are held up before us while we are asked to equate their function with the personal experience of a sound or a flash—something we naturally can't do.

If I was to re-write the earlier paragraph by substituting 'member of an alien race' with 'human', at what precise moment would readers lift their hand and call out, "Stop! We've got a grain problem"? As an educated guess I would say, as soon as we try to associate the wave-like excitation with a subjective sensation. It seems the semantic threshold that needs to be crossed consists in managing the conscious awareness of a particular dynamic (non-personal, electro-chemical, quantum-physical even), with the articulative arsenal of an automatic sensate experience, when neither its origin nor its intrinsic mechanism are accessible to its reach. They cannot be accessible because it is those elemental processes which form its constituents—just as an eye cannot see its own lens, nor a stomach digest its own sac. At the beginning of this little diversion I qualified the description of conscious awareness as being largely a myth. Our means to overcome this barrier lies in the ability of the mind to form a sufficient abstractive distance; it is this which allows us to side-step the straightjacket of conceptual absolutes on a rational basis. Throughout history wise men have urged us to practice inner detachment through meditation. The essential reason is not escapism; rather the ability to shrug off the intense ties which bind us to what we consider to be the 'real'. As long as we understand a sensate experience as just that, namely a sensate experience, our affinity domains are free to unfold their potential.

We could apply the same angle to OtoomCM. It is evident that a 500-odd node matrix will not have much to offer except a few circles; yet at the same time there is a process going on which generates output. If we keep adding more and more nodes to the system until it is capable of language as well as contemplating itself, the steady evolution of node numbers on one hand and its intellectual capabilities on the other can be followed along. It becomes obvious then that as so many nodes are responsible for cognition per se, there is equally a given amount which participates in a certain articulation. Therefore, just as a subset of the cognition complex does not entail cognition, and the subset of the articulation complex does not entail articulation, an utterance about this or that is not comparable to the adjustment process traversing a few connections. Hence for the utterance to demand equal functional status with an affinity complex and vice versa does not make sense.

At the time the firing mechanism of nerve cells began to be investigated the image of dynamic systems had still been sketchy. Neural networks, against which the viabilities of brain theories were sounded, made more sense to mathematicians and computer engineers than biologists. When Vincent Kenny wrote his paper on Humberto Maturana in 1985 and how that man saw living systems construct their own reality, the notion of an interdependent continuum moving itself through different stages of complexity and thus definition presented a challenge to the mainstream [24]. From conglomerates of cells to organ development to cognitive processes, they all seemed to confirm the principle of emergence as a function of an autonomous system, that is to say a group of modules and sub-modules which drew their activities from a rule book of information that in turn got fed back into the activity domain. As the system grew in sophistication, so did the applicability of the rules, and likewise would a manifestation impact at different levels of suitability. No more profoundly was this observed than in the field of language, a product as much derived from instinctual sensing as from rational consideration. Kenny refers to Maturana's take in the following way:

> At this point it should be clear that for Maturana there is no objectively existing reality. Whatever reality we experience it is one of our own creation, i.e. we bring it forth through our operations of distinction. For Einstein, scientific theories were seen as the free creations of the human mind which we used to explain the world—but for Maturana, what needs explaining is precisely this "free creation of the human mind", i.e. the way in which the observer brings forth his world. Thus, central to Maturana's theory is the ontology of the observer. "Languaging takes place in the happening of living of the observer. To explain languaging, I must explain the living of the observer". Languaging is therefore Maturana's instrument for explanation and also his central problem.

The conceptual ambience created by paragraphs such as this one comes close to the mindset of philosophers who strain to import their ideas into the structural framework of everyday experience. Maturana himself took care explaining his theory of cognition by moving from one stratum of biological organisation to the next higher, eventually arriving at the cognitive processes themselves [25]. So powerful was the concept of interdependency for constructivists like him that the role of each separate element (eg, cell dynamics, thought patterns, word complexes) was relegated to one of an almost coincidental participant. In his introduction to the article *Brain, Language and the Origin of Human Mental Functions* he emphasises that mental phenomena

> do not take place in the brain, nor are they the result of a unique operation of a human brain, but arise with the participation of the brain as it generates the behavioral relational dynamics that constitutes language. [26]

The development of language (as one example of a biological manifestation) in tandem with its environment provides for an interesting perspective in terms of the functional broadband of evolution. As Maturana says elsewhere in the paper, the

> internal dynamics of that nervous system will give rise to a succession of actions that cannot but be logically or coherently connected between them in the context of the historical circumstances of the realization of the living of the animal.

A view that is shared by the biologist Jakob von Uexküll, although an additional constraint is added in terms of the organism's subjectivity. Thus activities such as perception, food intake, etc, are performed according to their usefulness to the entity. Alexei Sharov, from the Entomology Department at Virginia Tech in Blacksburg, Virginia, finds that "Umwelt [von Uexküll's German word for environment] is not a set of objects in the environment but rather a system of signs interpreted by an organism" [27]. His text has been written to show how the model can be applied to semiotics, a field that deals with the role and interpretation of signs. The context is significant in itself, because the notion of a 'sign' does not exist for the purpose of identifying an item as such; rather it represents the object through the purpose an observer has designed for it. If one's environment therefore is a symbolic representation of one's subjective interpretation of the world, permitting nothing further into our perceptual reach than what our constitution condones as useful to us, we could summarise the perspectives of Maturana, von Uexküll et al, as a model of the world which serves our need to have a model of the world.

From the jarring images of the grain problem, to the embedded harmonies of Maturana's constructs, to Sharov's valuable signs, we can detect the results of self-serving affinity relationships. Their product, a complex of similar phase states, acts upon its domain because this is where the congruence works. Whether it ultimately makes sense to another formation of abstracts which in the end operates in terms of its own propensity, is neither here nor there as far as either standpoint is concerned. The experiment with a split-brain person mentioned earlier was only a dramatic instance of something which in principle happens all the time: there is an affinity-seeking complex of abstractions which is in the process of generating a congruent subset; in the process other domains are being accessed in terms of their affinity potential such that their content serves to enhance the initiator; if in due course the additional content settles within the potential of the existing affinity relationships the complex will grow, should it establish a congruence with some of its sub-domains at the expense of the whole the original complex will be diminished and assume a modified character. In Part I we referred to such developments as horizontal and vertical PAT chains around illusion trees, giving rise to i- and c-PAT-ls.

The scenario painted here may sound as if our mental lives are procedures that are predetermined by a master algorithm and therefore predictable. However, one need not necessitate the other. As OtoomCM has shown, despite AL's formalism the outcome across any given set of integers cannot be foreseen, especially should the number of elements climb towards the proportions of biological systems. For humans, not only does the network's complexity assure a virtually infinite range of outputs, but the selective input of consciousness heightens the optionality of bifurcations even further. But more of this later.

And in any case, not all speculations about the origin of life include cross-level abstractions. Liane Gabora from Brussels Free University presents one conjecture which handles the functional scales with care. In her paper on autocatalytic closure she starts with the previously advanced idea that the origin of life is based on the self-organization of autocatalytic polymers. She delineates from there:

> Much as catalysis increases the number of different polymers, which in turn increases the frequency of catalysis, reminding events increase meme density by triggering symbolic abstraction, which in turn increases the frequency of remindings. [28]

'Meme' in this case refers to any single idea or a train of thought one might have. As the memes begin to proliferate through continuous access they undergo a phase transition which enables the mind to create pathways to their associations. She makes a distinction between episodic and mimetic memory (which she turns into "memetic" to signify its connection with memes), two versions of data retention which are linked to the complexity of the nervous system. The former is restricted to a representation of events, whereas the latter is capable of abstracting from such representations and putting them to more sophisticated use. Her description of the functional scale in terms of the abstractive processes that early humans would have gone through is interesting in the light of our by now familiar nomenclature:

> Arguments for a reconstructive view of retrieval notwithstanding, highly abstract concepts that have been used thousands of times, such as 'space' or 'equal' or 'is', would be unlikely to emerge from memory retaining the associations of any particular usage. Thus it seems reasonable to begin with the working hypothesis that subsymbolic processing predominates for low-level, parallel, automatically-generated cognitive phenomena, and that symbolic processing provides a satisfactory approximation for many high-level, serial, consciously-directed aspects of cognition. (Creative processes may draw heavily on both.)

In our terms, the deconstruction of complexes, spurred along by the potential of affinity relationships, will sooner or later retrieve a useful sub-domain which then gets associated with another complex and now establishes a domain of its own. In Gabora's words, when a member of some tribe sees a "rotting, striped, bumpy, red gourd" it reminds her of the "striped, bumpy, yellow gourd her brother used" and so she proceeds to put it to use. When she discovers a leaking hole in the gourd and at the same time encounters the intestine of a recently killed water buffalo, she makes the connection between gourd = container and intestine = container through the transfer of abstracts from one mental representation to another (I have shortened Gabora's account somewhat).

In the article the basis for the entire development are the polymer molecules in neurons. When polymers act as catalysts they attain the ability to replicate. In an autocatalytic system the replication process is induced by its own molecules, and once each polymer's replication is catalysed by another member of the self-same set we have catalytic closure. If we assume, as Gabora does, that input is represented as a particular state of a neuronal complex, which in turn is a function of a certain protein formation derived through catalytic closure, then it can be shown that the probability of generating new complexes increases faster than the number of domains generated. In the case of complexes (the result of polymer catalysis), the lower their dimension (ie, the higher their generality), the greater the number of retrieval paths. To illustrate the exponentially increasing potential for successful access Gabora uses a fully connected system, and compares such a network to an n-dimensional structure (being mindful that memory is in fact not fully connected). Viewed as a trend however the results are still interesting: the more lower-level abstracts there are accessed by the mind, the more abstracts will be evoked, and the more active a domain happens to be the more likely new abstracts will form in its vicinity.

We have seen a similar situation with OtoomCM. Due to its depth-first organisation access between matrix nodes is similarly restricted, although a change in the connectivity and modification of depth traversal are directly related to the degree of diversity in output. While the nodes do not function in terms of chemical dynamics their integer ranges in conjunction with the adjustment algorithm produce particular trends towards phase states which either enhance or diminish potential affinities with each other.

In the computer model the limit to the realisation of affinities is based on the values of the integers which get collected along the node traversal; it is their respective distance to the reference value which determines their movement. For Gabora it is the probabilities inherent in the complexity level of the polymer conglomerates and their abstractive progeny. On the cognitive level the ultimate constraint is the applicability of affinity complexes, a situation we have already touched upon in Part I when we talked about latency within the context of the effectiveness chart and more generally earlier on in this chapter. Now, with more talk on abstracts behind us, it is time to briefly return to the topic.

In the early days of computing the mathematician Alan Turing came up with the reason why computers cannot calculate everything. Known as the 'halting problem' it centres on the fact that for any machine to know beforehand when to stop a series of computations it would need to know the results of such computations [29]. But this is impossible since the series has not finished yet and so any instruction regarding the end point cannot contain the knowledge necessary for its definition. Consider the relevant abstracts in detail: if there is a loop which keeps adding the number '5' to the previous result then it will go on forever unless something tells it to stop. For a computer that order could be "stop when the result has reached '100'". But I am only able to write such a command because I know that sooner or later the algorithm will come up with answers that approach and surpass the value of '100'. In other words, although the computer is doing the calculating, my mind has predictively evaluated the situation and acted accordingly. In cognitive parlance we are dealing with two levels of abstraction. The first and lower level concerns the parameters of the operation, and the second and higher one encompasses its rule set, its principles.

Suppose we heighten the complexity of the task and use an algorithm similar to AL (see the Appendix). Although simple to execute and returning predictable results for most integers, there is the odd occasion when this will not be true. In the first case I can collect the relevant numbers into a set and apply the same abstractive levels of articulation as has been done with our additions just before; that is to say, L_1 (abstraction level 1) deals truthfully with the numbers, and L_2 with the rules and their outcome. In the second case however I need an L_3 to describe the situation for those numbers where the results are periodic. Just as L_2 has been the administrator for L_1, L_3 is now organising L_2 and indirectly L_1. Yet periodic sequences are once again a subset of outcomes which also contain chaotic cases, and in order to include them I need a still higher abstraction L_4. Note that none of the expressions describe an actual value; they all refer to types.

Is it possible for a system to contain only L_1 and implement L_4? Clearly not—there is nothing in a random bunch of integers which speaks of the kind of scenario that produces a line such as in the third graph of Fig. 14-5. Let's turn this around: suppose we know L_4, would we be able to put together a collection of integers such that a result corresponding to the rules at L_4 emerges? At first glance the answer might seem 'yes', but the halting problem should tell us that it is not so. In principle (!) the problem is this: there are two types of abstractions, where type 1 represents the parameters to type 2, and type 2 represents the rules applied to whatever type 1 stands for. In the halting problem type 1 are the numbers themselves, type 2 the rule 'add x to y' ('x' and 'y' being members of the set expressed under #1), but the command to stop the loop would be of type 3 since it is another rule that controls the adding process; without it the computer cannot halt the loop. Thus type 3 is a rule which has type 2 as its parameter, just as type 1 represents the parameter for type 2. For a system to apply a rule productively (ie, the rule is activated and something definite happens as a result), it needs a set of parameters in addition to the control mechanism. To return to our question then, having something like L_4 could only be made to work if we had access to something like L_3, and for L_3 to act like a guiding principle we would need an L_2, and so on.

In the halting problem L_1 is expressed as the data set and L_2 as the rules applied to the set. Words such as 'data' imply parameters of concrete dimensions, but when transposing the issue into the context of abstractions such a label can be misleading. Once we are dealing with affinity complexes there are no values in measured terms, only relationships that either get realised or not. In Liane Gabora's description of emergent thought processes one thing led to another, but considering Turing's point there are certain restrictions. Using OtoomCM as a means to follow the transmission of data (easier here than in a biological network) we can observe the creation of an affinity complex when a set of integers becomes modified through AL, somewhere at the periphery of a virtual connection tree. Having attained a certain degree of stability, the adjustment process keeps the number range across the complex within the bounds of the domain's latency. Via the links of the connectivity plan further associated complexes are built which now co-exist with the established set. *If* there is now input which can be translated as affinitive to those representations already present in the system (for that the environment has to be capable of a meaningful response to the system's output), new complexes will come into being which can and probably will be associative with the current state but, provided the network possesses sufficient capacity (ie, enough number of nodes, enough connections per node, plus enough depth for traversal) they will exist as separate entities because their data transmissions followed slightly different paths. At this point their status is comparable to that of their predecessors, but suppose the process of adding new complexes in terms of existing affinities continues. At some stage the affinity relationship changes from an interaction between so many domains of a current order to one comprising an affinitive domain with a *set* of domains of the current nature. The relationship can still be evoked, but the format has turned from 'many to many' to 'one to many'. Note the importance of affinity relationships; once that connection is broken the relevance from one set to another becomes lost and the process turns unproductive.

To summarise: complexes are formed through the creation of affinity relationships. If they are stable enough they attract other complexes which form related domains. Provided the system has sufficient capacity and given a meaningful response from the environment even more complexes are assembled which form pools that allow a selective apportioning of affinities. Once distinct domains emerge from a repeated exposure to relevant input and its consequences, they can become affinitive to existing sets. With reference to its relations, such a domain acts as their abstraction.

If these sentences sound too trivial, consider that in the case of humans we are talking about a neuron mass in the order of 100G. That same mass requires years of constant tuning to achieve the abstract formations we acknowledge as 'common sense'; and for wisdom decades are needed (if it arrives at all).

The descriptors 'domain', 'complex', 'set of' etc apply to types of phenomena only, and their effective definition concerns the functions they perform. As long as the criteria at a particular level of enactment are met, the operation is successful. It will continue until such basis falters to supply the necessary ingredients; from then on another configuration will take over. The actual content in all this is quite irrelevant. While content might be seized upon by an observer (and thereby acting as input to that other affinity-constructing system as well as possibly providing some feedback), to the system under focus it will only then become an issue if it can give rise to a derived functionality which is able to influence the dynamic. In the chapters to come we will have ample opportunity to confront the sheer inevitability of this state of affairs.

Over and over again we have come up against the appearance of stable regions which served either the development or deconstruction of affinity complexes. These regions are characterised by a degree of permanence which sets them apart from temporary clusters, and throughout time they have been labeled as 'memory'. From our point of view their presence is not so surprising anymore; the formation of functional islands is something to be expected by now. Yet their very mode of existence has certain ramifications which should be listed, especially since in many cases they are anything but obvious.

The distinction between short- and long-term memory has always been made and for good reasons. Test runs of the computer model demonstrate the difference between short and prolonged input and also between a lower number of repetitions of depth traversal vs a higher one. Clearly, the longer the process of input embedding lasts, the more significant the modifications along the traversal routes are going to be and the more profound their subsequent influence upon future imports. To turn this around, for data to be incorporated such that their retrieval can occur effectively later on, the installing process must be given a certain amount of time. Since the crucial factors are the affinities met along the way, the traditional view of set time scales for short-term memory on one side and its counterpart on the other do not hold water, as contemporary research into memory consolidation has shown [30].

If the embedding of a particular representative complex depends on its associative properties with neighbouring domains, not only will the result reflect such properties but any access later on ought to occur under their auspices as well. This is important: just as an incoming data stream gets dispersed in some representative form amongst the clusters of associations, to retrieve those representations a way must be found to get close to the clusters. Since the system operates through affinity relationships, the greater the affinity of the seeking complex is to the target, the closer it will get. Note that the seeking complex is never synonymous with the target, whatever a close relationship may imply (hence I won't be trying to remember that I left the keys in the *car*—I know they are there). Refer to our remarks about abstraction levels not so long ago; for an abstraction to be productive there needs to be its parameter and its rule. Thus a memory recall functions either through the use of a parameter looking for its rule, or a rule looking for one or more of its parameters. Both are linguistic articulations based on the same functional contingencies of the moment, so the difference is superficial.

What is less superficial are the modifications the surrounding domains can undergo once a retrieval is in progress. Studies in memory recall have furnished surprising results. In an article about memory and its role in the making of analogies, Boicho Kokinov and Alexander Petrov (then both at the Central and East European Center for Cognitive Science in Sofia) describe some of the instances that accompany a conscious access to stored information [31]. They compare the operation to the reconstruction a paleontologist might perform in order to put a shape to an animal long gone. There are pieces of skeleton, extraneous information as to how these could fit together, and of course the inclination on behalf of the paleontologist to arrive at a certain form. They write,

> The paleontologist might make a second re-construction which could be different from the first one since something new was learned in between, or some fossils have disappeared, or new ones were found.
>
> The empirical question is whether such phenomena happen with human memory, and the answer is yes.

Even in what has been termed "flashbulb memory" the recall was not as complete as might have been expected. The name derives from the experience we all have when something really significant happens, for example the day John F. Kennedy was murdered in Dallas. Most Americans have some vivid memory about the circumstances they were in at that time. Nevertheless, even in those cases mistakes are being made when the scene is revisited from memory. Another example,

> Nickerson and Adams (1979) tested the memory Americans have for a commonly used object such as a penny. It turned out that on average each element was omitted by 61% of the participants. Some elements, such as the text *Liberty*, were omitted by 90% of the participants. Others, such as *United States of America*, *E Pluribus Unum*, and even *one cent*, were omitted by about 50% of them.

More dramatic (at least considering the possible repercussions) are false memories based on the blending of episodes. This occurs when a certain event that has been stored in memory is being recalled against the background of other, conflicting information. Say, subjects watch a slide show and afterwards answer questions during another session, but at the same time items from the show are being misrepresented. The authors report, "With a little help from misinformation, subjects have recalled seeing stop signs when they were actually yield signs, hammers when they were actually screwdrivers, and curly-haired culprits when they actually had straight hair". It also became obvious that people sincerely believed their false memories were true recollections. Such findings have been replicated many times over and are no longer disputed.

The above refers to relatively large scale instances, that is they relate to episodes or, as we would describe them, PAT chains of abstracts. That same phenomenon also occurs on a much smaller scale where almost singular representations become substituted as the occasion arises. For instance, the word 'land' was constructed out of 'lane' and 'sand'. On a somewhat higher level, when asked to memorise a number of faces the subjects where then shown different versions which contained features of the former. Many of these new faces were recognised as authentic. In another trial by Nystrom and McClelland conducted in 1992 sentences were blended and 10% of misjudgments were based on the mixing of parts of one sentence with those of a second. Again, those who made the errors rated their confidence very highly; in this case 40% of the false recalls "received the highest possible ranking". A statement by McClelland was cited directly:

> One particularly important observation that McClelland (1995) makes based on a simulation of these data is that "Intrusions from the other sentence rush in when the most active trace provides no information" (p.78).

In the split-brain experiment mentioned earlier on we saw the readiness with which the mind replenishes missing information. Words such as 'replenish', 'missing' etc need to be understood within the context of perception, the functional requirements imposed at the time of the associative process. If a domain of abstracts exists for which the affinity potential of another domain could fulfill the role of attractor, then these dynamics spring into action and the potential will be realised. In our 30-cycle test run with OtoomCM (see Fig. 14-12) in which a particular colour arrangement put up 'blue' and 'yellow', half-way through the process some affinitive input produced 'green'; when after such input the display returned to showing the former yellow spot then the representative format of the current phase state was changed towards the earlier configuration, although slightly changed. This is one reason why events such as these first and foremost ought to be understood in terms of their functionality and not their content. It is not the exact content which is reproduced here, but a replay of an earlier affinity relationship which, given the medium of representation, happens to be a similar version of coloured blotches.

Regarding the human environment, the same associative reorganisation occurs with entire thought patterns. Generic knowledge, that is memory holding blocks of past information, also becomes subject to alteration if a subsequent situation evokes the need to fit both into a harmonious whole. Studies have shown that people "systematically reconstruct their past in order to fit into their current self-image schema". For example,

> Sulin and Dooling (1974) had their subjects read a brief paragraph about a wild and unruly girl. Then in one of the conditions they mentioned that the name of the girl was Helen Keller, whereas in the other condition they called her Carol Harris. Later on, they tested the rote memory of the participants for the sentences of the story. The test demonstrated robust false recognition of a completely novel sentence—"She was deaf, dumb, and blind"—in the first condition but not in the second. This intrusion obviously came from the generic knowledge the participants had about Helen Keller.

A similar pattern can be found when it comes to problem solving. Since the discovery of a solution depends greatly on the ability to mine past experience and use it constructively, the nature of such experiences and the way it has been accessed will influence its informative quality.

Generally speaking we can say that any abstraction complex which happens to find an affinitive counterpart will, in the course of that very process, cause an alteration in the composition of both. The inherent nature of the dynamic ensures it. The above samples of how this happens suggest an essentially error-inducing, even dangerous, mechanism. This may be so in some circumstances, but there is an even more basic interpretation: what we are witnessing are the fundamental characteristics of learning. Without the capacity of the system to develop additional, associative, abstraction complexes we as cognitive organisms would not be able to view a scenario and become reminded of similarities with something in our personal past. Our recollection may be faulty, it may have been reconstructed according to some whim or fancy, but any attempt to implement the newly-found thought will quickly demonstrate its ultimate usefulness. Feedback reiterates the entire process and coerces an incongruent phase state towards something more productive. Every teacher, indeed every—honest—student knows perfectly well that the more comprehensive an experience can be made, the truer the recall. And they know it *perfectly well* because their life had presented them with just such information over and over again in so many ways. In their minds the principle of it all may have been stored abundantly, but the principle's representational versions were certainly not. That's why they could cross-check each other and finally emerge as correct. Over two thousand years ago the Pañcastikayasara of the Jainas exhorted us to check with reality, no matter how convincing our senses or our gods might appear. The constant referral to what is rather than what we think it is makes for a rational life.

As we have seen, the process of association endures with or without our conscious awareness. If we know what is going on, and especially if the outcome is a positive one, we label it 'learning'; if the result is negative, we tend to speak of something like 'false memory'. In case of a subconscious acquisition the term used is 'implicit learning', and research suggests that the neuronal circuits used for the latter type are not the same as those used for the former [32]; something we can expect by now. In a wider sense we are subjected to this form of moulding throughout all our lives, otherwise cultures would not have come into being. Since no culture provides for the entire set of eventualities this universe is capable of, the propensity of the mind to fill the gaps between the supply of daily encounters and their abstractive potential creates a self-serving mosaic of interpretations. The emergent picture swings between the demands of the central container (the identity) and the possibilities suggested by its abstractions (the ideal)—the result is religion. Thus one could say if there must be a god, let it be latency.

When traditionalists claim that nothing will ever replace the human mind, one needs to analyse what exactly they mean by 'human mind'. On the other hand, once the massively distributed, multi-module version of an affinity-based tera-scale network has been switched into the Grid, the question becomes: what features set the wet-ware apart from the System? Leaving aside the non-cognitive administrativa, one major difference would be the existence of bias-inducing vectors, acting as a kind of catalyst on the formation of complexes. Semantically our language refers to them as 'emotions'.

A glance through those fields which deal with higher-level articulations (psychology for instance) reveals a rather vague set of definitions; the focus tends to be on the effects of emotional influences and what they mean under particular circumstances. Neuroscience, always preferring precision, looks for certain biological factors in patients (sometimes through their absence). From such research an entity called the limbic system has been identified, but even here not only is the word 'system' not clearly restricted to one distinct physical area, its operational relevance is spread across such phenomena as emotion, learning, and memory. Therefore the trend exists to move away from such a singular nomenclature [33].

For us the neurological basis for emotional activity is a side issue. While interesting in itself, it does not provide a useful contribution on the functional scale we require. As far as Otoom is concerned, the following is significant:

there are complexes which are being established because their data enjoy a degree of affinity with already existing formations;

every complex thus formed can be viewed from a historical perspective and its degree of closeness between it and the original formations can be identified;

therefore, every abstraction complex can be expressed in terms of its correlation with the identity of the system;

Turning our attention to emotions, we can add:

there are dynamics which act in a catalyst-style manner on the formation of affinity complexes;

while the associative quality of the data set is still relevant, the presence of an emotion, depending on its contextual source, will either support or hinder the ongoing formation process.

The neurobiologist can point to chemical substances which enhance or inhibit the emotional state of a person, and so 'emotion' has a concrete basis in matter. Similarly, for the cognitive scientist and/or psychologist there are thought processes which either enhance or inhibit the severity of some attitude. While we have seen the effects of affinity relationships, in conjunction with the findings from neuroscience we can safely incorporate the influence of another source in the picture. In other words, if there is a TS_1 which acts upon a TS_2 to form TS_3, and if certain identifiable factors which are not TS_1 or TS_2 affect TS_3 anyway, it can be assumed that these factors are an additional participant in the formation of thought structures. The question is, what *type* of participant.

When we observed that a complex emerges through its associative properties with another domain, in a fundamental sense we are placing the effect under a Plus or a Minus. Earlier on in this chapter we talked about the simple role of the 'not' but also the possibility of more sophisticated structures of negation. Depending on this type of quality, a denominative thought structure will therefore influence the meaning which can be derived from its neighbour so that we, the beings who articulate all that on a much higher level, talk about "supporting a view" or "dismissing a view" or whatever is acceptable in the current demographic. We also know, again from a more fundamental point of view, that the Plus/Minus quality of a TS can be traced back towards its container and from this domain further backwards until we arrive at the identity of the person, that is the container playing the final arbiter. Despite our perspective right now all this of course evolves from the inside out, beginning at the birth of the brain and continuing throughout its life.

Therefore, since the positive or negative quality of a TS is a function of its status in relation to its progeny (this status being ultimately related to the mind's identity), and the enhancement of that status (in a positive as well as negative sense) is a function of what we circumscribe with the word 'emotion', we can say that the dynamical basis of emotion is to act as catalyst for acceptable or unacceptable TSs in relation to their owner. Note that under the definition of catalysis we include not only the possibility but also the acceleration of a process.

On a personal/psychological scale this can be shown (for this exercise it is advisable to restrict yourself to something of low intensity unless one's situatedness in the world is assured). Take any thought which has a generally positive connotation; deconstruct its content so that its constituent parts can be identified (eg, washing machine → time saver, household implement, status symbol); now associate as many negative values as you can find with each of those parts (eg, time saver: creates more haste overall, prevents physical exercise; house hold implement: uses up space, costs money; status symbol: only for others, pressure to follow trends). As these steps are taken (and provided you are being sincere), observe the gradual change in the original item. Suddenly the fact of owning a washing machine appears less attractive than it used to (you might reverse this process unless you don't mind getting rid of it at that point).

Given the underlying principles it could be said that any TS can be subjected to such treatment. If there is an argument about one's personal willingness towards the extreme end (ie, the closer we get to the core of the identity container), it can only be valid in terms of two factors: the existence of a strong enough counter force, and the availability of connections to carry out the necessary associations. The possibility of people going to certain death for a cause answers the former; and the obvious intelligence of those who managed to distance themselves from their cultural narrow-mindedness relates to the latter.

It is difficult to examine the precise distribution of emotive influence on thought processes when the focus is on the emotions themselves—so far the actual amounts of chemicals involved can't be measured. However, some studies have tried to approach the problem from the other direction since a person's cognitive output can be analysed under controlled circumstances. For example, Jane Richards and James Gross from Stanford University tested for a possible decrease in memory performance if subjects were given the task of suppressing emotion [34]. When shown a film with negative emotion-inducing scenes, participants were asked to suppress their feelings so that no-one watching them could discern what they were viewing. In another test subjects were asked to view a series of slides depicting individuals in various stages of injury; the idea was to suppress one's facial expression and/or reappraise the pictures from a professional's point of view, that is concentrate on the technical/medical details only. It was found that the effort of suppressing one's emotions contributed to an impaired memory performance, although it was the verbal memory that suffered. Overall the results would support our own perspective, which is the interdependency between the emotional contingent and the representative complexes it is attached to. The difference between verbal and non-verbal memory performance is significant; since verbal articulation in general is the result of conscious processes, the researcher's view that emotional control carries a cognitive cost in those terms would be confirmed. At the same time, since non-verbal dynamics occur mostly in the subconscious realm, there either is no bearing on the result or the cost is distributed across so many complexes that accurate measurements so far further down the line are impossible. In addition the very effort of exercising one's consciousness must become visible somewhere, no matter what the specific circumstances. On this note it is also interesting that reappraisal, ie emotional control *before* a particular TS complex has settled into some state, carries no appreciable cost whereas any controls applied *after* the event needed a considerable effort.

What these tests did not relate to were the links between emotional intensity and intellectual capacity. Although the description of the test groups supplied a breakdown according to ethnicity (percentages of African-Americans, Asian Americans, Caucasians, Hispanics and Others were given), demographic effects would largely ameliorate such definitions and make the groupings rather meaningless. More interesting was the comment that the average age was somewhere around twenty years in all the tests and the participants received either course credits or were given monetary compensation. 20-years olds do not necessarily supply reliable standards on common human behaviour, unless the focus happens to be on that age group itself. Moreover, from the kind of rewards offered it can be inferred that most group members were university students, another statistical stratum not necessarily representative of *all* humanity; and the remark "monetary compensation" may or may not refer to individuals who were non-students and possibly from a socio-economic group which gets attracted by the odd job offer to while away the day. The paper offers no further details.

The evident production of substances which enable an affinity complex to become more significant together with their possibility of getting redistributed across other domains suggests a generally steady supply of these substances and their purpose of stabilising the system. A redistribution can only occur if first of all the physical attributes of connectivity and neuronal mass are sufficient and secondly the awareness of an alternative exists; even then it can only be applied if the mind can avail itself of the brain's inherent capacity. Otherwise, the fixation of a particular complex in terms of its value to the identity will be enacted, and only a still stronger structure will be able to dislodge its position. But it can only do so if the basic conditions are met, which means the challenger fulfills the prerequisites anyway and if it wins then the conformity with the self remains satisfied. As such the occurrence of emotions can be seen as the forerunner to the association-building dynamic, in other words a primitive form of intelligence. When observing behaviour forms in various demographics around the world, the inverse relationship between the emotional energy supporting some cultural act and the availability of reasoning faculties for its maintenance becomes obvious (refer to the study given above). That is to say, the less capable people are to reason themselves out of a cultural norm, the more intense the emergence of emotion at any attempt to challenge such norms. The scale can be seen again, from the individual to society at large. In the case of a single person who has limited intelligence (ie, low general mass and/or low connectivity), an issue is not decided on the basis of its abstractive delineations but in terms of the theme's emotive quality. Trying to persuade someone of this nature turns into an exercise of finding the correctly tagged pieces in the sense that the tags are going to be read according to their emotional relevancies. A misjudgment will be responded to with a direct articulation of fear, anger, etc; a good decision evokes pleasure, love, etc. Note that the actual content takes secondary position to the emotional affiliation; for outsiders this can lead to confusing, even dangerous situations.

At this stage in our evolution no demographic is completely immune from such dependency, and nowhere is emotion handled as a source of entertainment only (as certain foods or drugs would—or should—be enjoyed for their momentary subjective value alone). Instead they are used to administer our value system. For example, consider these four items:

A. cooked snake meat
B. raw duck blood
C. an ant species
D. gilparka

- now, follow your inkling and determine which of them is,

1. good for your gastric juices
2. bad for your wallet
3. good for your libido
4. good for your sweet tooth

Write the combinations down and reflect on your tastes, your life style and your heritage. In the majority of cases the choices would have been made according to what your emotions determined; even seemingly rational decisions would not have been entirely devoid of their influence.

(The answers: in certain demographics snake meat is considered an aphrodisiac; A-3. Some culinary cultures treat duck blood similar to soy sauce; B-1. If you're ever in a rain forest around the Amazon and want your damper taste like cinnamon, sprinkle it with these ants; C-4. Gilparka is a big yellowish fish living in the waters of Guyana, tastes lovely but it costs you; D-2. Honestly now—who got C-4 right?)

There used to be heart-chilling adventure stories which told of strange countries and fierce customs. In many of them featured the piranha. Enter a fresh water somewhere in tropical South America and you walk back out a skeleton. To how many people is that fish synonymous with gruesome death and atavistic nature? Below are two photos, one of a specimen caught in a Guyanese river (they call them pirays there) and the other of the damage the fish can do to a knee.

Like most stories, they had an element of truth. But for the people living and working along the river banks of that region they are no worse than rattle snakes to Texans or red-back spiders to suburban Sydneysiders. Although someone returning from there would enjoy basking in the glory of having faced ferocious beasts en masse, the fact is that on a hot afternoon there is nothing like a splash in the black waters of Guyana (you can even drink it—with or without the ants). But accidents do happen, and whatever qualities the boy in Fig. 15-5 might have, his encounter with a piray did not make him a local hero.

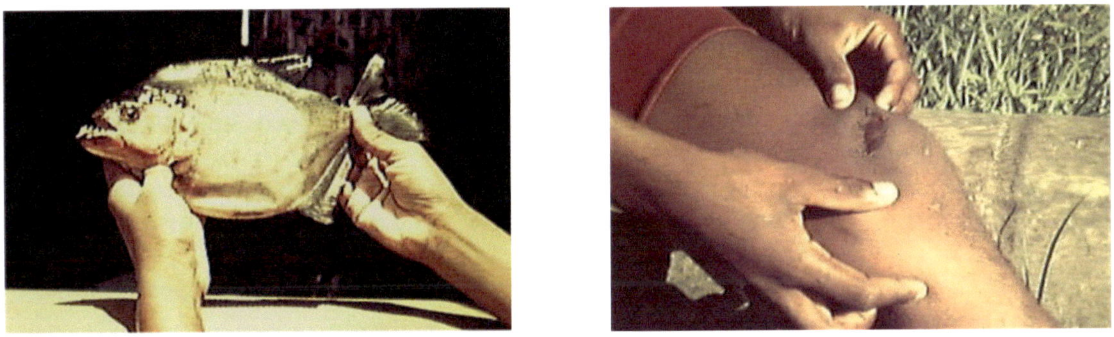

Fig. 15-5: A piranha from Guyana and its effect.

For the reader who has to swallow twice at the thought of falling off a boat on the Essequibo the emotional quality of particular TSs assumes significant proportions. Their link with the identity, grown through an evolving situatedness in a certain culture, predisposes the person to one type of reaction and not another. To the teachers who send their young charges on a regular swimming carnival in the same waters such sentiments would be utterly alien. Neither understands the other, but both know exactly what is 'right' because both are acting out their affinity relationships which had been enhanced by their emotional catalysts. Although a rational assessment of their mutual positions would shed some light on the nature of their feelings, there is a major obstacle that has to be overcome. The embedding of domains, their relational positioning, and so their eventual effect on the psyche occur largely hidden from our awareness. When we react we mostly do not know why, and any analysis is done after the main event. That what sits aside the "hidden" and the "mostly", and what is left over after the "main event", is consciousness.

There is no doubt that the investigation into human consciousness represents one of the most frequented fields of intellectual endeavour. Whether addressed directly or through reference to its many manifestations, there is hardly any other issue which has occupied religionists, philosophers and scientists to such a degree. Indeed, the very act of asking questions about a problem for which even its descriptors invite dispute hints at the profound mystery awaiting the curious. The laying down of centuries and centuries of perceptions, verbalisations, imagery, and hypothetical structures has created an interdependent lattice work of notions like a tightly woven fabric that by now has become more representative than its fibres. And yet it is those strands we have to examine if we want to achieve anything beyond a mere echoing of past sentiments, regardless of their roots in mythology or more recent rationalisations.

Just before I have linked the act of asking about consciousness to the import with which it gets treated. In earlier times the concept was related to identity. Recall how Aristotle referred to body and soul, and how the latter ought to rule the former so that one's life might be in accordance with nature. What he defined was a set of values, and those values in turn defined the man. A 'good' character exercised control over oneself in order to lead a 'good' life. The consciousness which administered the thought processes in ancient Greece did so under the auspices of a person's identity, manifested through their behaviour.

About six hundred years later St. Augustine wrote on the same topic. But now the assessment is not really about some particular act, but what that act represents in abstract terms. The 'truth' in one's demeanour derives from its meaning, and to understand it one needs to be apart from the immediate sense. Augustine's mind availed itself of higher abstractions and his consciousness, having received its offerings, returned them to their owner.

Another eight hundred years on and St. Thomas Aquinas abstracts further. 'Proportion' becomes 'proportionality' and he saw no reason why, while climbing the ladder to his god, the hierarchy of relations could not be extended to infinity. And his consciousness, responding to his fervour, complied.

By the 17th century we are in the Age of Reason and John Locke asks us to discern the faculties our mind provides for us. Through them we elevate an observed quality to an idea of it, and he cites a letter from an acquaintance asking him whether a blind man would be able to recognise anything later should his sight have been restored. The Age of Reason opened opportunities for consciousness to which it availed itself with gusto.

And so in the year 2000 Shaun Gallagher from the Department of Philosophy and Cognitive Science at the Canisius College in Dublin returns to that problem once again, the famous Molyneux question [35]. At this stage in our evolution he is able to articulate the concept of 'sense modalities', perceptional structures in the mind which relate to a specific sense and one only, although cross references are possible, in fact common. An inner image, derived from one modality, does therefore not necessarily relate to one caused by another—at least that's what Locke and the empiricists would have us believe. Gallagher then presents studies which show that newborns are able to distinguish visually between objects about which they have had only tactile information (they were given two differently shaped pacifiers). Since they were evidently successful, their internal imagery was able to relate to an object even though the previous experience had been conveyed through a different sense modality. Yet it is also true that individuals who had been blind for a long time and given sight experienced confusion and in many cases preferred a return to their sightless existence. The apparent inconsistency, so Gallagher infers, is due to an incorrect interpretation of the questions. He cites evidence of cortical developments in children and adults which point to the need for sensory experience in order for their respective functionality to unfold. If the visual area in the cortex, linked as it is to optical pathways, does not receive relevant stimulation its inherent capacity to represent visual images will deteriorate. On the other hand, representative complexes from various sensory input can and do interact to furnish an overall view of the world. So, in terms of visual acuity and interpretation, the Molyneux patient would indeed have difficulty identifying a cube or a sphere. In terms of adequate intellectual development, even a young child would not have such problems.

Several points stand out: the studies meant to confirm or otherwise the players' expectations as they centre around the activities in the cortex; the act of engaging in such an enquiry is a manifestation of consciousness; and the phenomena we have used to illustrate the issue are the product of a cognitive circuit going from sensory input to representation, to conscious awareness and back to re-representation.

Before we concentrate on the conceptual realm of Otoom we should pay a short visit to its neurobiological base. First, if we want to define consciousness it helps to look at its borders, and one of them constitutes what is described as 'minimally conscious states'. They are the threshold before which we can safely ascertain the absence of consciousness, and beyond which anything perceivable can be ascribed to its existence. Douglas Katz from the Boston University School of Medicine discusses various clinical features, from coma to the vegetative to the minimally conscious state [36]. In a table listing the anatomical and clinically-pathological consequences of these categories, the corresponding levels of consciousness are labeled "crude", "phenomenal", and "access", reflecting their functional capabilities. When it comes to their anatomical characteristics, the sequence becomes "brainstem-diencephalon", "partial cortical" and "integrated cortical". Further in the paper he explains that an important distinguishing mark between vegetative and minimally conscious states as well as higher conscious states is the evidence of cortical activity. Note also that such evidence is gained from testing patients in terms of their ability to respond to outside stimuli (eg, light, verbal commands, pictures), something which requires the integrity of cognitive circuits involving the rest of the brain too.

Secondly, the anatomical configuration of the cortical layer is rather special and for this reason has fascinated researchers for quite a while. Much of the investigations have focused on the cerebellar cortex in the rat's brain. The mammalian cerebellum is an important site for motor learning and, although its output is not quite at the same level of sophistication as other higher cognitive functions, it can nevertheless be considered an important feature of conscious feedback in general. Its cortex consists of three layers, the lowest containing granule cells, the next higher one Purkinje cells, and the topmost being referred to as the molecular layer [37]. Purkinje cells receive their input from deep inside the cerebellum and send their dendrites into the molecular layer where they interact with parallel fibres. They act in a modifying manner on the cerebellar output. Another form of transmission into the cerebellum occurs via so-called mossy fibres, nerve cells which synapse with the granule cells in their layer.

Fig. 15-6 shows the arrangement between the mossy fibres, the granule cells, and another type, Golgi cells [38]. The data coming from the mossy fibres are transmitted to the Purkinje cells after having been encoded in the granular layer, with the Golgi cells providing inhibitory input to the input layer, and basket and stellate cells performing a similar function for the output layer. The entire system is a combination of excitatory and inhibitory feedback loops with output to the Purkinje as well as granular layers and input coming from the white matter of the cerebellum plus the parallel fibres going to the Golgi cells.

Despite the interest shown the exact details of the neurological mechanism underpinning the cortical dynamics are not yet known. Certain points can be made however: the area draws information from the rest of the brain associated with higher-level cognitive functions, a complex system of feedback loops enables real-time modifications to the data, and there is connectivity back to the cerebrum (neuronal pathways between the mossy fibres and the cerebral cortex have been identified).

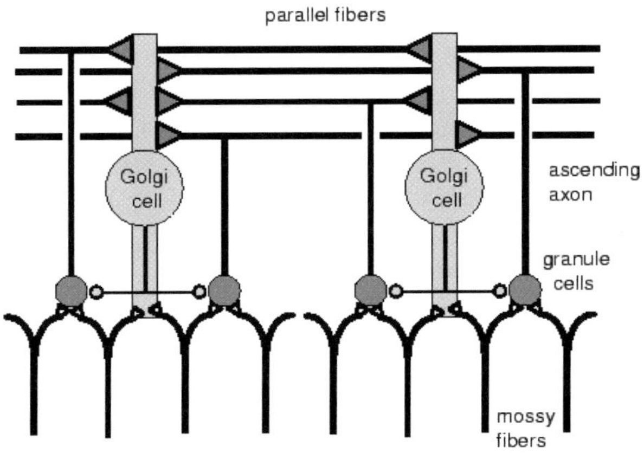

Fig. 15-6: Schematic representation of the organization of the granular layer of the cerebellum, transverse view. Mossy fibers originating external to the cerebellum excite both granule and Golgi cells, granule cells excite by their long parallel fibers Golgi cells, and Golgi cells inhibit granule cells. Each granule cell receives about 4 mossy fiber inputs and about 10 inhibitory contacts but it is unclear whether these come from different Golgi cells or not. The number of parallel fiber contacts onto Golgi cells is not known. *From:* Coding in the Granular Layer of the Cerebellum, Progress in Brain Research.

Relating this to our perspective we can say the following: there is a definable part of the brain which is associated with higher-level cognitive processes in the conscious state. Its anatomy suggests the existence of physical attributes which allow a certain amount of modifying control over its dynamic, where the dynamic concerns the processing of data received from the rest of the brain.

This leads to certain ramifications. Only those data can be modified that are contained within the cortical area, and the degree of modification depends on the locale's general capacity. Since biological systems work under the principle of feedback, and feedback is communicated via sensory data, the feedback processes are based on the interpretation of sensate events. 'Interpretation' being a cognitive process itself and thus relying on the interdependence of data transmissions, the modification of data within the cortical region is therefore a function of sensated experiences during the state of consciousness. In line with the overall functionality of the system, modified data sets can again be seen as ongoing input, entering the cycle once more.

Extending our reference even further, we can include the affinity-seeking properties of thought complexes which attract associative domains to the transmitted packets coming from the domains under consciousness. The subconscious structures, which we named sub-TSs in Part I, become the input to the containers dealing with their conscious counterparts (we named those con-TSs). The similarity between con- and sub-TSs led many to view consciousness under the umbrella of an emergent, complex system and ascribing to it the dynamics generally related to self-organising systems, as for instance Arnold Tannenbaum from the Department of Psychology and Institute for Social Research at the University of Michigan does [39]. It must be said however that the interest of researchers was first of all directed towards the experiential side of cognitive phenomena which excluded subconscious states; any inference to the wider system was therefore almost coincidental.

The difference between sub- and con-TSs lies in the latter's property to sensate. A sensate experience is an innately organic event, but we can only express that view because we are sensate beings. Our consciousness enables us to articulate an experience in terms of its sensatory quality (apart from our ability to abstract such things of course). Because our awareness allows the experience of touch, smell, taste etc we can point to these and declare them to *be* of a tactile, olfactory etc nature. Therefore the process of articulation goes from the experience to cognitive processing to conscious awareness to articulate output. In this case consciousness is the administrative agent making it happen.

The problem with the identification of consciousness lies in the immediacy of the functional participants of such events. Whereas before our agent delegated our sensate experiences between in- and output, with both being subordinate to the dynamics of the agent, this time it is the dynamics of the agent themselves that are supposed to articulate the experience—but the experience *is* the agent. Just as a film cannot carry an image of the film, just as a camera cannot take a picture of itself (and no, holding it up to a mirror is a different thing entirely), so consciousness cannot articulate itself in the manner we are accustomed to. Thus to the question, "What is consciousness?" the answer is: "When you experience a flower, consciousness *is* the flower; when you experience thirst, consciousness *is* the thirst; when you experience happiness, consciousness *is* the happiness."

Sometimes, for a thought to make sense we have to let it melt on our tongue.

Here's another way of describing the process. Take the toy in Fig. 15-7 displaying a face. We can see it is one because we know what they look like. Of course, it is not *really* a face—the surface made out of the heads of many nails has been warped by someone pressing his head against the other side and so each nail was made to move accordingly. Now imagine that the board holding the nails in place possesses some kind of sensory apparatus which enables it to identify the position of each nail in relation to its vertical plane, and let's assume it can somehow communicate this information to itself. If we were to print the data onto a sheet of paper the output might be something like this:

nail 1:	23 mm
nail 2:	24 mm
nail 3:	16 mm ...

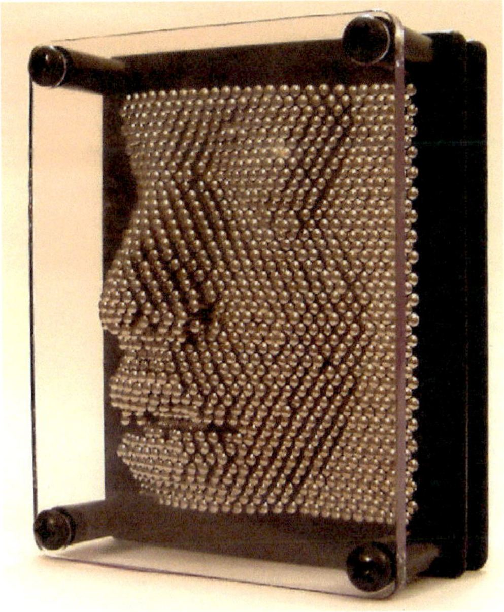

Fig. 15-7: Is this a face?

Any time that board would need to tell what it is doing, there emerges a string of numbers corresponding to the positions of all these nails. Would such a sequence sound like a face to us? Of course not, but if we were to translate that series into a suitable geometric framework so that a point on a display would be plotted exactly according to its integer value, the resultant whole would once again look familiar. This procedure is nothing out the ordinary; radar screens for instance do it all the time. But suppose now there are other boards, and they have no need to communicate with us but are quite content sharing their information amongst each other. To them the string of numbers makes perfect sense, and a certain sequence could be met with, "Ah—I know that face!"; another with, "what's that car?" and so on, each word of course being another set of numbers. How does the board know it has a car on its slate? It doesn't really, but it *thinks* it knows because whenever the number sequence happened to be just so it was accompanied by confirmatory events which told it "car!". Eventually this particular pattern became familiar and sub-patterns accentuated which allowed groupings. Hence there is a set of sensations (human observers would call them a series of numbers) which to the nail board means one definite thing and not another.

To complete our metaphor we attach little muscles to each nail so that any position can be modified and the new setting communicated back to where the numbers came from in the first place. At this point the boards could happily talk to each other and share whatever it is they represent with their nails. Their messages would be understood because every one knows 'what it means' when something like 23 24 16... comes in. With their muscles they could do things like, "Are you sure it's not 23 **25** 16...??" and a lively debate ensues. Until one day, a particularly picky board asked, "But what does it actually *mean* when I say 23?" And life has never been the same...

Nigel Thomas from California State University argues that the problem of consciousness only appeared around the 17th century [40]; the ancients weren't bothered and there are no words for it in Latin and Ancient Greek, nor indeed any other non-Western language, such as Chinese. However, it is not that simple.

For example, in Latin there exists the word *conscientia* which translates into something like 'knowledge' or 'understanding' but comes close to our term when expressed as *magna conscientia* (self-awareness), although there the meaning leans more towards the notion of 'conscience' [41]. As for Chinese, the word *yishi* (itself a composite from *yi* - 'meaning, to mean' and *shi* - 'to recognise, acknowledge') is commonly used to articulate the opposite to 'unconscious', 'to be awake', but also the idea of class consciousness and identity. The language is semantically layered to a high degree, and in the case of *yi* a reasonably comprehensive dictionary offers 'to stand upright', 'sun' and 'heart'; for *shi* the complements are 'language', 'speech', and 'only'. A certain alternative to *yishi* is *shing*, meaning the heart of something, its nature. Then there are contextual neighbours such as *jingshen* relating to the aspect of matter vs consciousness, *zhijue* referring to 'recover' or 'loose', and *juewu* having more to do with political consciousness.

Some reflection on these variants will illustrate the ambiguity of the concept because of its very nature. It also teaches us that language, a result of referential articulation constantly being tuned through feedback from the outside and inside, is largely a settling process that never fits precisely what it is meant to represent but can only ever be a compromise. Both points make sense if we consider the essential dynamics of cognitive processes in conjunction with the impossibility of dismantling a thought structure and expecting its loose parts to form into something meaningful again. But most of all, such semantic lattices demonstrate how selective a component consciousness is compared to the entire theme it confronts. The window of awareness is a small aperture that permits the organism's mental senses to view what the mind at large contains.

What the mind contains is the result of experience, and what consciousness does with it depends on its aperture and the dexterity of its 'little muscles'. The 'aperture' is a function of dendritic connectivity and relates to the individual's overall intelligence. It is known that children afflicted with mental retardation show a significant change in their dendritic structure; the spines are fewer and they are long and thin [42]. Dendrites are responsible for synaptic input, and genetic and/or prenatal deformities in this area make for a low IQ. As for the 'little muscles', they are a metaphor meant to circumscribe the type of wilful act which accompanies focused thinking, or rather the absence of it if we want to test its efficacy. The experiment is easily performed provided the person has a modicum of control over their thought processes. Those affiliated with religious and/or phantasy-laden concepts might choose not to go ahead for reasons which will become clear shortly. The idea is to be comfortably at rest and then proceed to close down one thought after the other. In the normal conscious state there is always something that flits through our minds, whether we look for it or not. In this case we focus on the 'hold' we have on a thought and then relax its 'grip'. If this has never been attempted it may take some tries to synchronise this particular act of relaxation with the effect of that thought actually disappearing (my use of inverted commas signifies the inherent ambiguity of my words—but it is the best I can do under the circumstances since we are dealing with highly subjective states). I say "synchronise" because what one understands under 'relaxing the grip' may not be the act which affects the thought—for instance, we might mistakenly relax some real muscle instead and the thought remains.

After a while the act of letting go and a thought being gone become more accomplished, but now we experience another difficulty. The *thought* of letting thoughts go is in fact an act of consciousness in its own and as long as we retain it our minds won't be completely cleared. For that to happen we need enough practice to perform the trick without, well, thinking. It can be done, and once it occurs what is left is a grey state of dim lucidity.

So far the experiment has demonstrated the possibility of control one can exercise over one's conscious imagery, and the existence of wilful interference with the appearance and disappearance of ideational events; none of which should come as a surprise. At this point one of three outcomes is possible: (a) a sudden itch or some such disturbance yanks us out of our tranquility and we return to everyday life; (b) we fall asleep and wake up some time later; (c) we almost fall asleep in the sense that we loose control over our modification faculty but not our sense of awareness. During the last option the body has relinquished its conscious control (which includes the limbs as well as those 'little muscles') but we remain lucid nevertheless. It is a state which some, if not most, people mistake for a real-life experience no matter how bizarre its contents may be (this is the reason for the warning above).

Considering the functional dynamics of consciousness we can readily identify the players: affinity relationships are being generated throughout the sub-TSs, con-TSs are actively mining those parts that can be reached, yet the person has no control over the kind of combinations that eventuate. The result is a lucid hallucination that comprises the individual's propensity. Its nearness to ordinary conscious states makes it very persuasive, and due to its underlying associativity the experience may appear to be meaningful to the holder. Note that all the attributes of consciousness are present bar one. However, due to the perceptual intensity even the loss of control may not be interpreted as such, because the affinity-seeking properties of these virtual con-TSs ensure a conceptualisation which is commensurate with the overall tone of the event. For example, a person may think he encounters higher beings who take him to a place of authority, where the notion of 'higher' and 'place of authority' denote the control-less state of the individual himself. Or someone might think she has been transported to another place or time, and scenes which seem entirely real confirm the perception. Since with most people such states are entered only very rarely by coincidence but rather through a predisposed religious tendency prompting them to link to something 'higher', the results will reflect their preoccupation with mythical themes. We also know that one's emotional energy affects the creation and stability of affinity relationships and that their ultimate character is a reflection of such TSs' congruence.

Tying these and previous points together we can say the following:

affinity complexes are grown from the inside out, that is they represent an organism's identity due to the temporal sequence of affinity layers;

affinity relationships are enhanced or suppressed in terms of their emotional content, such content being measured against the organism's identity;

within the biological framework, con-TSs are a subset of cognitive processes in general and in addition are subject to wilful interference which has its basis in similar affinity-seeking complexes;

the resultant event is termed consciousness and forms a sensate experience which, because of its representative nature, equates with the contextual format of its con-TSs;

an extraordinary subset of conscious events consists of a sensate experience but without the usual accompanying availability of interference.

In Part I we have followed the steady development of congruent TSs and compared them to the situation in wider society. At the same time the unfolding of abstractive ability caused the creation of incongruent domains, their groundedness in question. Thus there was a time in our evolutionary history when the confluence of general abstractive ability was tempered by an as yet insufficient mastery of rational conceptualisation of input. That is to say, there were people who had the intellectual capability to reflect profoundly upon themselves and their environs, but lacked the necessary depth to articulate their insights towards anything other than ego-centric notions. Their contemporaries understood them on the commonly shared level of interpretation. It is no coincidence therefore that the world's major religions emanated from that juncture. Without a mind to imagine, without the will to believe, and without the strength to follow, no people can make their belief system truly dominant. The generations to come rationalised over the raw spectres, making them palatable to their own intellects. But never again would a human collective be so enthralled by its own phantasms.

It was not so long ago that a genuine emotion radiating from a human face convinced any onlooker of its truthfulness no matter what the message might have been—at least in Western society. The person doing the articulating conveyed the meaning his or her conscious mind projected, and there was no doubting its realism; what consciousness expressed must therefore be true. "*I have seen it with my own eyes!*" the man shouts, his hands trembling from the impact, and the sheer honesty of it all stirs the soul of everyone. Across the globe we still find many demographics whose members live in that same state; it is a wise and prudent traveller who watches for the tell-tale signs of such emotional intensity and begins to tread very carefully once they have become visible.

Yet the inherent dynamics are with us whatever the culture, and the exclusivity of the source from which we draw our information about the world cannot be challenged. The picture that con-TSs manage to portray is the one we find valid because it is the only one which is able to be sensated. Whatever it contains becomes the purchase we have on reality—nothing more, nothing less. This includes the perspective we have of our own selves; after all, it is nothing else but the sensated result of an affinity domain transported into a con-TS currently active.

For example, how many people would consider themselves to be on a par with bats? These animals orient themselves through analysing the sound which is reflected from surrounding objects. It may surprise many that the human brain can and does do likewise. Repeated tests have found that blind as well as sighted people (after some training) locate obstacles by interpreting the reflected sound they make; thus echolocation as an ability is not in dispute [43]. What has been disputed by a considerable number of people, notably those who use echolocation to compensate for their impaired vision, is the fact that it is the sound through which they acquired the ability to locate. Eric Schwitzgebel and Michael Gordon, both from the University of California, reported that one nearly blind subject for instance insisted that he detected objects through pressure on his face; another thought it was "cutaneous sensations on his forehead" that enabled him to avoid collisions, and only after several experiments was his conviction overturned. Subjects who ascribe their ability to a sensitive skin claimed to actually "feel" the pressure on their faces, although tests clearly pointed to the real means for orientation.

Just like the split-brain case from a previous section the above demonstrates the readiness of thought complexes to enter into affinities. If for some reason the person is inclined to consider himself to have a skin with extraordinary faculties, the mind will comply and consciousness will create the sensated reality for him—even to the extend of providing the appropriate 'feel'. While all this may sound rather distressing for the realists amongst us, we need to consider the situation from a pragmatic point of view. Here is a system—biological as it happens—which receives input via a sensory apparatus. Such input needs to be processed for the system to operate productively (ie, to live another day and reproduce). It can only do so if the result of such processes is commensurate with the reality interpreted in terms of that system's character. It will not be productive if the reality is understood in a more objective manner; for example, if a fish identifies air currents as barometric pressure differentials rather than whirls on the water's surface. For the fish to know one pressure from another is useless, but to recognise the approach of something towards itself is vital. By the same token, for the shaman of a tribe to be taught about high voltages during lightning may or may not sound fascinating, but how would this fare against his skill to placate the thunder god and so calm down his people?

The organisation of suitable TS complexes and their arraignment before our consciousness takes time. The moment an event becomes re-represented within the feedback circuitry of the cortex we will know about it, but no sooner. Obviously, the actual occurrence has gone before and so we can expect a temporal mismatch between the two. However, once again consciousness prevents us from becoming too distraught about it and so obfuscates the event. A long series of experiments tracked the circumstances surrounding these temporal shifts until finally the situation became clear [44]. What follows is articulated under Otoom's perspective.

Suppose something confronts us which requires a response. The necessary TP is formed through the realised affinity potential of representative domains which at this stage includes sub-TSs only. The content of sub-TS clusters furnishes the information that is deemed relevant for the response (this is all they can do since there is no other reference). These clusters link with pathways connected to the rest of the organism to actualise the response. Through the connection between sub- and con-TSs the response-TS is sensated to consciousness and we become aware of what we are doing. For the system to work two points are important: the created affinity relationship ensures that what eventuates will be in line with our expectation (after all, this is what our expectation gets built upon in the first place); and consciousness satisfies our need to comply with reality by placing the sensated event backwards in time to concur with the *real* response triggered by the sub-TSs. Note that our "need to comply" is not a demand made by some stern objectivist; rather, it is our emotive self which wants to feel comfortable in the knowledge that reality is taken care of—according to *our* way of seeing reality. Benjamin Libet who performed a major part of these experiments is quoted as saying,

> This leads me to propose that the performance of every conscious voluntary act is preceded by special unconscious cerebral processes that begin about 500 ms or so before the act.

Suppose we would not be 'tricked' in this way. Then just about everything we do, based as it is on some kind of reaction to what happened before, would be met with a significant amount of stupefaction: we would constantly find ourselves observing some act of ours *after* we are had initiated it, wondering what on earth is going on—but the very act of such recognition is yet another act in response to the first one, which in turn occurs after a similar delay; and so on and so on. Surely, madness would result. Hence it seems reasonable to assume, given the instances witnessed so far, that consciousness, in order to stall any emerging feelings of consternation, would readjust itself to such a degree that our sensate experience syncs with our subconscious activity, or at least the results of it. Since there is no mechanism to modify the sub-TSs, the target are their conscious counterparts.

The experiments by Libet and others used the electrical action potential of nerve cells to identify the onset of activity. But with a bit of luck it is possible to catch the system out when entire trains of thought are involved. A few weeks ago this is what happened:

I had just completed a series of test runs with OtoomCM. As is usual during the making of this book my mind was preoccupied with its details, and sleep is no exception. At about half past five that morning I woke up, triggered by my subconscious which proceeded to bring into awareness certain points I had missed on the previous test runs (nothing unusual so far).

Lying there in bed I realised that (a) I could make some changes to the setup and run another thirty cycles, (b) at that particular configuration it would take roughly three hours for the run to be completed, and (c) being that early in the morning the test could be done by about half past nine (by now it was almost six), which meant that I could catch some more sleep and have the results ready for the day (I am a night person). By the way, the program emits a beep once a test run is finished.

At 06:14:12 the program had completed its first cycle (every screen shot carries a date/time stamp), although I did not know this at the time because I was asleep again. A while later I woke up because in my sleep I perceived a beep; my mind was on the program with its current technicalities, and the sound came within the context of the test. A few moments later I was fully awake and a glance at the digital clock told me it was 09:34. It was at that moment that I—now fully conscious—heard the beep again.

Loading the screen shot from the last cycle told me it was saved to disk at 09:33:22. As it happened, only a couple of days earlier I had synchronised all clocks and watches, including the computer clock. Allowing for some delay caused by clumsy fingers, the time span between the various clocks' units could not have been more than one second (even that would be overly pessimistic; one second is a long time to hesitate before pressing a button). It can be assumed therefore that my glance at the clock did indeed occur at 09:34 (plus some seconds maybe), and the sensation of the beep being either at that moment or shortly after. This means that my subconscious, unrestrained as it was by any conscious activity, received the sound at the time it was made at 09:33:22 and thus at least thirty seconds earlier, and had passed the entire package to the con-TSs which then roused me from sleep. But my cortex did not spring into complete action until 09:34, by which time the earlier sensation had slipped through. From that moment however the modifying faculty was up and running and it brought me the event, not realising I had received it already. Note, firstly, my preoccupation with the event and therefore the particular contextualisation of the sub-TSs; secondly, my light sleep mode predisposing me towards receiving input from the sub-TSs; thirdly, the discrepancy between the half- and fully functioning cortical activity leading to the same experience but separated in time; and finally, it was the effect of the former which preceded the effect of the latter.

The quote above has Benjamin Libet referring to "special unconscious cerebral processes" when compared to the activity of consciousness. It is a certain slant which hints at the predominance we give to our conscious thoughts. Based on what we have learned so far one can argue however that it should be seen the other way around. What is special are the sensate experiences that accompany the rest, the vast reservoir of ongoing dynamics which is the main system; it's just that we don't 'know' it. There is also the historical perspective: just as all the other organic processes can be grouped in terms of their functional sophistication and thus be sequenced on the evolutionary time scale, so it is with the neuronal mechanisms. The differentiation can be observed throughout the growth of the central nervous system. From the neural tube at the earliest stages of foetal development to the accentuation of vesicles which become the fore-, mid- and hindbrain, to eventually the fully-blown organ comprising its vast array of features, there is a systematic build-up of one functional layer after another [45]. Naturally, cognitive dynamics follow in tandem.

If we break down the functionality of consciousness we find two main composites, the facilities to sensate and to modify (the experiments listed above suggest as much). In terms of sophistication, to sensate would be on a lower rung than the other, given the apparatus necessary for either. Hence it can be argued that during the evolution of the human mind the ability to sensate came before the ability to modify the thought processes within the dynamic context of that newly acquired stage. Note that we are not referring to response mechanisms per se; the focus is on a wilful interaction with sensated re-representations of neuronal data, the tips of the mental icebergs as it were.

Tying this together with the phenomenon of 'lucid dreams' (the kind of experience which can be generated through the thought experiment mentioned earlier), and we have a likely basis for the mythologies of ancient lore. There was a time when human beings were in the process of consciously awakening. They sensated thought events, but were restricted to interfere with them to various degree. Those thought events were the results of affinity relationships, configured according to their identity containers, stabilised through self-referential emotions, and conceptualised by the limited abstractive capacity of their bearers. From lizards there emerged dragons, from thunder there came the voice of god, and life and death, sorrow and happiness, lust and love, greed and contentment, all become interwoven with representative articulations that projected their images into a hitherto passive cortex. Only gradually were the forms moulded into something more realistic, but by then the prototypes had found their place in the collective memory of the group. Notions of 'old' and 'dim past', of 'power' and 'magic', of 'high lords' and 'monsters', of 'true love' and 'deep wisdom', but also of 'murderous hell' and 'paradise', they lingered on and provided their material for ongoing conceptualisations.

While this may be conjecture (although its components are true enough), there is further confirmation, this time found in the present as well. If such 'half-consciousness' gives rise to specters which are largely beyond the control of the mind, the inability by their owners to modify does not mean lack of awareness. Therefore, once a particular power structure has emerged in the process of societal rearrangement and that structure owes its status to the imagery of its mythological base (as it invariably must), the mechanics of affinity relationships ensure that defences will be erected to safeguard against any competitor; the fundamental principle for it can be found anywhere in the biological realm. An analysis of religion-based demographics will show that there exists a direct relationship between the degree of phantasy inherent in the semantic structure of the belief system and the severity of the social mechanisms designed to uphold that system. In other words, the more likely it is that members of a society will come up with ungrounded interpretations, the stricter the control over those that already exist. Since any act, whether by the individual or the group, is mostly derived from sub-TSs, the entire societal dynamic functions in a space that is barely accessible to grounded abstracts. The effects can be observed whenever an outside influence is perceived to be a challenger. We will examine the consequences in the next chapter.

At this moment it may be appropriate to reappraise our concept of the 'self'. As long as we derive the image of ourselves from the sensated experience of a conscious act, the result will be problematic. We won't *know* it is problematic unless we have access to higher-level abstracts which allow us to redirect our thoughts to those thoughts already held. Thus people in antiquity did engage in debates about the self, but their conversations were bereft of the kind of complications their successors in more modern times had to put up with. Once a mind could formulate the question of "Who am I" where the "I" stands not for the persona but the functioning entity of the questioning act, we have more or less arrived at the limits of conceptual deconstruction. To get any further we have to divert along a different route. When it comes to assertions of a representative nature regarding the physicality of the self therefore, an articulation synchronised with the conscious self will suffice. For anything beyond we need to reformulate ourselves into an abstracted object with characteristics confirmed through grounded observations along the same lines (if in fact we *are* interested in grounded observations). So you could say that the search for a soul in Man is like trying to find acceleration in a car.

The central nervous system determines the status of its owner. Its faculties supply the organism with actions and responses commensurate with their functional scale. At any given level the output can only ever be articulated in those terms—interpretations that go beyond the dynamic locale lead to errors. In an interview neuroscientist Rodolfo Llinás of the New York University School of Medicine said when asked about 'consciousness',

> I think consciousness is the sum of perceptions, which you must put together as a single event. I seriously believe that consciousness does not belong only to humans; it belongs to probably all forms of life that have a nervous system. The issue is the *level* of consciousness. [46]

It is that level which becomes honed to the degree we are accustomed to in our social environment. Take away some of its components and the result will be diminished. In his book *Savage Girls and Wild Boys* Michael Newton writes about the boy from Aveyron in France, who was caught there in the year 1800 [47]. No-one knew how he had come to be in the wilderness, but it was obvious that most of his young life was spent away from human society. He had no language, rudimentary concepts of hygiene, and his consuming interest seemed to have been food. One sentence is most telling:

> On being given a mirror, the boy could make nothing of it, looking at once behind it, to try to find the child that he had seen within its surface.

Evidently the first-level perception of the image in the mirror is intact. His inability to go further stemmed from his ignorance about himself as an entity in its own right. While it could be argued that the absence of self-recognition was due to him having never seen a mirror and therefore being unaware of its reflective properties, the core issue remains. To recognise the synchronicity between his own movements and those of the image would not have been beyond his intellect, although untrained in more formal exercises; without a basic pattern recognition facility he would not have survived on his own for so many years. His fault lay in the lack of sufficient abstraction to identify the image as himself as the object, a second-level conceptualisation or PAT^2. Apart from his infancy (one would assume) he had never experienced human interplay, and when his mind organised itself around the perceived input on a daily basis there was never any prompt to refer to oneself as 'another'. There are wider parallels with demographics around the globe in which the degree of group-identity is enhanced in favour of the individualism we are used to in more industrialised regions. A level of industrialisation (particularly if it had been generated by the people themselves rather than merely imported) engenders a more complex interaction with one's environment and therefore the push towards higher and higher abstractions.

Another case comes from the United States. A girl had been imprisoned by her parents for nearly twelve of her thirteen years, locked in a room, tied to a chair, force-fed, only able to move her hands and feet. She was beaten when she called out, never talked to, the only form of communication a growl or a dog-like barking from the father [48]. When found in 1970 she was thought to be only eight. Despite signs that she was not retarded her behaviour lacked the situatedness of those who grow up in a more benign environment. Her impersonal interaction with others, her habits of spitting out food and vomiting, her remoteness from a whole host of social detail the rest of us take for granted, they all indicate the disrupted and askew internal representation of what was her reality. As difficult or impossible as it is to understand the world from the perspective of someone like her, a true realisation could only come from a position that has at its disposal a similar set of frameworks. Only then could one follow the point-by-point moves of every thought, inflection, and response.

The emotions which abandoned children evoke in normal people ('normal' as in 'norm') are a response to the discrepancy between their own reality and what is perceived to be that of the child. To be sure, physical suffering is felt by anyone because it doesn't take a wider reference system to feel pain. The mental states however are a different matter. Do well-meaning officials *really* know what it is like to hunt in a forest, to feel the wet grass on your skin and smell the wind as it comes through the trees? Can the solitude of a night under the moon be equated with sitting by yourself at a party? When concerned members of the middle class are aroused by the plight of others who do not conform to the standards of Western home life, they do not act on the desires of those they see; first and foremost, they react to their inner representations and decide on such terms. The context is not the other, it is their own. Individuals who practice the detachment from mental frameworks are likely to sense the circumstances of subjective assertions. Phouttasone Thirakoul, in an essay on Buddhist thinking, expresses it in these words:

> Experience is only meaningful when events are placed within a context; if novel events arise, in order to truly experience them (that is, make sense of them), one must reformulate one's constructs so that the novelties are not merely foreign events, but systematically explicable under a new framework. [49]

The enveloping atmosphere of social commonality leads us to the belief that our perceptions are shared by everyone. That assumption is not without foundation; the constant exposure to the acts of others, the daily opportunity to see our own gestures bounce off the members of our group, provides us with the context of human society. Our minds adjust the re-representation to suit the context, and the context is constructed from the multitudinous articulations of minds. Together they weave what we call the 'social fabric', and consciousness convinces us of its truth. As long as these external and internal systems are able to act in tandem the world is a safe place, but it takes only a small rupture and reality begins to fall apart. The collective familiarity with the customs of one's group is the only protection an abstracting mind can have against the unraveling of contextual connections. Thus the ultimate abode of homo sapiens is not a cave, a homestead, or an apartment. It is the however limited range of ideations, compiled within the pool of expected abstractive potential. To step outside those boundaries opens the mind to a higher clarity, yet deprives it of its home; stay inside and your vision sinks to the ground, but you'll be safe.

The subconscious and conscious means by which humans make themselves feel safe can be summarised with one word: society.

16. Setting the scene

Society

In the beginning was the herd. - -

Would an alien intelligence, analysing primitive organisms on a younger planet Earth, be able to forecast the emergence of a central nervous system? If it identified the nerve cells in those organisms and inferred from there then yes, it probably would. So let us re-phrase the question: would aliens, analysing protein formations on our planet, predict the existence of organic modules which are able to carry data specific to their host?

Logical purists might have some problems with these questions. Firstly, aliens which are capable of investigating Earth would most likely know a few things about themselves and therefore have certain expectations of how life forms in general are organised; but more importantly, they would understand the principles underpinning a dynamic, self-regulating, pattern-seeking system that is situated in the world and interdependent of it. Either way, we would have to answer in the affirmative and leave it at that; we would also miss the point.

Perhaps we should frame this differently still: if there is the question, "Is it possible for nervous systems to emerge from the current protein formations on Earth?", would the clusters of enzymes give us the answer? I can see the purists raising their eye brows again. For this time, who exactly is asking the question, and how can a bunch of chemicals articulate a response? In other words, there needs to be a system which has a knowledge base about nerve cells, it must have the ability to formulate a cohesive string of utterances in relation to the data, and such data will have to be understood by the proteins. Even assuming our system is the perfect translator (like the Babel fish in *The Hitch Hiker's Guide to the Galaxy* [1]), it could not transmit the concept on a level that is acceptable to the organisms. Since they do not have nerve cells, let alone an apparatus to receive and transmit messages, the intent would be entirely lost on them.

We see that our question is trivial. Not because nervous systems are in dispute one way or another. The point is, at this stage in our evolution we do not need to ask since the ability to ask *is* the answer; and if we really did not know, we could not ask.

So, back to the beginning—the herd.

When a sensory datum moves along its path from interface to brain tissue, it traverses a structure which performs the function of a carrier; the gradual extension of the organism has made this necessary. The carrier, growing into a many-stranded net, is now the recipient of pulses converging on its centre. These pulses are no transients; they do not enter at one end and leave by another, and similarly they do not wash backwards and forwards between the terminals either. They cannot do this because as the functional development of the structure responds to the transmission, which in turn represents the continual effect of the organism's environment supplying the input, such input will always be the stronger factor. The environment exists and acts, regardless of the organism's state. The organism *has* a state because it is dynamic. Eventually the system reacts to the supply of data on one side and the impossibility of deflecting them to somewhere else. The result is an emerging region of pure reception, with the sole purpose of those cells being a catchment for modified states, each one representative of some input. The dynamism of the module ensures the adjustment moment by moment, and the compositional multitude of its elements (the cells with their internal configuration) makes it a self-regulating one. The possibility of data transmission from the outside inwards predisposes the system to respond to its dynamic, self-regulating core by enabling its states to be transmitted outwards. As the physical movement of a hair can trigger a change in the chemical potential in a cell, so can a chemical potential be translated into the physical movement of a hair. The internal states of the module, representative of the outside, transmit their responses back to the environment which then in turn becomes the source for yet another input. As the cycle continues the potential of the system synchronises with the type of input it is capable of digesting, and so it reacts accordingly. Hence we arrive at a dynamic, self-regulating, and pattern-seeking system. Its basic engine is the central nervous system.

Let us be clear on the components involved: (1) an entity which contains the capacity to change its internal state and drawing on its internal state to do so, (2) data which are being carried to the entity from the outside, (3) a mutual synchronicity between the internal state and its environment to the extent that the data representing one are responded to by the other.

Liane Gabora (we have met her in the previous chapter) sees the evolution of culture similar to the chemical composition of matter as it forms and reforms itself to produce a different level of manifestation [2]. This occurs against the background of the possibilities inherent in that particular environment, but she makes the point that there is no clean trace from one type of mechanism to another. That is to say, the evolutionary elements surrounding the creation of cellular organisms for example are not synonymous with the progression observed in cultural developments; although one can and does affect the other:

> Some of the ways in which biological information gets tainted with cultural information seem relatively inconsequential, such as the trimming of hedges, whereas others, such as dog-breeding, have a long-lasting effect. (In fact, one could view dogs as the consequence of a memetic trajectory that was launched by the need to protect property.)

She continues by drawing parallels between the genetic concept of 'allele' (a version of a gene) and 'meme', an informational element comprising ideation and its implemented articulation. She admits however that such a comparison lacks finesse on the side of the meme. While the combinatorial potential of genes is huge, their one limiting factor is the discreteness of these units.

On the other hand, a meme (to remain with the above definition) cannot be deconstructed into standardised categories. The quality of a particular sample is not an absolute category but an amalgam of situational effects, brought together by the conceptual ability of the observer. Note that what we call the 'observer' does not necessarily imply the role of 'participant'.

For example, Gabora mentioned dog-breeding as a co-existant with human evolution. One could interpret her statement as saying that dogs were bred in order to protect property (ie, "was *launched* by the need to protect property"), something we could agree with when considering the mechanisms of human settlements in general. In terms of the evolutionary perspective however where the focus is on the *emergent* properties of cultural behaviour, such a view is misleading.

In an interview Iain Davidson, Professor of Archaeology at the University of New England, Australia, talks about the gradual domestication of dogs by humans [3]. Australian Aborigines, whose traditional culture is that of the hunter-gatherer, attracted the dingo because a tribe would be a collection of people who produced scraps of food on a regular basis, and that would be nourishment for the dog. There was never a proper domestication however because the tribes never settled and so a genetic modification based on human selectivity never took place either—the dingoes just hovered around the periphery of the camps. Yet where settlement did occur elsewhere dogs became a more constant member of the group and people started to select the animals in terms of their own preferences; the genetic changes in dogs can be referenced to such periods.

The scenario painted by Iain Davidson would suggest a more or less coincidental, tandem development rather than the targeted, intentional one we usually tend to infer. The reason for the latter's preference is our status as abstracting observers, where information is perceived from the outside but then is interpreted according to the already existing knowledge base in our minds. Therefore, we know that dogs are being bred; we observe earliest historical evidence of dogs having existed in the vicinity of humans; so we assume that humans appropriated the dog in order to extract a service from the species. Eventually of course this is what happened, but it does not describe the origin of the human-dog relationship. At that time, the idea of breeding dogs in captivity because they just might be useful to protect property could not have been present because that particular meme did not exist. For the dog to become useful there must have been some sort of settlement, the relationship between dog and humans would have to have the opportunity to develop further so that dogs would be drawn to protect not only their own territory but—as it happened—that of humans too, and the connection would have to be made on the human side to recognise the usefulness in this displayed ownership. In principle the situation is comparable to the emergent affinity relationships between TS complexes due to the presence of appropriate content; or to the production of relevant colour patches in OtoomCM by just the right input.

Since we referred to OtoomCM's output in terms of stable, periodic, and chaotic states, it is interesting to read Vladimir Dimitrov's perspective of meaning when it comes to the dynamics of our mental space [4]. He sees an idea as an attractor which operates in the mental space of an individual and performs according to the contingencies of chaos, that is amid the fitness landscape of phase states leading to either consolidation or entering zones of instability. Although references to chaos theory are useful for the mathematical framework they provide in relation to the dynamics played out by the elements in that space, they do not in themselves tell us anything about the elements' ultimate potential for creating more complex clusters.

The causes for this kind of productivity have always fascinated thinkers. One might say, they have also fascinated people in general, but for different reasons. We have seen the interdependence between in- and output, the connectedness between information contained in the environment and its representative form once perception has occurred. Without the environment making sense to the observer, and without the observer providing meaningful input to the environment, no emergence of any kind will take place. What differentiates the basic central nervous system from its later descendant so dramatically is the latter's additional depth, its ability to abstract its clusters of representative states in order to form re-representations. The organic medium with its hundreds of millions of neurons supplies the capacity to actually realise the affinity potentials in its network; it is the total organism which in turn articulates the result. Thus the functional space is no longer the ground on which homo sapiens walks and finds his food and makes his tools, it is the mind which deploys a world entirely on its own. In this new world the laws are those of its owner, the possibilities those of its inhabitants. There lies the fascination as well as the terror.

A landscape featuring natural components such as plants and animals, and supplemented by other physical objects such as the artifacts manufactured by its human inhabitants, demonstrates a certain degree of associativity. Trees and fruits, water and fish, axe and weapon, they all lend themselves to combinations of one kind or another. But it takes the abstractive ability of the mind to connect a tree with a deity, or water with a border, or to turn an axe into a symbol of the chief's power. And once done, what next? As the tribe reverberated the notions of gods and their earthly representatives, could not a creative younger member reassemble the conceptual elements so that not the chief but he himself became the focus of glory? Most certainly he could, and there is no doubt that history, whether recorded or not, is full of these candidates. Just as in OtoomCM the arrival of some new input was being tested against the affinity potential of the existing clusters, so does the group of humans confront a new idea for the same purpose. Except that now the mechanism is not a strictly configured one (albeit subject to the unpredictability of complex systems) but consists of many players who represent what we might call super clusters of affinitive entities, in other words who possess brains with neurons, and where these brains again form modules which are part of the group.

The affinity potential of OtoomCM's matrices could be expressed in terms of the product of the matrices and their elements; the resultant latency defined the functional space in which the system could operate without running into constraints. That space is finite, and as we have seen through the effectiveness chart there are two—positive—solutions to a progression lock: either increase the space itself, or increase the abstractive ability of its elements. The former entails adding more matrices, the other assigning more elements to each matrix as well as improving the connectivity. Transposed onto the human level this becomes a higher number of group members and/or a heightened intelligence on behalf of each member.

In a herd the functional space is created by its members and their respective capabilities. The members, via their numbers and their know-how, form the herd; in turn, the herd is the reflective reservoir of the members' identity, at once its sounding board and its silencer. For an individual to live, their articulations (ie, their output) have to meet the affinity standards of the system (ie, the group) so that the system can respond in kind. The ramifications are obvious: the larger the group, the greater the chance any of its members gets away with something; the more intelligent the group, the greater the chance anything different will be met with tolerance. Due to the interdependency of affinity relationships such ratios can be turned around quite readily: the longer a thief can follow his trade and still be free, the larger the territory must be which he inhabits; and the more outlandish the statements of an iconoclast are with the person still alive and well, the more knowledgeable and insightful his community will have to be. As far as novel ideas are concerned then, the response by the group will be one of fascination if its functional space allows it, and one of terror if it does not. Luckily for us, Western culture does contain those resources.

Which is why Antonio Preti, an Italian psychiatrist and psychotherapist, can discuss genius and madness in his article *The Gift of Saturn: Creativity and Psychopathology* [5]. While there are many factors leading to the ascertainable presence of mental disorders, one common denominator is the heightened predilection for conceptualising "opposites in a free-response situation", termed 'Janusian thinking' here. At first glance my last sentence may appear to be upside down (an example of an 'opposing free-response'?), but considering the functional dynamics of cognitive mechanisms it should make sense. If the degree of conceptual associativity reaches random proportions, the result will be a predictable judgment concerning disturbed behaviour; that judgment is passed back and used to categorise the person. Likewise, congenital defects or subsequent chemical imbalances in the brain also affect the associative potential of the neuronal network and hence the mind; the result is the same.

Preti uses Kekulé von Stradonitz our chemist from chapter 10 who discovered the ring-like structure of benzene. Our analysis showed that in this case the usurped format of traditional thinking was not without foundation; rather, one set of rules got replaced by another one. Therefore the novelty consisted in the juxtaposition of conceptual formats that so far had not been contemplated, and von Stradonitz's achievement consisted in his mind's willingness to admit unsanctioned but logical thought processes. The arraignment of rules on one contextual level and the placement of another set on a different one, with an accompanying contemplation as to the usefulness of whatever transference between these sets, should not be confused with the mixing of contexts outside any grounded references performed by schizophrenics. Nowadays professionals in the field tend to relate a heightened degree of associativity more to bipolar mood disorder (its older name manic-depressive illness), as Preti writes.

But note the element of randomness, not only in the sense of Preti's associations in unusual minds, but also in the entire process of the interplay within that triangle of individual, group, and surroundings. In OtoomCM the integers within the element matrices were not important, but their mutual relationship was; likewise the constituents of an environment are not absolutes, but eventually acquire their importance once the flow of information amongst the participants has started. What emerges is a specific culture, and what the self-reflecting players see in its nature is their own identity. To the dynamic it did not matter whether the landscape consisted of sand or snow, forest or taiga; the principle of affinity relationships supporting and/or dismantling the shifting mosaic of clusters could operate provided their system had the space and connectivity to support them.

Since the abstracting powers of this functional space need data to perform their transformations, in that secondary sense the characteristics of the environment do matter after all, but the significance lies in the contextual format, not the functional one. Hence a tale of love rings true for anyone, whether the setting happens to be a mountain village, a kingly palace, or even an imaginary scenario of talking sea creatures. At the same time, the ideas coming to the spurned protagonist will reflect his situatedness and so his adventures will not be random.

Throughout our discussions of TS dynamics we distinguished functional and contextual formats. In Part I the difference had been demonstrated through the examples enabled by this or that philosophy; so far in Part II we mostly concentrated on the fundamentals. It is time to accentuate the connection between the two.

In a herd the contextual pool consists of the environment, the members, and the representative counterpart of their environment. As soon as cognitive abilities increase, its representative component experiences a rise in complexity which has certain ramifications. If in OtoomCM we could follow the creation of specific clusters that hinted at the first instances of primitive memory, and if specialists of the 21st century can articulate societal patterns in terms of memetic vectors, then somewhere along that path there occurs an ongoing differentiation of objects on one side and functions on the other.

The pointy end of a spear is an object, but without an accompanying function which allows its owner to speculate on a slightly different format so that it becomes a needle, neither will come into existence since both are arbitrary instances placed somewhere in time. In terms of thought structures, the object as well as its functionality are abstracts and accordingly we denoted them as OAs and FAs. The flint found on the ground is a PAT^1, and without the mind's capacity to adjoin with a PAT^2 the stone will not be shaped nor will it be placed at the end of a stick. From that—particular—initial abstraction many branchings can be produced and all with the same functionality. There comes a time when the functional space of this OAs-FA combination is exhausted, and we have reached a functional progression lock. Progress is only made once a further abstraction can be produced and so a new space is created to be explored again. For example, using the flint-stick as a spear will be effective, spawning numerous advantages for the tribe. Sooner or later the damaging, hole-making properties of the instrument will have been explored, and the cognitive energies employed up to now will become idle and restless. The act of pushing a lance through a hide and so leaving a hole, and being pushed through another spot and leaving a second hole, could spawn abstractions which concentrate on holes → spaces to be filled → object connecting all the spaces → etc and eventually arrive at 'sewing'. Similar processes run their course in other minds, at other times, for other occasions.

Recall the 3-dimensional effectiveness-chart from chapter 10. There the red lines marked the number of products, the green ones the realised latency and blue stood for the degree of applicability inherent in the entire resource space; the yellow lines represented its degrees of complexity.

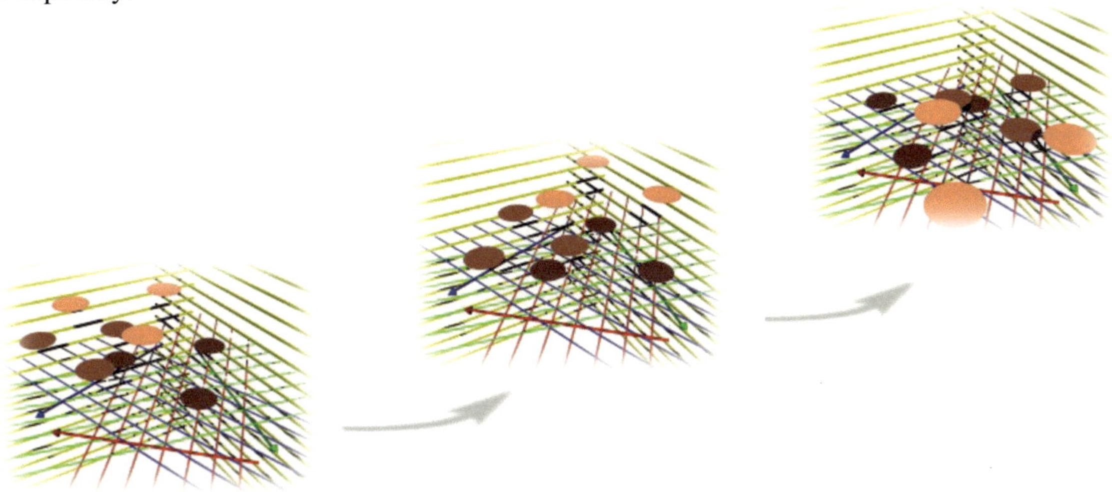

Fig. 16-1: Three e-charts, each one representing a particular region of abstractions. The resource space on the left gives rise to the next (centre), and this in turn leads to a higher region still.

Fig. 16-1 illustrates the setup once more, but this time the resource space is open-ended to illustrate the flexibility of the system, and one context can be placed next to another if the overall latency permits the evolution of conceptualisations. The sequential placement of neighbouring resource spaces makes their backward dependency obvious. A higher abstract cannot be entertained unless its foundation has been created. The relativity of it all is equally evident.

Despite the possible creation or destruction of singular entities within that space, at any given level of complexity the affinitive nature of the participants ensures that there is a certain homogeneity in terms of their mutual interplay. Just as in the single mind the input is measured against the pre-existing formations, so will the dynamic of the resource space evaluate any newcomer. Note that the elements on one level become the basis for the next. Since the potential of the entire space relates to the product/latency/applicability set, and therefore an expansion is only possible in terms of an elevation of the set's members, the long-term viability of the system depends on the actualisation of its inherent OAs as well as FAs. Progress means the attainment of yet another level; stagnation implies exhausting the available space with nowhere else to go; and degeneration happens when its elements are being dismantled.

The environment changes due to geological reasons; the climate fluctuates on various time scales as well; and it is being changed by its inhabitants, be that plants, animals, or humans. Groups grow or decline; they conquer or become defeated; and their culture undergoes modifications. Finally, individuals are born, conduct a life, and die.

Given the interdependency between the environment, the group and its members, any one of the following combinations is theoretically possible:

	no change	change
environment	3	3
group	3	3
individual	3	3

Up to now the relationship could apply to any organism, but humans are different—they have the capacity to re-represent. Once this type of perception is possible, an object is not merely a 'thing', it is also seen as a thing which does something. Although in a technical sense the internal representation is the same (ie, protein clusters, integer matrices), *what* gets represented matters to a system which can dynamically integrate the result and use it as another input. If a rock can be picked up the resultant complex contains OA(rock) and FA(pick up). Because the act of seeing a rock is functionally less embedded than picking one up (there is less engagement by the whole body), the associated TS complex of the former is more malleable than that of the latter (we have discussed the issue of stability in chapter 14, see also the Appendix and its description of the computer model). Therefore the initial instance of the lifted rock will enable more versions of different objects being handled similarly rather than the rock being used for other purposes. Archeological finds would support this: one has only to consider the range of throwing, cutting, hitting tools and compare with the variants of those objects in terms of usage (eg, at any given time there were many items which were used for throwing at something, but only a few were used to *build* a structure, that is to place one item above another in a certain way; for that more resourcefulness is required).

To return to abstracts in general, we have an additional set which exists in the mind, and it is the abstracted version of the above which ultimately determines the viability of the system. No matter what the environment may be like in factual terms, for a human society the dynamic fashions itself according to how the world appears in their heads. This applies to all the elements. Even before considering any detail, the complexity of the whole is set down by the inherent dualism the environment, the group, and the individual incorporates. We can say then that the fundamental resource space of any human system consists of the dualistic nature of its three elements, environment, group, and individual, and at any given stage its status is a function of the overall potential residing in each of these elements. As we have seen, the overall potential can be broken down to the smallest scale; in OtoomCM it is the realm of integers and the adjustment algorithm, in the bio-version the protein formations and autocatalysis. As to the higher end of the scale, we leave that for later.

At this point the reader may reflect on our criticism of the mind-body dichotomy in chapter 2 and what we are saying right now. Does not this dichotomy precisely reflect the dualism in our resource space? Actually no, it doesn't. The mind-body relationship as interpreted by the ancients and others later on is a product of their re-representing mind; it is the qualities they ascribed to their physical vs the intellectual selves which formed their views towards themselves and their world. The overall system still contained the physical body per se, with its own organic structure, its parasites, and its mechanical effects on everything else. Long before the real purposes of the organs were known, long before microbes could have been discussed, and even long before the physics of lever and fulcrum made sense to anyone, the human entity interacted with its world—and had its very own version of it. The mind-body dichotomy resulted from the re-representative quality of the mind, but its abstract manifestations remained in that particular resource space. Not until the mind was able to abstract sufficiently to allow PAT chains affinitive with its identity container *as well as* others that had equal depth but were grounded, could the mind-body set be regarded as a separate complex existing alongside many others. Before that it was the sole referent for anything to do with the self√world juxtaposition.

When in the previous chapter we placed the subconscious-conscious interplay within the context of myth creation, we can now do the same on the next higher level. It is no coincidence that every major religion (my own lack of anthropological expertise in ancient mythologies prevents me from simply writing 'every religion') refers to a dim past when its people existed in a semi-conscious state, a 'dream-time', or 'paradise', where things were taken care of by the god(s) and Man lived in naïve happiness. Eventually, so the myths tell in one form or another, that satisfying state was lost because humans 'questioned'. Note the sequential placement of the story's parts: it is the—now conscious—humans who describe the scenario, referring from their current point in time backwards. The myth can only be told because people have the capacity to articulate their self-perception; at the time of their espoused 'happiness' they could not have done so. Thus the tragedy of Man exists because we know—and if we didn't know it wouldn't exist.

For a mind to hold this perspective, especially in terms of its qualitative connotations (and so labeled with descriptors such as 'paradise') its resource space must be large enough to hold the abstractions necessary for self reflection. Yet at the same time they cannot be large enough to facilitate the creation of a further abstraction complex which permits them to see themselves as they see themselves, and therefore recognise the ultimate constraint their view evokes. The result is the relegation of relatively sophisticated thought structures to the subconscious, the attempt to limit relatively simple considerations to the conscious, and so to ascribe that major part of their resource space to the god(s). In the next chapter we will explore how this attitude plays out in the real world, but for now let us just point to the general effect: a resource space which is largely governed by the sensed but un-seen, and its potential which is restricted to those few instances of consciousness the self-created system allows.

Since the backward dependency of abstraction levels occurs in the resource space of a single mind just as it can be observed amongst people, it becomes clear that the conceptualisation of a notion represented by a PAT^3 for instance requires the presence of affinitive PAT^2s and lower, whether in the person or the group. The word 'affinitive' is ambiguous however; it merely asserts the appropriateness of one TS compared with another in the technical/functional sense. It does *not* denote any higher validity of the concept as such. For example, if the idea of 'expensive car' has evolved from an experience spanning lots of work, money saved, and diligent financial planning to take care of its use and maintenance, we can define the conceptual resource space as comprehensive to a substantial degree. On the other hand, if someone is given a sudden choice of owning a vehicle without understanding the ramifications of ownership, that person might well be inclined to get one. As can be guessed, the outcome will not be happy. In terms of any affinity relationships however the choice may not be unreasonable; there could be knowledge about this model, appreciation of the status that goes with it, a clear idea regarding how people could be impressed, and so on—what matters of course is the correct affinity once the complex is placed into the here and now.

Even in a single mind the TS complexes can be described as so many resource spaces defining their potential in terms of the parameters known from the e-chart. A PAT chain can be congruent, it can be ungrounded, it can lead to an illusion tree with its own arrangement of vertical and horizontal links; each one of these sectors can be expressed through the types of potential we have associated with the red, green and blue grids of the resource space. Basically these interpretations are merely another way of referring to the abstractive potential encountered at the level of OtoomCM, but as we move through ever wider dimensions of scale the focus turns from the matrix or neuronal cluster to the greater arena of where such states are being played out. A slight change in conceptualisation is therefore warranted.

Still, having become familiar with the underlying dynamics of TS formations, we can understand now the technical detail of such things as contextual associations, resonant and denominative structures, default abstracts (especially in the context of cluster stability), homologous and synonymous TSs, and of course illusion trees and association complexes. They all are variations on the theme and result from an increase in complexity of affinity relationships conducted under various circumstances, where the 'circumstances' are nothing but states within that selfsame system—autocatalysis writ large.

Our example of the 'expensive car' above described the viability of PAT chains, particularly in the face of possible interruptions. They can be horizontal or vertical ones, depending on the type of associativity employed. As we have seen, the contextual association that emerges may not be part of a congruent structure—all it needs are some affinitive elements. This holds especially true if the path crosses different abstraction levels, because the trigger may be an OA arrangement that looses contact with its FA component (recall the greater stability of OA complexes compared to their FA counterparts). To remain with the car example, the ideas relating to the actual performance of maintenance payments can be lost amid the imagery regarding status, wealth etc; where the tedium of financial responsibility outstrips the intellectual resources required for entertaining the exuberance of having such a car, the result is an illusion tree (in more ways than one!) of incongruent TSs. It also means there are abstractions which, linking as they do across incompatible levels, raise the possibility of attracting other TSs homologous with them but carrying various contents, again leading to incongruence (eg, to impress with the car a life style is adopted which is even more precarious).

Although the foregoing implied a more or less circumstantial development (something happens and the mind's dynamic responds in a certain way), the presence of sub-conscious as well as conscious TSs leaves the owner not entirely helpless. We defined 'social horizon' as the awareness about one's actions in relationship to society at large. On a smaller, more personal scale we can now add the concept of *conceptual radius*: it is the distance between the centre of an idea (its core) and those concepts which are seen to be relevant to the core by the person. Since the distance is measured in terms of PAT chains, we can express the possibility of a congruent path being developed through the extent of the complex's conceptual radius. If that range stretches as far as other TSs which provide congruent links, the result will be a congruent PAT chain; if not, groundedness will prove elusive. Furthermore, if the solution to a problem relies on the associations with abstracts at different levels (such as with the benzene rings), the conceptual radius will be an important determinant as to that solution's probability. On an individual scale the radius is a function of the system's connectivity, the availability of abstracts, and the emotive stability or otherwise of its formations. On the societal scale the factors are the same in principle: communication between people, the presence of artifacts and/or tools, and the rigidity of the culture. In turn, those factors are interdependent with the system's resource space, hence its products, its latency, and the applicability overall. And so we arrive at another definition: a society's resource space is defined by the parameters of the e-chart implemented across its social horizon, and represented by its members through their conceptual radii. Given the re-

representing nature of the system, to the members the same phenomenon will appear as their identity.

We can observe the role of conceptual radius at certain age levels. For instance, take a mother with a four-year old who is crazy about computer games and both know that store XYZ has them. Discussing with another adult where to go next and remain sane she says, in front of the boy, "If I tell him there is no XYZ store he won't pester me." An older child won't be fooled by this; but to the younger the possibility of being told 'no ...' precisely because there is '...' does not occur because the abstractive link is not made—the conceptual radius is too short. Similarly, if the emotive quality of the TS 'she told me there is no ...' is so strong that no other options are even considered, the result will be the same. Or, if the connectivity is so low that the combination of 'tell him' and 'no ...' is meaningless to begin with, no response will be made either.

In the world of adults the same criteria apply. While the senses convey the information about this or that situation, the subsequent process in terms of the individual's and their society's resource space defines the outcome. Because in all situations emotions participate in the forming of TSs, the degree to which they exercise control over the distribution of content is influenced by the distances involved. The more extensive the entire complex is, the less direct influence they have over any specific part. If the number of abstractions is low, if the content's representation is relatively simple, then the emotional quality attached to the participating modules will be rather intense. It follows that the con-TSs, that is those TSs which form the visible contingent of the particular dynamic, are accompanied by a forceful display of emotion of whatever type it may be at the moment. Fig. 16-2 shows two examples, one taken from the field of sports, the other from politics.

Fig. 16-2: Emotional displays in a low abstraction, tribal environment—in sports (left), and politics (right). [6]&[7]

In both cases the participating TSs are not very complex; in sports the issues are quite simple, and in Papua New Guinea the alliances of tribes equally follow straight lines—there are not many shades of grey. Note that in either case the peripheral issues can indeed be complicated, such as planning a strategy for the game or manoeuvering one's position between the pitfalls of thorny relationships. Nevertheless, the deciding gesture is derived from the ultimate meaning the outcome has for the owner, and it is that context which draws the emotions.

In contrast the presence of a high number of abstractions tends to distribute the content and therefore heightens the associativity of the module overall. There are several consequences: emotions are not as concentrated (the opposite to the cases above), the possibility of connections to other contents is increased, and the chance of grounded structures getting involved in the affinity dynamics is greater too. The second item relates mostly to problem solving since it is the degree of realised potential which enables useful conceptualisations to be raised. Thirdly, a larger area of involvement means the threat of isolated PAT chains is lessened and this in turn places the content within a more effective framework of cross-references. Hence intelligent people are more likely to be diligent and careful to the point of being perfectionists (this does not necessarily mean that anti-social behaviour is limited to a low conceptual radius, but more of this later).

So direct is the relationship between high abstractive potential and participative content that one of the characteristic signs of a child with above average intelligence is its obsessiveness with certain subjects (or even a single one), plus the desire to be neat and have things organised just so, as Jane Fynes-Clinton writes in an article on gifted children [8]. The obsessiveness is a term we use—unfortunately in my opinion, especially in the current context—to describe someone who explores a subject to the tiniest detail. The fact that a subtle nuance here or there can mean so much to a gifted child is indicative of the connectivity underlying the relevant TS. For instance, to the average kid playing with dinosaur figures their shapes and colours may be of superficial interest, one item may be significant because it's rare amongst collectors, and so on; but to the connoisseur one head shape means 'meat-eater', another means 'plant-eater', one set of legs means 'fast-moving', and a certain tail evokes visions of a very particular defence mechanism. When such a child sits amongst its 'toys' (I am using adult nomenclature here) these are all images which perform a fascinating dance inside the head—while the rest of the world can get lost. With all these details having a special value, their treatment must reflect such properties. Therefore meat-eating dinosaurs will be packed with other meat-eaters and *not* plant-eaters, and the book will be placed on *this* shelf and not *that* one.

As the child grows older, other representations will form in the mind and so gradually the universe expands. Naturally, the attention per time value given to a single item will be reduced and other considerations will be judged on a more extended basis. Hence a similar degree of 'obsessiveness' in an adult will justifiably be met with concern. However, the attention to detail in the context of a particular purpose remains a hallmark of intellect, because in the end it is the mind of the individual that decides the importance or otherwise of all that it contains and is capable of containing. Moving on to the larger scale again, the appreciation by others matters therefore in terms of the feedback dynamic of the system, whether that be expressed through the individual or society at large. For a child who gets laughed out of the group because they do not grasp the reason for a question (to them it is useless and hence 'stupid') the lesson is 'be careful with what you say' if it happens to be already confident enough, or 'oh no, I made a mistake' if it had not been given the opportunity. In both cases, the destruction of already existing TSs and the constraint imposed upon the abstractive potential of the mind is considerable. For that reason intelligent children seek out the company of those older than their immediate peer group, but even adults can fail to understand what they sometimes call the 'precocity' of a youngster. If an adult happens to be above the rest, it becomes a measure of their society's standing whether it suppresses or supports such difference.

In this book I have used examples from architecture before to illustrate a point; here is another. Fig. 16-3 contains a building from a society which appreciates intelligence and one from another that does not. None of the samples are from a particularly rich or poor area, measured against the overall context of their domain. One is from Paris, France, the other from Banjul, Gambia. One displays a framework of orderliness, of precise configuration, a dedicated placement of purpose and understanding. The other tells of negligence, of unwitting randomness, and a shallow level of care. While it can be argued that the owners' or builders' financial resources would have determined the outcome, this is only partly true. Regardless of the amount of money available, there has to be a mind which considers the effect ballusters would have on the parapet, or the cornings behind the console, or what type of semi-circular arch ought to rest on top of which pilasters. Indeed, without a readiness to embrace such concepts there would not *be* a pilaster or a parapet.

Fig. 16-3: Architectural styles representing a society's particular appreciation for detail.

In the multi-variance of life the resource space of a society is an amalgam in which the functionalities of TS formations go through their paces. The product of a person who delights in the subtleties of a scenario needs the abstractive ability of the group to accept such an item and value it. In turn the person will be encouraged to continue through the contributions made to them. In the absence of appreciation the input will flounder and with it all the potential and so possibility of future enhancement. Naturally this translates into the big picture: in a fact sheet on Africa its total gross national product for south of the Sahara is listed as $204.7 billion in the year 1991 [9]. "This was only slightly higher than the gross national product of Belgium", the article notes. In comparison, Belgium has 10 million people, Africa about 500 million, and for Australia the gross national product in that same year was $295 billion.

An agglomeration of representative clusters can only occur if its base has had a chance to emerge from the affinity-seeking dynamics of its progenitor. We have identified the parameters of the resource space and so can say that the process of articulation requires the presence of all of its participants, notwithstanding the impossibility of a perfect balance (in fact, in the absence of assessable alternatives it would be hard to identify such a thing in the first place). The considerable extent of a Western society, given its own numbers as well as its connectedness to the rest of the world, contributes to the formation of domains which need not interact with the rest as would otherwise be the case. The very idea of openness implies the room to move (in more ways than one), and it is a hallmark of civilisations that their path to greatness leads through the well-traveled roads of commerce and intellectual interchange. Isolated communities on the other hand do not generate the same diversity within their own resource space, nor are they able to relate meaningfully to intruders from the outside, as inevitably happens. They may evolve a sophisticated culture if their resource space allowed such dimensions, but no matter how refined their locale, the rest of the world is always bigger (the histories of China and Japan are two such examples).

If the identity container of a society is constrained by the lack of conceptual radius and emotional intensity, the system becomes self-adjusting in terms of those qualities. Since this large-scale phenomenon is a manifestation of the mental dynamics in its members, similar affinity relationships exist there. Hence such a society will relate to a concept represented by a short PAT chain rather than a longer, just as its members will be unaffected by considerations of a wider reach. By the same token, its products exhibit characteristics which will be understood according to their faculties by those whose own mindset works in a comparable fashion. In a world as interconnected as today (via the means of highly abstracted Western technology) the opportunity exists for many to avail themselves of a conceptual smorgasbord from around the globe; the only limitation being their personal space. So it can happen that someone who at the time was a client services director of an Internet connection company in Brisbane, espouses the view that the connection between the right and left brain is determined by race and gender; in indigenous races who have a Dreaming mythology it is developed to the highest degree and so these make best use of the brain; and within races it is women who have the best connections [10]. Presumably it was the use of her brain which enabled her to acquire her current name derived from a belief in 'Spider Grandmother', a 'North American creatrix older than God who spun the world' (in the interview she declined to give the one she was born with). This person, who it is significant to note happens to be white and female, was invited to outline her theory at an Australian Institute of Management women's business breakfast. She also was a Queensland finalist in the Australian Businesswoman of the Year award in 1996.

Several points can be made. There is a member of a society (*her* society) who, despite being involved with some of its more sophisticated products on a daily basis, can nevertheless come to the conclusion that a hunter-gatherer life style is the product of superior minds; that intelligence (as defined by herself) is race and gender based; that to present herself in the world a name hauled from distant myths is the thing to do; and there are demographics within Queensland, Australia, which see a purpose in aligning themselves with such a mind set. As such it is a good example of seeing affinity relationships, ungrounded abstracts, and illusion trees in action on every scale.

At least her social environment seems to have been convivial. When groups of unequal characteristics compete for the same space, the outcome is determined by the overall qualities of the participants. Given the types of resource spaces involved, we can expect differing reactions to the 'other', depending on the direction of view. The more compact mentality will distribute the perceptions across relatively few conceptualisations, emotions tend to be more intense, and any solution to be regarded as viable would have to meet the restrictive ambience of a small resource space. The more extensive mentality will view the situation in a larger context, emotions will be measured, and possible courses of action will be drawn from a much wider reference to be juxtaposed at a considerable depth. Since the material on neither side will resonate with the other, the first viewpoint will see the problem as due to insincerity and obfuscation, cold-bloodedness, and being too abstract to be viable. To the other side the difficulties are an overly simplistic attitude, hot-headedness, and a black-and-white narrow-minded mentality. In functional terms the outcome is predictable: if the antagonist happens to be the more intelligent entity, that is to say the other operates in an environment which is at a lower level, it is up to the antagonist to exercise restraint (after all, that's where the understanding lies). In its absence the animosity can reach murderous proportions—the short conceptual radius sees to it. Should the antagonist be the less intelligent side and be faced with a more effective opponent and environment, the result will be more complex (as is to be expected). In terms of realised potential the resource space of the environment perceives a vacuum which will be filled, resulting in activities that tend to take advantage of the other's limited capacity for defence. Or, should the higher intelligence entail a dose of altruism (of which there is no guarantee), actions of a paternalistic nature will be taken (which also could be seen as filling a vacuum). In both cases the initiative lies with the host.

Throughout history societies have responded to such situations within the range outlined above. From ancient times to today the options went from slavery to affirmative action and anything in between. It should be noted that these relationships are of a relative nature; the expressed values on either side are a function of perceptions that had been re-represented under the auspices of the receiving mind. To return to the topic of gifted children, here is one type of person who had to suffer the consequences of being different even within an otherwise homogenous demographic; in addition they upset the power relationship between mother and offspring. Consider the manner in which they are mostly portrayed in stories—as dangerous, calculating, cold-hearted monsters. In the American film *Village of the Damned* for instance women of a small town become pregnant after suffering a mysterious blackout caused by unidentified aliens [11]. When the children are born they turn out to be so intelligent that even as toddlers they know how to spell their names and can soon speak in proper sentences. Any antagonism on behalf of the humans is met with deadly mind force—the persons targeted are compelled to kill themselves by whatever means come handy. In the end the hero of the piece sacrifices himself by blowing up the children with a bag full of dynamite. Except for one boy—he is saved to be accepted as human because he 'learned' to feel emotion. The lessons for the audience are manifold, but they all fit the template of equating anything higher than the average as a threat to be destroyed by any means: (1) the very reason for the children's existence is a rape by dark creatures; (2) the unsuspecting victims are good small-town folk who go to church and live simple; (3) if a child displays abilities out of the ordinary the expected reaction by the mother is disquiet escalating into fear; (4) never mind if the animosity by their parents forces the children to react, when they do so it must be murder, no less; (5) no deeper understanding is offered, but the response by the community is the full force of the law which in US parlance translates into a contingent of armed soldiers accompanied by air power; and finally (6), the destruction, when it comes, has to be complete.

Another film of that genre is *The Lawnmower Man*, although in this case it doesn't involve a child but an adult whose intelligence gets jacked up through a computer/drug combination [12]. Anyway, it all gets too much for the protagonist and the movie ends with explosions all around; the preferred solution it seems for being different, especially when a touch of superiority is included.

While these can be dismissed as 'entertainment', their tone is significant. Firstly, no movie is made without a careful glance at the type of audience it is made for. The scenes can be exaggerated, the plot a phantasy through and through, but the principled manner in which the narrative deploys itself gets tailored according to the sentiments of its host culture. Thus there can be 'typical' French, Indian, Japanese, and also American films. The functionality, not so much content, can serve as a window to its society. And secondly, a repeated format of a certain behaviour pattern does influence its audience. Provided there are no clashes with other, competing, values, a continuous expression of fear and the demonstration of belligerent responses will in time taint the source of such arousal. At the same time, the general antagonism towards relatively higher intelligence is also a hallmark of religion, which never sat easily with probing minds. From the very beginning the US has been a very religious society, but we will deal with that issue in the next chapter.

In Australia the situation is not that different. Helen Dowland, Gifted Children's Coordinator with Australian Mensa, describes the lack of provisions for gifted children in the schooling system as an "atrocity", "even in many schools which do claim to provide for them". The problem was identified in most Western countries decades ago, she writes, and it continues to be identified in every study of that nature, but no action is being taken here [13].

The issue is not limited to helping children as such. In the e-chart we have seen the important role latency plays, and with OtoomCM the abstractive potential of a given cluster is a fundamental function of the system. Without the cluster's opportunity to unfold into something more substantial there will be no resource space—end of story. Considering the respective role each parameter plays in the entire dynamic, the actualisation of latency underpins the viability of the rest. Transferring this to the wider scale of society, we can say that despite individual differences in physical attributes of the brain, they have to be minor compared to the structural quality of the brain's environment, in other words, the culture. If there is such a thing as inherent mental capacity (and to a certain degree it must exist, otherwise we would not be able to observe various depths of understanding in people even against a normalisation of extraneous circumstances), the major drawback for any individual is the society they inhabit. In a diverse, benignly challenging, supportive and reasonably tolerant atmosphere the mind will use its inherent curiousity and absorb data to be stored and distributed in multi-faceted forms; if it is stark, dim, antagonistic and restrictive the inner world will be image-poor and beset with inertia.

In chapter 15, in the context of autocatalysis, we discussed the exponentially rising probability values for the creation of abstractive complexes due to the availability of access. Moving up through the dimensions of scale we see the accelerated diversification in a culture that has managed to move along the path of ongoing development of its artifacts, where 'artifact' means everything from infrastructure to technology and the arts. The most profound example in modern times is Western culture, but the principle can be observed in every country who has decided to open up and become prolific. Conversely, a culture which refuses to relax its constraints will maintain its base for as long as such strictures apply. On this planet there are representatives of either kind. On one extreme end of this range an indigenous culture is one which has preserved the values of its ancestors dating back thousands of years. Some people see that as an advantage (inside *and* outside such demographics). In terms of resource space however the matter becomes one of limitation coupled with the pressure to uphold it. For a community to live according to specific rituals, customs, and morals for literally millennia, a combination of two factors need to be in place: there must be a certain incapacity on behalf of individuals to think of alternatives, and/or there must be an effective authority within the group that manages to prevent the implementation of alternatives. As to the former, its extreme form of acceptance would mean that in the entire history of that tribe there had never been anyone with a better vision—clearly a ludicrous assumption, but keeping in mind the interdependency between an existing knowledge base and its potential not entirely without foundation. But as for the latter, confirmation for just such an arrangement is not hard to find.

Every culture seeks to impose a degree of conformity upon its members. In practical terms this may or may not be perceived as a restriction. If I enter a jurisdiction in which there are one hundred varieties of running a business but definitely not more, I will experience less trauma than in another which offers only two, with a similar rider attached. Indeed, there comes a time when the offers are so manifold that I simply could not exhaust the possibilities because I would sooner exhaust my own. Thus in a low-abstractive environment with a limited social horizon the point at which any member will run up against some form of opposition in their attempt to do something will occur much earlier. Accompanied by emotional intensity, shallowness of understanding, and little physical as well as intellectual room to move elsewhere, the result is a brutish existence spanning over long periods of time. Note that such words are *my* interpretation, not that of their members. For them the identity is preserved through self-referencing feedback which excludes the outside—the very reason of course for its existence.

The control is strict, and inevitable. In a more gentle form it manifests in the simultaneous insistence of one's own ways and the rejection of the other, as Murrandoo Yanner, an indigenous spokesperson and political activist from Australia's north, demonstrated in a speech he gave at the sorry day commemorations at the Sydney Opera House in June 2002

[14]. Holding up the rights to live as his ancestors did for thousands of years, he declared, "So society has got to accept us, not change us". He then gave an interesting example of corralled exclusivity, so characteristic of inert societies:

> I said 'I can't use a gun or a boat while you give all of your, all of your navy, air force, army give all the gun powder you own back to the Chinese, 'cos you whitefullas stole it from the Chinese in the 16th century off Marco Polo', is that a fact, or what?

To witness expertise somewhere else and import and improve on it would indeed be considered in the negative by someone in whose perspective the societal domains of diverse communities are barriers to be obeyed. Talking about education he said,

> But if the whitefullas can technologically adapt isn't it racist to say no other culture can? I'd love to use a spear and a dugout canoe but all them skills of how to do that, or make that have been taken from me, that's the other point I told the high court, you can't take a bloke from the bush, force him into pre school, school, all this crap all through his life, up his parents, send them to jail if they don't send them to your to school and learn your crap.

And, in the next sentence,

> All that time that's happening he's not out bush learning what he would have been learning in the old days. So you're responsible for me losing those skills and you're racist to say I can't technology adapt like you whitefullas have with gunpowder and stuff. we were always technologically adapting, any culture, any race, blackfullas including.

In the rest of the country a report found that education places taken by Aborigines are disappearing at a rate of three thousand a year [15]. An indigenous officer with the National Tertiary Education Union is quoted as saying that "lack of participation in education was contributing to genocide" (space prevents me from entering into a debate on the use of the word 'genocide'). In an article on brutality at a Youth Detention Centre in Brisbane a meeting of indigenous elders and agency workers protested against the treatment which allegedly involved beatings of handcuffed children [16]. Referring to detention rates the reporter mentioned that "Aboriginal and Torres Strait Islanders comprise more than 55 per cent of Queensland youths in detention, despite making up only 3.2 per cent of the state's population". When it comes to placing endangered children in foster care, the rate of representation is similarly lopsided. In the indigenous community the need for foster carers is particularly acute as 18.3 per 1000 children are in want of outplacement, compared with 3.6 per 1000 elsewhere [17]. In tedious succession one report after another finds the conditions in Aboriginal communities unacceptable for a First World country, but the solutions offered are repeats of the same old theme: unemployment, alcohol, reliance on the welfare system. Nowhere are the cognitive necessities for society-building touched upon.

Just as narrow views prevent the recognition of general solutions, so do they promote quick fixes on the small, that is domestic scale. The rates of violence in isolated communities are by far the highest in the nation. In the year 2001 one in twenty people living on Palm Island in Queensland's far north received an average of $30,000 in taxpayer-funded compensation after being raped or assaulted [18]. In the year 2001-02 $1.3 million was awarded to 37 victims of crime, and in the year to November 2002 almost $1.5 million was paid in criminal compensation to 48 victims there. The population of Palm Island is just over 2000. Its inhabitants, the article reads, were outraged when in 1999 the *Guinness Book of Records* "named it as the most violent place on earth outside a combat zone." In the meantime the Federal Government has deprived the Aboriginal and Torres Strait Islander Commission (ATSIC) of its funding powers and insisted on a separation of governing faculties, as is the norm in Western-style nations [19]. This was prompted by investigated allegations of misuse of funds, where commissioners had the power to use moneys at their own discretion. ATSIC's annual budget is $1.1 billion and to be elected as one of its functionaries the person has to be of indigenous descent.

For the general citizen in a modern society the expression 'tribal culture' is largely meaningless beyond a vague concept of smallness. In cognitive terms there is much more to it than that. A member of a tribe will see and interpret the world through the lens of a quite specific, strictly configured and bounded set of perspectives. Every value that is placed on anything will be referenced to its importance relative to the group. This does not mean that he or she cannot interact with outsiders; but the moral bonds which in one way or another guide us through the decision making process of daily life only hold inside the group, not outside it. The often subtle nuances of right and wrong, of 'yes' or 'no', have no purchase beyond the familiar. When such people are forced to react to the rest of the world, the standards are bound to go haywire. For this reason someone in the Apunipima Cape York Health Council has seen it necessary to apply for a $30,000 grant to investigate turning usually black-coloured condoms into culturally acceptable, "local products" (the funds are not meant to cover the actual manufacture etc) [20]. The money would be spent on procuring artwork to be applied to the condoms, such as the name of the community or totem symbols. Moreover, the word 'condom' would not be used at all but instead whatever the local word was for 'protection'. The writer of that article reported all this in a somewhat humorous fashion, but in terms of TS complexes he is missing the point: tribal cultures really do have problems with anything foreign to them. For them it is a serious issue and it demonstrates the general state of confined conceptual radii.

Short abstractive spans mean straight justice. Variants on the theme are uncommon because there is no conceptual room for them. Customs reflect this. Marriage rules in tribal societies do not allow for multiple input, and transgressions are dealt with swiftly and severely. If a promised wife runs off with a younger man (mostly the preferred option would be a younger male in place of the chosen older) the punishment when caught could be death [21]. In Jordan, another demographic not noted for its expansive attitudes, so-called 'honour killings' are practiced by Muslims and Catholics alike; it's part of the Arab culture, the best friend of a woman who has met such a fate tells us [22]. She is now a campaigner to stop the custom. Back in Australia the punishment for desecrating a culturally important image is spearing, according to Dula Njurruwuthun who, being a tribal member herself, should know [23]; and when men suspect their wife or partner of infidelity the traditional response is disfigurement so she would not be attractive to another suitor [24]. That particular article is accompanied by pictures of women who have lost their sight in one eye because it has been knocked out of them "by broom handles, sticks or other weapons".

There is an opinion, held by certain sectors of black and white society, that violence in Aboriginal communities reflects their dispossession and similar standards elsewhere. Not so, argues criminologist Paul Wilson and cites anthropologists who have documented that "huge levels of violence existed well before colonisation" [25]. The statistics show the average violence rate among today's Queenslanders is 819 per 100,000 population; in the indigenous communities of Doomagee and Kowanyama it is more than 12,000 and in Aurukun it reaches 15,420.

Recall the need for continuity when it comes to the deployment of TSs across several abstraction levels (particularly in terms of the $L_1 \rightarrow L_2 \rightarrow L_3$ etc links mentioned in the context of Alan Turing's halting problem, last chapter). Relating this to a perspective either from the low abstractive environment to a higher one or vice versa, the prerequisite for any grounded evaluation is the presence of the appropriate parameter and rule set under its auspices. Hence illusion trees can also be expressed as the result of a missing L_n in the strata of abstractions from the current rule set on downwards. For someone who is situated in a modern society, the reference markers leading to a proper conceptualisation of restricted resource spaces have usually been lost (these would be the L_n). It takes the experience of actually moving within the grounds of a world that does *not* encompass the alternatives we in the West have come to expect to understand what such an existence entails on a day-by-day basis. Tourists who throw themselves unprepared into an adventure somewhere in the wild only to run foul of its dangers are one example; commentators who opine about a rendition of traditional culture are another.

In a review of *Skins: contemporary indigenous writing* Anita Heiss enthuses about the existence of such a work (no problem so far), but then goes on to exhilarate about some of the themes found there [26]. Here is how she presents one segment dealing with revenge among the Inuits (keeping in mind the reviewer's positive attitude throughout her text):

> "The shaman took hold of the adulterer's penis and swung him around in mid-air ... his penis and testicles were ripped right off his crotch."
>
> That's swift justice!

One might also detect a touch of feminism in the above but more on this in the next chapter. While a considerable amount of literature introduces the ordinary Westerner to life in the arctic North I do not recall encountering its finer detail among such sources. No doubt there would be many, young and adult alike, who at least to some degree romanticise about the white wastes under the midnight sun and all that; how many school children would know that one of the options consists of having your genitals torn off?

One might add: how many teachers would dare setting the record straight? But never mind that now.

When a newspaper article appears telling us about yet another massacre the facts as related are certainly horrific, but for most people they remain the distant abstracts they have become in the process of global transmissions. The same goes for entertainment based on this reality. To be sure, they can lead to meetings, to debates, maybe even to dedicated get-togethers such as radio and TV shows. There interested parties pronounce their views on the artistic merits of a particular camera angle that shows a mutilated stump just so, or a scene which is 'so terrific' how it introduces us to gang warfare in the favelas (applause to the director!), or how some wretches do their movements on the screen in such wonderful measure of profound meaning that many suburbanites cannot help but wipe a tear from their eyes.

The civil war in Congo has led to a recent slaughter of nearly 1000 people who were shot or hacked to death; the attackers "slit their throats and slashed many on the head. People were killed like cows" a local chief is quoted as saying [27]. There is no need to repeat the particular configuration of TSs here—we should know this by now. What matters for the course of life as it is lived is the scene in that locale. Replace the electric light with kerosene lamps and burning timber and the night is instantly transformed into a mysterious world, where you can see as far as you can spit and beyond that there are only sounds. What you hear is what you can make of it, and one's inner imagery rules supreme. As you stare out into the blackness, so someone else watches you in return. What you see are the shadows of huts, posts, rocks, leaves and trunks; what the other looks for is a light-flecked shape of a body that moves about the periphery of his own self, outside the bounds of any kind. The tribe in his head is his home, the machete in his hand the power, and the creature before him a territory to be taken. When his blade springs open that flesh the gushing blood is the tell-tale reward of conquest. Lust is sweet.

Meanwhile the ordered universe of street lamps, fashion sense and office mission statements allows its citizens to walk around its artifacts and ponder through their personal priorities. Here is one of them:

> It is with the utmost disgust that I must condemn you for your photograph of Nicole Kidman complete with deadly fag stuck in her mouth (in a film role, Page 3, C-M, Saturday). Surely there must be thousands of other frames from this film (sans fag) that you could have chosen. Young female smokers will see this picture as an excuse to continue their filthy and deadly habit. [28]

A TS domain which acquires stability through its links with the identity container of its host is set to influence a considerable range of relatable complexes—the familiar scenario of denominative thought structures. Its interdependency with the inner plus its similar relationship with the surrounds produce a mentality that is uniquely configured to manipulate perceptions in terms of the identity. Although beset with illusion trees and incongruent PAT chains, these are only descriptors useful for our analytic purposes; to their host they are virtually meaningless since they *are* the reference. In a sense the phenomenon occurs all the time, especially when the system's conceptual radius and thus the resource space is rather large. For example, it is possible to involve oneself deeply in some problem and becoming so lost among the lattice work of multiple detail that being interrupted by some words renders them completely unintelligible for the first few moments. The TSs for language and speech patterns need to be reset for them to make sense again. If this can happen with a momentary complex how much more effective would the result be in the case of long-term deposits of experiential layers?

Obviously the daily wash of communication would preserve the faculties for language. But even they must relate to the object- and function-related structures of the mind and through them connect to the containers. Thus perceptions from the outside are configured in terms of the conceptual realm of identity complexes and the articulation derived from such domains will reflect this. At the 1994 congressional hearing in the US about the influence of tobacco companies on the health of the nation, a woman declared that her exposure to passive smoking caused her tears to be brown from the nicotine she was forced to inhale [29]. The committee members nodded gravely and thanked her for her contribution. As far as the hearing was concerned, the issue of commerce vs public health was added to by somebody's experience, the addition being another component which the health authorities and the company executives would have to deal with. For the woman the event constituted an opportunity to air her emotional life and include her persona in circles normally removed from her demographic. The societal, commercial, and political environment represents a high-abstraction complex which suddenly was made to connect with an entity on a much lower level. This does not necessarily mean that the input has to be useless or inappropriate; it does however involve the possibility of mismatch so that the dynamics of one system grate against those of another, basically unsuitable one.

We can describe the situation in the following terms: there is an abstraction complex of a certain degree which is exposed to another of a dissimilar degree, with an outcome that depends on the type of mismatch. If complex A is more highly abstracted than complex B, then the result is an interpretation in terms of A if B happens to be the addition. If A is less abstracted (and B is the addition) then the interpretation will also be in terms of A but will reflect the lower potential of A. In practice the interpretation in the first case tends to read into B something which isn't there, and in the second the perspective will be an oversimplification of an issue which should warrant far more depth. We can call the general event a *conceptual intersection (CI)* to reflect the interaction between two dissimilar conceptualisations where the result is a summary re-representation of both. As one part is considered being stationary and the other moving towards it, the scenario is akin to the role the resident and reference values in OtoomCM played in the adjustment algorithm. Here the entity is either a high-abstract complex (C_H) or a low-abstract complex (C_L), taking on the role of resident and/or reference value. In principle the only difference is the scale of the enterprise and as such once again demonstrates the scalability of cognitive dynamics. The two types of intersection are therefore $CI_{L\rightarrow H}$ and $CI_{H\rightarrow L}$.

In the foregoing example the committee represented C_H and the witness C_L. If the initiative stays with C_H the woman will be considered under the mantle of bureaucratic and political expediency and her emotional state becomes a player on that stage.

Conceptual intersections are a common occurrence. Note they define a functional aspect of society, not a contextual one. For example, the education system in a constituency dealing with a truant; a navy vessel being confronted by peace protesters; a prime minister having to answer a question about a local park; and so on. All the foregoing have the same high → low direction in common. From our point of view it gets interesting when the situation is reversed, say the courts in a fundamentalist autocracy are called upon to judge a dissenting intellectual. In terms of sheer weight the legal system with all the resources at its disposal would be the C_H, but the dissenter could have the superior intellect. It is a truly philosophical mind that can observe the contorted perambulations a challenged authority undertakes and find nothing but amusement in its stilted suffering with the only purpose of finally breaking the usurper's back—and all this when any street thug could have performed the ceremony with a swift stab in the dark for them. The fates of Galileo Galilei, of Giordano Bruno, and the many others who ever ran foul of institutionalised stupidity, they represent the conceptual intersections to which the wealth of Europe owes a debt.

And what about an overall positive ambience—when the contact between a C_H and the C_L is made in a state of grace, and the one reaches out to the other full of goodwill? The priest blessing the congregation, a king addressing his people, and a passing-out parade fall into this category. Again, it is the functional aspect which is significant here, and the cognitive and emotional strength of either participant determines the outcome so that the event could be a contest leading to destruction or an advance that goes down in history. On an individual level a well-used example is the placebo effect, whether by putting a pill into the mouth of a trusting patient or placing a hand on the head of a fervent believer (who in due course, now exercising conceptual intersection of the low → high variety, proceeds to elevate their sudden spiritual dizziness into cosmic emotional significance).

An opera singer (whose name I unfortunately cannot recall) had once been asked who is the more accomplished artist; one who can sing an aria or the other who sings popular music. To which the answer was, "Simple: how long would it take the opera singer to learn a pop song, and how long would it take the other to sing an aria?"—CI in action.

There are occasions when it forms the deception we use to enrich our lives. There is a series of photographs doing the rounds on the Internet, each one a brilliantly captured panorama (such as an autumn forest, snow covered fields, distant fjords, or sun-lit land against a black storm sky). The last picture ends with the caption, “Who said heaven is not on earth?” There is no doubt the images are wonderful; they cause us to assign something like ‘grandeur’ to these scenes. And yet—the significance lies not in those landscapes, it is in us. Just as the snowy white looks so enticing when it covers huts and trees, the same meteorological forces can produce freezing sleet, hypothermia and frost bite that results in foul-smelling agony. Thus the low-abstract object (the landscape) becomes transformed through the coincidental arraignment of a few characteristics into a high-abstract representation of what some humans refer to as heaven (which is of course another product of high abstractions).

The types of interaction mentioned above are yet another source of input for the single mind and society. They form the matrix of perceptions with an uneven distribution of abstractive depth across the public and private domains. The larger the system, the more diverse these domains can be. In a tribe the laws are homogenous with no exceptions allowed and none missed. At the other end of the scale a Western democracy provides a sufficient number of niches for the emergence of entire subcultures which may remain unrecognised for a considerable amount of time. There is the balance: the group possesses less depth and therefore less scope to address an eventuality of behaviour, but those it does know about are handled deftly; the million-fold society contains the depth but is overwhelmed by the multitude of instants which stretch its administrative resource space to the limits and whatever it does gather becomes abstracted to the nth degree.

Consider the use of recreational drugs. Initially hidden behind such euphemisms as ‘demi monde’, ‘café society’, ‘youth-’ or ‘counter culture’, the world of alternative consumers at once participated with the rest yet preserved its own secrets for decades. Arguably it was not until the democratisation of diverse drug use that the mainstream awoke to the sensation of sharing its comfort zone with alien habitudes. There are still many who do not recognise the full range of semantic options in such descriptors as morning glory, black fruit, purple people eaters, ice, California sunrise, bat man etc etc, not to mention the local idioms in every corner of the globe. In public places a nod or a wink would be as obvious to the cognoscenti as carrying a big sign on your forehead, and against a context as rich as the market can bear the slightest twitch could say it all.

In Europe before the second World War, that is to say before the reform agendas of American purists began to influence public morality there (extreme upheavals such as the wars themselves or the Third Reich notwithstanding) one’s indulgence in substances other than alcohol and nicotine was either more or less tolerated or wilfully overlooked. An Arthur Conan Doyle could imbue his Sherlock Holmes with superior clarity of mind and still present him with a cocaine habit (indeed, the two are in fact linked more than once), and this without being castigated as a corrupter of souls [30]. Coca Cola, the all-American drink for children and adults alike, used to contain cocaine and was marketed as a therapeutic agent during the late 19th century [31]. And until the 60s it was possible to order doses of LSD via mail from laboratories in Europe [32].

It could not continue. The tendency towards prohibition in American culture draws from several sources, with religion and its accompanying wariness of the physical being a major contingent but not the only one. In the context of its societal dynamics in general the duality of individualism (ie, distance from authoritative norms) and the need to reinforce a state of nationhood on an immigrant mixture leads to a sharpening of forces which could coexist more peacefully elsewhere. Recall the relationships of affinitive complexes played out within the resource space as defined by the e-chart. A system can only be viewed in its functional entirety if its subdomains are somehow relatable through their affinities with their neighbours. While horizontal connections can be disrupted anyway, the problem becomes especially serious should the vertical links be broken. The vertical links are important because without them any subdomain cannot escape a progression lock towards a higher abstraction level. We have seen the effects of illusion trees and incongruent PAT chains on the validity of their representational content: myths and phantasies abound. To a wider extent the same danger presents itself when entire domains are prevented from accessing higher or lower abstraction levels. Since a domain defines itself by its framework of containers and their identity-building influence, on this scale (we are not talking about TSs anymore but demographics of people) the domains are turned into islands of collectives, each with their own substrata of identity, customs, and preferences. It does not take much to severe the links with the outside, especially if the outside is yet another domain obeying the same principles of formation-building and so creating more clusters of similar trends.

Some of the most obvious effects are seen in educational standards, a classic variant of the conceptual intersection of the high → low kind. A recent report for the federal Department of Education and Science in Australia identified an increasing number of first-year university students requiring remedial courses in English and maths [33]. In the US, about 30% of first-year college students need the same kind of assistance and it costs the country $US1 billion a year; in California the ratio rose from 38% in 1989 to 43% in 1996. In the UK Prime Minister Tony Blair declared truancy and behaviour problems in schools a priority issue [34]. Every day more than 50,000 students stay away from school and according to their Education Department 40% of street crime and 25% of burglaries on 10 to 16 year-olds is to be blamed on these children. In Australia the violence exhibited by pupils in the classroom is also a growing problem [35]. The type of acts teachers are subjected to include throwing projectiles, kicking, biting, spitting and punching. In another set of statistics supplied by Education Queensland and covering the period from July 1999 until March 2001 over one hundred teachers were injured by students in state schools; the students were known to be violent but were included with the rest anyway [36]. Then French Minister for the Interior Nicolas Sarkozy introduced a law under which children as young as thirteen can be jailed for abusing teachers [37]. What these statistics demonstrate is a considerable defect in the $CI_{H \to L}$ relationship which via the feedback process becomes a $CI_{L \to H}$ type phenomenon.

Citing figures about problematic educational standards is not a divergence from the drug issue; neither is it meant to indicate a superficial link through stoned kids on one side and unruly behaviour in the classroom on the other. Over the past few decades a broad range of interests have claimed ownership of the issue until it has become a bizarre hybrid virtually unidentifiable even to the participants themselves. It is a typical case of mismanaged identity dynamics churning through ungrounded conceptualisations.

In order to address the players we need to move away from the customary approach which categorises drugs according to chemical type. While there is an obvious need for this kind of classification, within the context of societal dynamics it leads to confusion due to mixing of metaphors. I use the word ‘metaphor’ on purpose because in current debates such descriptors as ‘cannabis’, ‘cocaine’, etc are usually not traded in terms of their chemical composition, nor their psychopathological effects (as ill-understood as they are in any case), but are meant to convey a broad concept of social and legalistic negativity more than anything else. The words have become icons.

Let’s be clear on this. There is hardly a tribe, community, culture, race or society of humans on this planet that does not use drugs in one form or another to cement its identity. From the shamanic rituals in the Amazon rainforest to the catechism in a Catholic cathedral, mind-altering substances have been and are being used to facilitate the individual’s escape from reality. Nor are these substances treated as the chemicals they are. Notwithstanding contemporary knowledge about biochemistry, the products of such plants are referred to in terms of the mythological context in which they are seen. Amerindians chew coca leaves to give them strength, but the strength comes from the spirits, and when they inhale Yopo powder they are in communication with their ancestors; cannabis is one the sacred plants of India opening a gateway to the gods; and what is termed the holy communion in the Christian church utilises red wine to make it easier imbibing what otherwise would have to be taken as blood from a corpse (albeit a special corpse). Naturally the noble habits filter down towards the general populace and more common tastes are born. In principle however they serve the same purpose: to create an identity relationship between individuals through the momentary by-passing of grounded PAT chains. The reason is found in the need to balance cultural strictures.

The prescriptive nature of the social framework has already been mentioned. It applies to smaller as well as larger social horizons and lays down the rule set for its members. However, just as the need for a common law is recognised, so is the occasion for dealing with exceptions because it is through these exceptions that the more honest version of one's character can be glimpsed. "Get him drunk with wine and observe how well he handles himself" is one of the nine ways to find out what a person is like, advised the Taoist author Chuang-tzu over two thousand years ago [38]. Its modern form will be remembered if not rued by anyone who has ever sat down to a round of business over a course of rice wine in Chinese settings. The Western-style 'business lunch' runs on similar ideas, and there is always an added dimension to friends who are willing to engage in repartee made more convivial by a few glasses, joints, tabs, or whatever. "A friend with weed is a friend indeed" says the old sign in Fig. 16-4, but it means more than applying a selfish bias towards one's acquaintances.

Fig. 16-4: A politically incorrect but socially necessary definition of the term 'friend' [39].

The gradual emergence of TSs during the course of life are not only shaped by outside input; they are equally the result of reprocessing of such data. Recall that the definition of 'input' in relation to an affinity complex is not limited to sensory information from beyond the system, it also covers neighbouring formations inside the system. Since any data will only be recognised as a valid conceptualisation if it had undergone further processing (and hence be articulated), we can expect what is called a person's 'nature' to be representative of those various structures. Psychoactive substances modify the affinity relationships in general together with the regulatory feedback system of the cortex; as such they affect the sub- as well as con-TSs. *How* the effect manifests in a person's behaviour is therefore significant.

For an organism to be endowed with consciousness the price is a greater optionality when it comes to interaction with others. While some may prefer to use the word 'advantage' instead of 'price' our current focus is on the objective assessment of the situation. In any given group a member's ultimate validity is defined by the contribution they are able to make to their social habitat. Since the identity of the group is a function of its environment, and the character of each member a result of internal exposure and adjustment, we can say that the perceived value of a member's contribution reflects the general contingencies the group is subjected to. As such the regulatory framework of the group will have formed in accordance with those contingencies. Therefore, the greater the degree of alignment between the behavioural format of the individual and that of the group, the greater the stability of the whole. Deviations are responded to in tandem with the conceptual radius the members are capable of. In animals (from our point of view an extreme form of short radius) a newborn that is deemed unsuitable is abandoned or gets eaten; in indigenous societies the options include infanticide unless prohibited by modern-day governments [40], and we have seen an example of punishment for breaking the rules. Current levels of modernity facilitate the productive inclusion of otherwise handicapped people through technology and its sophisticated sense of the self, although remedying anti-social acts are less advanced. The 'price' then is the effort it takes to stay within the bounds of accepted responses despite the potential alternatives available to the mind—plus the cost of associated transgressions.

Still there are some regulations which will always be necessary because they inject a degree of reliability of expected behaviour into our affairs. To willingly enter a phase where such conditions are relaxed implies a certain amount of trust on behalf of the participants. In traditional acts of drug taking, that is to say during events that have been organised and integrated within the wider framework of cultural markers, the time and place are chosen to reflect the extraordinary nature of the occasion. Whether they are the more formal versions, conducted as religious or commercial ceremonies, or less so designed as general get togethers, the dynamics are similar. Regardless of the type of substance used, problems arise when the boundaries of freedom and/or responsibility are disregarded.

So much for the societal dimension; what about the mental side? The outputs from OtoomCM have shown us how varying configurations of element matrices produced the effects of specific colour patches and associated memory. While the actual integers assigned to each element are significant for the processes of the adjustment algorithm, the ramifications at the higher levels of interpretation render them meaningless. It does not matter whether an incursion of purple into a green area was based on '24-67-112..' or '75-9-61..' or whatever, what counted was the initial re-representation on the screen, the response it invoked, and the accommodation of display vs feedback from then on. The open-ended specificity of neuronal states in the biological brain suggest a similar mechanism. Protein formations in nerve cells are not inherently predisposed to represent a datum as it gets interpreted by another organism or the host itself. The conceptual qualities which emerge are a product of cell states and the response they invite in terms of ongoing affinity relationships. Therefore, as far as the representative quality of neuronal states as such are concerned, the particular underlying configurations are not relevant. For example, although the initial exposure to alcohol will modify the mental state of a person, a prolonged exposure to the drug will ameliorate the effects because the mind has adjusted itself to dealing with in- and output under new conditions. Note however that such a statement is made with reference to Otoom and not the physiological necessities of cell dynamics. In the above example the alcoholic will be in a better mental state with alcohol in his blood than without, but the physical costs are still being incurred. In principle it is the same with any other variety, but representative states being reliant on the cells' physical properties as they are, what eventuates in terms of the individual as a whole is a mind trying to make sense of modified perceptions against the background of an altered body.

Thus the consciousness which results from the ingestion of a mind-altering substance derives its nature from two levels: the physiological base belonging to the biological mechanism, and the perceptive/conceptual realm of representative states. Since human beings rely on both to *be* human beings, the effects of one will be played against the other's, with the general framework of society acting as their sounding board.

If alcohol is held as the cultural mediator, customs will be built around its use to establish acceptable limits for behaviour. The process begins early in life, particularly in demographics where historical links with the past are maintained such as in Europe. Twelve-year olds would not be allowed to go out and buy themselves a bottle of beer, but at a special dinner with the family a few sips from a wine glass can be permitted. From the age of fifteen or sixteen years onwards high school students may participate in student organisations where the drinking of beer is not only allowed but in fact required to conduct certain procedures at formal gatherings. As these are only held two or three times a year, the result is not constant drunkenness but a gradual education in the effects of alcohol on one's mind and body. By the time the teenagers are ready to leave school they would be unlikely to cause the kind of upheavals as their counterparts in Australia, where an event called 'Schoolies week' stands for several days of abandon at the end of the school year in November; at least so we are told. Held at the Gold Coast, Queensland, it regularly attracts comments in the press and is a source of concern for people in the area as well as the police who are called upon to put a lid on the assaults, damage, and general disruption. On the other hand, not so long ago it was impossible in that same state for a family with children to have dinner at a licensed restaurant because it was illegal to expose their youngsters to an atmosphere were alcohol was consumed, regardless of whether they drank nothing stronger than water.

When statistics are published they tend to come out of context and in a manner that raises public concern. Consider the following figures which appeared in the Courier Mail, Brisbane's broadsheet, designed to demonstrate how serious crime is on the Gold Coast when young people indulge in drugs and sex. Headlined *Extent of Schoolies shame revealed*, we read that the arrest figures for November and December between the years 1996 and 2001 were for, amongst others,

serious and other assault combined - 472

sexual offences - 30 [41]

Since the numbers cover six years, the average per year is about 79 and 5 respectively for those two months.

About a month later and in a different context the paper published another set of crime figures, comparing several categories across the states and Australia as a whole. For the entire year 2001 the crime rates per 100,000 people were for,

assault- Queensland: 557.4, Australia: 782.9

sexual assault - Queensland: 110.8, Australia: 86.4 [42]

(I have not used any other figures from these articles because their respective category labels are too disparate). Given that the Gold Coast has a population of 425,000 as of July 2001 [43], the figures of the first set, adjusted in terms of 100,000 population, are now,

serious and other assault combined - 79 / 4.25 = 18.6

sexual offences - 5 / 4.25 = 1.2

But because the original Schoolies statistics only refer to November/December, we need to multiply the above by 6 to arrive at a whole year: 18.6 becomes 111.6, and 1.2 becomes

7.2 (even assuming that these rates do apply for the entire year, which is obviously not what the article is meant to convey—we are supposed to believe that November and December are quite exceptional).

So, the pro rata statistics for assault on the Gold Coast are 111.6 vs 557.4 for the rest of the state, and for sexual assault they are 7.2 vs 110.8—keeping in mind that these results are on the pessimistic side since firstly, I have used the November/December set as indicative for the entire year and secondly, the figure for assault on the Gold Coast is the sum of two categories which possibly covers more incidents than the single category 'assault' used in the state- and nation wide statistics. On the other hand it could be argued that my adjustment from the Gold Coast population figure to the 100,000 pro rata value was unfair because it is the arriving students and not the residents who cause the mayhem; perhaps so, but in that case the bias would be offset by an overall *increase* in population which would cause the result to be smaller again. By the way, the Schoolies figures were reluctantly supplied by the Gold Coast police, so the article says; the second set has been quoted from the Australian Bureau of Statistics, and their values seem to tally with comparable findings published by the Australian Institute of Criminology for that same year [44]. Therefore, rather than plunging the Coast into darkness, in those statistical terms Schoolies week actually brings relief to its residents. While it is obvious that incidents do occur, one has to question the ambience created by a set of data in order to prove a particular perception.

When it comes to one's traditional drug culture the bias works the other way. A couple of years ago the Brisbane City Council instituted a new commercial sign policy and amongst many others it was the local brewery's 'XXXX' signage which attracted the attention of its officers. Belonging to the Fourex Brewery it consists of these red, illuminated, metres-high letters prominently displayed above its building on a busy road in an inner suburb. In an article about the ensuing stoush between brewery and council we are told,

Castlemaine Perkins public affairs manager Michael Blucher said the Fourex sign was the unofficial symbol of Queensland and deserved special treatment.

"(The red X's) are one of Brisbane's most recognisable landmarks," he said. "They have cultural—some would say spiritual—significance." [45]

Only a few weeks before those comments Paul Wilson wrote that over the past twenty-five years serious assault per 100,000 persons had "increased by more than 600 percent in Australia" [46]. By far most of the violence was caused by alcohol, and it occurred in the home, between neighbours, in pub brawls, but also on the street as completely unprovoked attacks by strangers. He went on to say,

> The youth of Japan and of other Asian countries also drink heavily during festive times but they do not engage in anywhere near the same level of violence as do our youth.

As a reality, this is nothing new. Back in 1986 the Royal College of Psychiatrists in the UK published a report (*Alcohol—our Favourite Drug*) in which they pointed out that alcohol brought more "deaths, suffering and mayhem than heroin and cocaine", and government should do more to bring home such "sobering facts" [47].

The Sydney Morning Herald also ran some news on a report about ecstasy use and its effects [48]. Several comments were made, such as brain damage in rats and monkeys, an increased degree of risk when used at rave parties but a reduced risk when done "sitting quietly with friends", and the problems of dehydration and overheating. The article had the headline *Research points to the agony of ecstasy* and the first paragraph read,

> People who take the illicit drug ecstasy risk poisoning, possible brain damage and death from heat stroke, British researchers say.

Then followed 21cm of column space containing information such as mentioned above. It was only in the very last paragraph that the whole issue is put into perspective:

The report noted that while five deaths from ecstasy were recorded in Britain in 1990, alcohol claimed 30,000 lives and tobacco 110,000.

We know how relevant and at the same time how irrelevant the integers in OtoomCM or molecular clusters are; it all depends on the scale of conceptualisation. In principle we can observe the same relationship at the level of societal dynamics. The effects of drugs are not considered relevant in terms of the actual damage they do but how they fit into the conceptual framework created by the community's identity containers. While the facts of the matter are the known data that can be accessed objectively, they can be compared to the integers in the computer matrix. As such they do not represent the output—they are the medium through which the imagery is being formed. When it comes to the result they are as dependent on the affinity relationships expressed within their domains as are the numbers in the element matrix, and they similarly produce clusters and subdomains that serve the existing containers. Their inherent value (whether it happens to be '24-67-112..' or '5 ecstasy deaths vs 30,000 alcohol deaths') is irrelevant to the greater dynamic.

Ironically then, the reason for drug use and its often controversial role in society are intimately linked. In chapter 2 we began to observe the attempts by the mind to construct ideations which fulfilled the needs of its abstractive potential. The mind-body dualism was a major consequence, as were the ideals produced by St Augustine, Aquinas, the Cathars, and so on. To perceive an object is not enough; we want to extract its innermost meaning from it, even if that means leaving the grounds of reality altogether and embarking on a journey into realms we hardly understand ourselves. How could we—if they are the effects of ungrounded PAT chains they are our own personal phantasy and that's that. What manages to reach consciousness is but a shadow of what the subconscious really contains. It is true that without abstractions we could never build a machine, but those abstractions themselves do not contain the safeguards to prevent their owner loosing touch with reality. Not only that, since as cognitive dynamics their structures obey the same principles of affinity relationships regardless of content, given the right opportunity they can relate to clusters or containers that have the potential to be affinitive. What determines the outcome is the availability of links, and it is those links which humans try to evoke.

Consider an individual who is dissatisfied with his identity and seeks an alternative conceptualisation of himself; the choice will trend towards something he can perceive as profound rather than trivial, and so he will undergo exercises that demonstrate to him his connection with a god. Or the person who brings meaning into his life by joining a political party with desirable goals or a leader who attracts. Or, on a more trivial level, someone who muses about a situation and dreams up non-existent alternatives. The common denominator is the same—there are two thought patterns, and they need to be joined together. Note the denominative nature of the event: there is an idealised version of some concept, and it needs to be paired with content on a less abstractive level. The higher TS can be defined in terms of its associative intensity towards the lower, and the outcome is a function of the associative resonance of both. Because the TSs occupy different abstraction levels the resultant association is a functional one. Both TSs tend to become resonant thought structures because firstly, the lower-level abstraction has acquired a greater significance from its association with something higher, and secondly, the higher-level abstraction is perceived to be of greater significance because it has gained a closer link with the mind's identity container.

Thus drugs are the means to facilitate associative resonance. Since the phenomenon is of a functional nature, that is to say the viability of its dynamic relies on functional characteristics rather than contextual ones, in terms of efficacy the contents of the exercise do not matter. In other words, if a certain substance allows the person to dream about powerful beings, it is the cognitive formation of 'self + power + idealised beings' which will result—whether these turn out to be loving angels or terrifying monsters is neither here or there as far as the neuronal system is concerned.

This is the reason why organised drug taking is performed in ceremonies, the evoked framework of culturally familiar and desirable concepts serving as the markers for a path over which the conscious mind has little control once the first steps have been taken. For the church ministers who warn against involvement with different spiritual customs their concern is only partly due to territorial jealousy (although it often gets expressed that way). When the mind undertakes a meditative experiment, in moments of stress the comfort that childhood images of trusted icons bring should not be underestimated. Conversely, to step out into the conceptual void on one's own and suddenly coming face to face with the sensations of subconscious turmoil can indeed destroy the network of identity TSs. At this point one flirts with madness unless one's own measures are in place—just as not everyone can climb a mountain.

To socialise is also an act of associative resonance. Making contact with strangers or reconfirming one's familiarity with those we already have come to know serve the same purpose of connecting one's self with the perceptions we have of others. It is no coincidence that the degree of intimacy inherent in a subject of conversation rises along with the consumption of whatever drug is used on such occasions. There is the common view that drinking a few glasses of wine for instance loosens one's tongue; this is only partly true. If the attempt is made (and more importantly, if it *can* be made) to observe the manner in which the communication between people proceeds, it can be seen that the participants spend a great deal of effort on accommodating their own views towards what they perceive the others represent. A phrase like "that's right" is probably the most often used acknowledgement of what has been said, and inserting slightly divergent points into the round demonstrates that people are willing to change their own views to a considerable extent so that the overall theme appears agreeable. Note however that the willingness to do so must be present in the first place. Also, because perceptions in general are being modified the accuracy of interpretation tends to suffer, and consequently the point at which disagreements emerge may be missed. By the time the communication flow carries disparate elements while still treating them as potential building blocks for denominative TSs, their destructive influence has had a chance to act. From such an entrenched position it is virtually impossible to retrace one's steps and the inevitable conflict arrives (once more it should be emphasised that the above refers to the cognitive layer of mental dynamics, and not the physiological events induced by the substance; the latter add their own mechanisms to the process).

To summarise: the human mind, as a cognitive system, has the inherent tendency to form identity containers which in turn promote the formation of associative clusters. On the societal scale the same happens in principle and so groups and communities emerge. Indeed, without this facility a society per se would not come into being. Due to the presence of cultural constraints the abstractive potential of the mind has to be harnessed to prevent untenable behaviour, where 'untenable' ought to be understood in terms of the individual's as well as the group's identity. Psychoactive drugs (by this we mean any substance which influences the mental processes, from coffee to cigarettes to LSD) are used to facilitate the establishment of links between otherwise disparate ideations, with the proviso that it is the structure closest to the identity which is being served and not the other way around. As such the identity is strengthened and its periphery aligned with the core.

Thus one could say that drugs help individuals to form groups—and groups help individuals to take drugs. The ceremonies in student organisations, the rites in the Christian church, the toasts at banquets and the sharing of needles amongst junkies all come under the same banner.

While the functional dynamics can be compared across the globe, their cultural settings provide the contexts in which such dynamics are being played out. The reinforcement of identity has therefore different results in high school students, diplomats, or slum dwellers. The associative resonance in the single mind follows similar tendencies. Since a person's consciousness is only a small part of what the entire mind holds, it suffers from two limitations: the wilful feedback mechanism of con-TSs is largely a linear system, whereas the rest of the neuronal network operates as a distributed one (for a description of linear vs distributed systems see under *Computer Model* in the Appendix); and cultural restraints make the existence of undesirable domains possible with the result that they are not being accessed by the owner's con-TSs. As the functional environment changes under the influence of a drug we can expect the resource space of con-TSs to mutate as well. To understand the nature of altered resource spaces we need to consider the effects of drugs at the physiological and the Otoom level.

The formation of affinity complexes within a neural network rests on the ability of cells to transmit data which are representative of their own states, and do so in a manner which induces relatable states in the next cell. Without the stability of these synaptic interchanges no 'meaning' can emerge because the relationship between one state and the next can never be referenced reliably. If the exchange of neurotransmitters becomes modified, the induction of phase states is now placed under a new regimen with no means to relate the current characteristics to those from before. In OtoomCM it would be similar to taking the integers in a set of element matrices and change them without due regard to their history, the reason why they exhibited the values they did after a course of adjustment processes.

Generally speaking there are two classes of agents which affect the activity of neurotransmitters, those that enhance their transmission and those that inhibit it; agonists and antagonists [49]. An example of the first type is nicotine, of the other atropine, a derivative of which is used by ophthalmologists to dilate one's pupils for an eye inspection. Another of the latter is curare, a far more powerful poison that Amerindians put on their arrows to paralyse their prey.

It gets complicated when the effect is more selective and targets some receptors but not others. Leaving physiology aside and moving on to Otoom, we have seen how the entire set of integers in an element matrix was modified in relation to another set of reference values in a way we could interpret as a specific response, even memory. We stressed that for the emergent state the actual values of those numbers were irrelevant since their meanings derived from the interdependent effects they had in conjunction with the response from outside their domain; that the correct level of conceptualisation of these two types of events (integer values on one side and what their entirety means in terms of complexes in another part of the system) is instrumental to an understanding of the phenomenon; and that an incorrect approach leads to such things as the eternal debates about consciousness and the grain problem.

Suppose in our computer model we have a test run during which one particular node in one particular element matrix has the values '... 645 1184 570 ...'. We know this because we instructed the program to print a list of matrix values. We also know from observing the output on the screen that in a corresponding area the colour green is giving way to yellow (these events are taken from the example listed with Fig. A-18, A-19 etc under *Computer Model* in the Appendix). Further, checking a purple spot somewhere else in the output we discover the number sequence '... 813 1150 2490 ...'. The figures '... 645 1184 570 ...' (A) are significant for us because they affect the colour patches we perceive in the output (but as pure integer values they are irrelevant). Likewise we associate '... 813 1150 2490 ...' (B) with purple. Suppose now during one cycle a computer virus would change the first set so it resembles the second. If A brings on yellow, and B stands for purple, turning A into B would certainly get rid of yellow, but it may not generate purple. Such sets cannot be simply exchanged with each other because what we perceive as 'yellow' represents the culmination of a specific node domain which, in conjunction with its functional neighbours (ie, those it is linked with under the connectivity plan of the matrix) produces an effect that has been processed by the output subsystem such that the result happens to be 'yellow'.

If sentences like the last one seem long-winded they do so because they describe a process which is anything but straightforward; we are faced with the dynamic of a multi-layered resource space, and our language needs to account for the changing levels of conceptualisation.

In any case, we have come back to the elements of our e-chart. For something like 'yellow' to become visible to us, there has to be a specific product (the integer values), with its potential being realised (the integers being adjusted in terms of their neighbouring matrices), and becoming applicable to us in a certain manner (what we are able to recognise as a yellow patch). Note the interdependency here: since the resource space relevant to 'yellow' is a concept for us and not a formally delineated region (which precise node can be identified as belonging to 'yellow' and which to the next colour?), it is as reliant on the dynamics of its periphery (wherever that may be) as are those of what is happening here. The nature of such dynamics are the basis for the affinity relationships that eventually cause whole events to emerge. Hence the resource space of a perceived event is a function of its affinity potential, realised within its neighbourhood—we are looking at the scenario depicted in Fig. 16-1.

Alter any one of the components in the resource space and the result will reflect the relationships as they had become established during the course of events leading up to that point in time. Therefore, if a conscious event is based on some configuration A and contains elements which have an affinity with some other configurations B and C, then altering A will create the possibility of having an affinity with a D or an F. It can only ever be a possibility since neither A_1 or A_2 is strictly definable in terms of an exact extent and so a precise influence; besides, a workable affinity needs *two* matching entities and there is no guarantee that the other will meet the criteria.

In the case of psychoactive drugs (keeping in mind our interpretation of the term) the influence ranges from a mild enhancement and/or suppression of synaptic transmissions to an actual modification of their content. As far as the potential of a TS complex is concerned, even a mere enhancement can make the difference between accessing a domain or bypassing it, as any late-night worker downing his espresso will know. Having accessed another TS can then lead to further complexes and entirely new trains of thought can eventuate; as such the effect is not trivial at all. In the case of a modification however the result is more profound. If the configurative basis for one particular ideation and therefore its affinities with other concepts is changed, not only will the ideation itself be altered but so will its affinity potential. Given the interdependencies within the resource space (at whatever extent we are willing to consider it), the new TS can align itself with structures it normally would not link up with. Note the type of formality: what emerges is not some random flickering of consciousness (for that to happen the physiological conditions need to be sufficiently severe) but an ordered display of re-representations—except that it is not an order its owner may expect.

In previous sections we discussed the formation of identity-related TSs as they occur during a life time. Since exposures to one's environment do not follow the dictums of culture and morals, at any one time there could be complexes which do not fulfill the imposed predicates. They are not deleted; rather, they are relegated to islands which in the ordinary course of events are being left alone and so don't bother their holder. They are left alone because the person's consciousness wills it so, and they are not a bother because we are not directly aware of subconscious states. They do however exist and they do participate in the formation of affinity relations with the only proviso that the results may not be accessed by the con-TSs. The more of these unwanted complexes there are, the greater the chance that any given PAT chain will touch upon one and hence the greater the possibility that it will be rejected by the 'law-abiding' con-TSs—it is the major reason why liberal environments are also creative ones.

Under the influence of a drug the affinity potential of TSs is changed and as such its regulatory powers become muted in relation to the usual set of rules. As the affinities still exist, connections can be made which seem disturbing to the person, but just as memories can be modified to suit the context in which they are accessed (see chapter 15), so do the emergent TS complexes reflect the circumstances of their evocation. Fears, phobias, unreconciled elements of one's persona, they all rise to the conscious surface from the subconscious deep. By the same token, positive sentiments also come to the fore and mutate into a contextually relevant imagery. In principle the same effect can be achieved by intentionally delving into one's self and so overriding any customary defence mechanism installed by everyday habits. Deep meditation by followers of a religion produces images relevant to that religion, or self-hypnosis calls up vistas of pagan scenery if one is so inclined, as the pictures of Rosaleen Norton, the 'witch of Kings Cross' as she became known, so insinuatingly demonstrate [50].

In his book *Human encounters with death* Stanislav Grof details his experience with terminally ill patients who underwent LSD sessions under his guidance [51]. LSD (lysergic acid diethylamide) was synthesised by the Swiss chemist Albert Hofman in 1938 and belongs to a class of compounds usually labeled psychedelics, which in itself is a construed descriptor and represents the ambiguity associated with these substances. The circumstances surrounding Grof's selections are telling in themselves. His patients were unsettled by the ominous advent of death, their religion did not provide them with the calm they thought they could expect, and they were willing to undergo an experiment which was explained to them at length. Most individuals benefited from the sessions, and their attitudes changed markedly towards the positive. Given the supportive psychological environment in which the experiment was conducted, the experiences turned into profound re-evaluations of their inner selves. The common denominator is significant: subconscious niches were reached that now were able to integrate again with the rest, and the persons saw themselves as complete wholes where death was a natural event in line with their human nature.

A more diverse view is offered by Aldous Huxley in his *Doors of perception* and *Heaven and hell* [52]. Glorious exaltation is mixed with abyssal terror, but in the absence of Huxley's personality profile and inner world it is obviously difficult to trace the sources for his images. It is interesting to note however that in the earlier stages of its history psychiatrists used LSD in order to uncover the subconscious of mentally disturbed patients [53], and the US Central Intelligence Agency together with the Army investigated its hallucinatory properties by giving it to soldiers and prisoners alike [54]. Presumably the often unpredictable consequences rendered it less and less attractive to formal treatments, since a more in-depth evaluation of the subjects' mind states was not available.

In the light of affinity relationships and associative PAT chains however it can be said that the cognitive effects of LSD and the other members of that family such as ecstasy, peyote et al are a function of that person's inner state. A well-balanced mind set with an appreciation of one's natural self and untrammelled by religious angst will use the existing TS contents and manufacture a fascinating, glittering world. While it is artificial and should be understood as such, in the sense of its conceptual engineering it is no different from sophisticated computer graphics, but in terms of actual content it is far more intimate than a third-party program can ever be. From this comes an honesty that often is too confronting for the world at large, especially since a society's standards are never set by intellectual exceptions but by the steady majority, its main structure. Fig. 16-5 gives an example of Rosaleen Norton's work (arrived at through meditation) and what the media around the middle of the 20th century made of it.

A cultural niche which has the energy and will to assert itself can be an incubator for perceptions that the mainstream is not capable of holding; neither its members nor its framework posses the adequate resources. Kings Cross in Sydney was—or still is?—one such niche, and during the years 1970-72 a group of artists came together and created the physical dimension to their own extended resource space. Known as the Yellow House it became a centre for exhibitions, readings, meetings, or simply events; a focal point for a type of culture only possible in a metropolitan environment buffered from the rest of the world through its multi-faceted faculties and the capacities of minds. One such perspective transposed into the articulation of a room (a space within the space within the space) is shown in Fig. 16-6. Its ambience could be called self-explanatory, were it not for the sheer number of possibilities a 'self' can contain.

Fig. 16-5: "Bacchanal" by Rosaleen Norton (left), and the artist's re-representation in society [55].

Fig. 16-6: "The Stone Room". Design Martin Sharp, completed by George Gittoes, Joyce Gittoes and Peter Kingston. A recreation for the Yellow House exhibition at the New South Wales Art Gallery in 1992 [56].

So context matters. In a Dutch study which investigated a random population sample over a three-year period it was found that there is an association between frequency of cannabis use and the presence of psychosis [57]. 'Use' was defined in terms of nearly every day, 3–4 days per week, 1–2 days per week, 1–3 days per month, and less than once a month; 'psychosis' was interpreted as "unusual thought content," "hallucinations," and "conceptual disorganization" and was classified in terms of "no need for care" and "probable or definite need for care" in relation to the symptoms [58]. The data suggest however that exposure to the drug and vulnerability to psychosis in the user together account for between 79 and 83 percent of incidence [59]. The authors note that the longer-term trend towards higher risk seems to be due to gradual physiological changes in the brain rather than the short-term effects induced by cannabis [60], and warn that although statistical relationships can be ascertained the substantial difference in the sample pool for all individuals (4045 subjects) compared to those who reported any incidence of psychosis (38, comprising only 0.94%) [61] should give rise to caution. The conclusion stating a "maximum proportion of psychosis outcomes attributable to cannabis use in psychosis-free subjects" being "higher than 50 percent" is preceded by the rider "If a causal relation between cannabis use and psychosis outcome is assumed.." [62].

For us two things are significant: (1) the identified level of cannabis use was based on the subjective assessment by the individual. In a climate of illegality this may cause under-reporting, but otherwise (personal use of cannabis is legal in the Netherlands) it may have the opposite effect, especially when it is clear that one's habit is linked to potential psychotic behaviour. Rightly or wrongly, some people tend to overstate a problematic situation in order to demonstrate their prowess. (2) notwithstanding the physiological changes identified in the brains of users and/or schizophrenic patients, the type of interaction with one's environment is as much due to socio-economic as to personal-psychological factors, and when the representative properties of the mind are modified (for whatever reason) the overall mental state can be challenged in ways ordinary circumstances would not require.

The very atmosphere which gives its elements the air to breathe also poses its demands. In principle the organised articulation of abstracts are replicated right across its expanse, but the conditions vary from one instance to another. The identity-building mechanisms of approved versions are subsumed into the wider framework of society where they are nurtured and confirmed, but for the cultural exceptions such a reach is not available. Yet the dynamics relating to identity containers and their affinitive properties still hold, and so in a functional sense the effects are the same for all.

When a shaman introduces the youngster to the spirit world where he is to encounter the forces which shape his and his tribe's existence, a successful return enriches not only the new man but also his community. The interplay between individual and society ensures a union that benefits all, and in the process both parties reinforce the status of their identity. The same goes for the rite of Confirmation in the Christian context, the bar mitzvah in Judaism, and so on. While for some the connection between taking the Eucharist and smoking a joint may seem somewhat tenuous, in terms of the societal dynamics the links exist. In both cases the persona is called upon to integrate with the group, create abstractions which serve as the group's identifiers, and confirm one's commitment to it all by participating in the sharing of ingestible substances. After all, how more serious and more honest can one be than being there with one's body, one's mind, and consume in a communal setting? As said before, without such ceremonies society would not have come into being.

Consider now the same interaction for a group which does not enjoy the wider support of the community. What would have changed? Looking at the sets of e-charts from Fig. 16-1 we can extend these images, select one somewhere at the higher end and move backwards to reach what we could deem a representation of a particular demographic. While there are changes from then onwards (our assumed situation), for this set anything below remains the same as far as the resource space dynamics are concerned. What *will* impact on that region is the upward-directed potential in each case, influenced as it is by the modified content coming from above. Therefore we can say that for any region within a set of resource spaces an influence from above downwards will result in certain contextual modifications, leaving the functional aspects unaffected. If the influence happens to be a positive one, the members of the uppermost set will enjoy an increased possibility to actualise their potential, and for anything lower the dynamics remain the same although an improvement could occur due to better products and their ramifications. If it should be negative, the uppermost set will experience a progression lock, and there is the likely chance that the products in those sets will diminish, again with relevant ramifications in tow.

A more general example can be seen in isolated rural communities some of whose children eventually make it to university in the big city. Despite the options available for a successful student, life in the hamlet itself will continue to follow the same rules; although one could expect that the presence of a university education will stimulate the standard of subjects taught at the country school. Conversely, a civil war in the city will drastically remove the options for rural people, won't change the habits there but would most likely dry up the income derived from city dwellers and so would result in hardship for the locals.

When it comes to the control of drugs however the dynamics happen at a far more fundamental level. Since we are dealing with identity-related formations any invasion into the dependencies of such subsystems will have profound consequences down the line. In a resource space the option to counter a difficulty will only be there if the space has the necessary capacity. As we have seen, that capacity draws on the elements existing at that level and/or the availability to move to a higher abstraction. The intolerance practised by cultural regulation acts on both ends; it attempts to stifle the confirmation of the target's identity as well as undermines its means to function in terms of its nature. The result is the evocation of a defence mechanism which seeks to protect the target while attacking the aggressor, all within the confines of its resource space—it cannot act otherwise. Again, as a phenomenon in itself this is nothing out of the ordinary; think of partisans in enemy territory, cells of an outlawed political party, prohibited religious sects, even children sticking together lest they be found out. The only difference lies in the capacity of the players and the permanence of their status.

Such fundamentals turn any efforts to curtail the use of drugs into a different matter altogether. In earlier times the general availability of abstract-forming substances was limited to agrarian know-how, and its specific versions to cultural demographics. Today it is possible to draw compounds from a variety of sources, and the focus is on effect rather than mythological significance. Yet the functional principles of the mind have not changed and as always they govern the ultimate outcome.

The response by the mainstream has not evolved either. The religious puritanism in the United States committed itself to change human nature by instituting the 18th Amendment to their Constitution and from 1920 until 1933 the prohibition reigned. As an entry in one history book dryly comments, it "fostered smuggling, gangsterism, and lawlessness" [63]. The affinity-seeking properties of demographic clusters induced like-minded citizens on both sides to reconfirm their identity, and the violent element in Christian righteousness found a ready sounding board in the Ku Klux Klan. It enjoyed a heyday during the 1920s because the sharpened atmosphere allowed it to campaign for morality and so against blacks, Catholics, Jews, and of course all those criticising the prohibition.

In 1937 a law was passed in the US declaring cannabis a narcotic [64]. Harry Anslinger had been appointed Commissioner of the Narcotics Bureau and articulated his antagonism towards the plant by calling it the "assassin of youth", informing congress that "you smoke a joint and you're likely to kill your brother". In 1961 he orchestrated an agreement amongst sixty nations with the aim to eradicate marijuana use within twenty-five years. Around the same time Washington was declared the crime capital of the nation because most offenders were heroin addicts trying to finance their habit. As an example of cultural parochialism a program was instigated which replaced heroin with methadone and within a year burglaries had dropped by 41 percent [65]; the difference was people were now getting their fix from a chemical supplied by the government but in principle it was the same. By the spring of 1971 a report found 10 to 15 percent of US military personnel in Vietnam hooked on heroin and marijuana use "was endemic". President Nixon felt something had to be done and started a programme to eradicate drug use once and for all. When cocaine came along in the seventies the Medellin cartel supplied the US market to the tune of around $100 million per month. The operation was compared to a Fortune 500 company and it spawned massive investments in the States and in Europe, from real estate to planes to boats to islands.

In Australia the Vietnam war brought R&R visitors to mostly Sydney and Kings Cross and with them hard drugs such as heroin, creating a counterpart to the cannabis and psychedelic movement already in place. Just over ten years later the Sydney Morning Herald reported that "drug trafficking in NSW in early 1984 was almost out of hand" [66]. In 1986 a joint Federal-NSW task force asserted that ninety percent of heroin users committed crimes to support their life style and the supply was in the hands of a "highly organised criminal conspiracy."

Statistics from 1988 showed alcohol to be the "biggest killing drug among young people in Australia", where it was responsible for "73 percent of drug-caused deaths in the 15-34 age group" [67]. Something had to be done, and so the biggest anti-alcohol campaign to date was launched. According to State and Federal police forces cocaine traffic was increasing that same year, and fifty percent or more of the staff from Customs, the National Crime Authority, and the Australian Federal Police were assigned to fight illegal drugs.

By 1989 we read, "Seven out of 10 crimes committed in NSW are drug related—and there is no solution in sight", according to a Federal parliamentary committee [68].

In 1993 the Advisory Committee on Illicit Drugs in Queensland published a quite comprehensive report called *Cannabis and the Law in Queensland* [69]. In its summary of inquiries regarding the recreational use of hemp we find no less than twenty-five attempts to come to grips with an obviously continuous phenomenon [70]. They start with the British East Indian Hemp Commission 1893-94 (which concluded that a ban on hemp drugs would likely do more harm than good), move on to an investigation into the Smoking of Marihuana among Military Personnel in Panama in 1932 (present military prohibition should continue, but no further legislative action to prevent the sale and use proposed), and other samples such as the 1968 Wootton Report in the UK (restrictions should be maintained), the Le Dain Report from Canada in 1972 (suggested a repeal of prohibition against the simple possession), the 1973 Shafer Report in the US (possession for personal use should not be an offence), the 1978 Joint Committee Upon Drugs in New South Wales (not enough evidence to remove existing restrictions), and the 1991 Select Committee on HIV, Illegal Drugs and Prostitution - Third Interim Report from the Australian Capital Territory (staggered restrictions in terms of quantity, with a lower base line of 'no offence' and reaching up to two years imprisonment plus a penalty of $5000). It also featured an estimate of the Australian market in 1988/9, namely $1,905 million [71]. Apprehended offenders are "overwhelmingly young, single, unemployed or unskilled males" [72]. Overall, drug offences constitute 5.4 percent of all charges before the lower courts, but the paper emphasises that other offences generally get reported *to* police whereas drug offences tend to be detected *by* police [73]. In the report's conclusion the drug's role in assisting in mind-altering experiences is noted, linking it to religious traditions which have concerned themselves with just such processes [74]. There are elements in mysticism which modern society does not address and the failure to enter into altered states deprives humans of their creative and cultural potential. The report ends with the words,

> Thus, we may understand the use of cannabis in society not only as a public health issue, but as a sign of a fundamental but unfulfilled human need, which cannabis users attempt to fill by use of the drug, albeit inadequately. [75]

When police arrested members of an "interstate drug ring" on the Gold Coast they found ecstasy, LSD, amphetamines, cocaine and cannabis [76]. As most of the drugs were sold in nightclubs the managing director of the Surfers Paradise Licensing Venues Association distanced herself from any "drug-related activity" on their premises, saying it "obviously detracts from their own business of providing entertainment and provision of alcohol in a safe environment" (note the demographic differentiation).

Given the continuing information from the media and the conservative nature of Australian society it is not surprising that parents are resorting to spy on their children, with some paying private detectives up to $20,000 to find out whether their offspring is using something illegal [77]. Since such actions are perpetrated by people to whom their own values matter to a substantial degree, on a superficial level it may seem ironic that family-oriented parents drive this kind of wedge into their relations. Considering the societal dynamics in force however we can understand the taking of drugs as well as the parental initiatives as being driven by the same principles. If both parties are prisoners in their own resource space neither has any choice.

At yet another national conference on drugs (this time focusing on jails) Queensland's phenomenal growth in the prison population was mentioned as a cause for concern [78]. Queensland Corrective Services Commission director-general Barry Apsey is quoted as saying,

> No country or state has experienced the rate of increase which has occurred in Queensland since 1994 where a prison population has risen from approximately 2000 to the current 4550 in April 1998.

There are several reasons for this with drugs being only one of them; we will examine the role of low social horizons shortly.

On the other side of the fence there are strident developments too. In 1999 the Australian Medical Association says that the prescription of mind-altering amphetamines for school children has increased 21-fold since 1990 [79]. This translates into six percent of children currently taking such drugs, although doctors "have reported more than 70 percent of children diagnosed with attention deficit hyperactivity disorder [ADHD] do not actually have the condition". At the same time Association of State School Principals president Tom Hardy thinks that instead of kids having ADHD all they really need is more attention at home. He goes on to say, "There are some parents who aren't relieved until they find a doctor who says 'this child has ADHD' and they give them the tablet".

Read the above carefully. Not only do we have the increase in prescriptions for mind-altering substances (the legal ones of course), we also have parents who actively push for their child to be included in the rush, plus we have *some* doctors who prescribe amphetamines to children but we also have *some* other (?) doctors who are able and willing to reassess such actions to the tune of 70 percent of the former, all that against the backdrop of parents wandering from one surgery to another looking for a fix by proxy. Somehow the numbers don't add up—or do they after all?

And every now and then we have a lone voice like the one of Australian Drug Law Reform Foundation president Alex Wodak who told an Alcohol and Drug Foundation conference in 1999 that it was the criminals who reaped billions from the drug trade and were propping up campaigns to keep drugs illegal, leading to unacceptable levels of drug-related deaths, disease, crime and corruption [80]. By the way, it might be interesting to note at this point that the just quoted article occupied an area of 22.5cm^2 in the newspaper, whereas the report about the 'smashed drug ring' on the Gold Coast for example in which the alcohol industry expressed its concern about competition was given 544cm^2 (5×4.5cm and 32×17cm respectively). But back to the story.

Whenever attempts are made to move the mechanics of supply and demand out of the hands of criminals into something more regulated and reliable everyone seems to get scared. In 2000 a Brisbane entity called the Fortitude Valley Community Consultative Committee representing police, drug support groups, government agencies and the Brisbane City Council, initially suggested a scheme whereby pharmacists would sell heroin for between $3 and $20 per dose [81]. Hearing of the plan Mark Le Grand, who used to head the Criminal Justice Commission and who is described as a "high-profile crime fighter", thought parents should be alarmed at the prospect of heroin injecting rooms being legalised. These rooms "were 'fraught with danger' for the youth of Australia". Brisbane's Lord Mayor could not publicly support the committee's idea because it went against State Government policy, the Opposition claimed the Labor city council was trying to push through a heroin trial against the government's wishes (which also happened to be Labor), the State Premier confirmed his hard-line stance against injecting rooms, and others in the city council feared the report would attract unwanted media attention. After the furore (judging by the article) the task force report had been "carefully 'sanitised' for fear of a media backlash", and now does not mention anything about trials and/or injecting rooms. Meanwhile, back on the streets, between 1997 and 2000 there had been a 130 percent increase in heroin overdoses.

A short while later the Year 2000 Illicit Drugs Reporting System found that the use of amphetamines in Queensland "outstripped that of any other state or territory", with their availability and quality higher and the age at which young people injected them lower with "people as young as 14 years old getting into IV drug use" [82].

At the Australasian Paroling Authorities Conference in 2001 a member of the Queensland Community Corrections Board, Ross Fitzgerald, put the number of inmates with alcohol and drug abuse problems at eighty percent [83]. He called the current rehabilitation programs "scandalous", suggesting complete abstinence rather than the harm minimisation models in use now. As it is, prisoners were "destined to re-offend".

Campaigns targeting cigarette smoking and alcohol abuse seemed to have an effect. A study by the Australian Institute of Health and Welfare found the national consumption of alcohol and tobacco had dropped over the past ten years but at the same time was accompanied by a steady increase in marijuana use [84]. With the problems of under-reporting in illegal environments being recognised, it is interesting to compare the Australian data with those from overseas: the declared marijuana use in Australia for 2000 was 18 percent, for the UK and the US it was 9, in Spain it was 8 and in the Netherlands only 5 percent. Since Australia does not suffer from the zero-tolerance policy of a United States for instance its data should be given more credence than the American figures, but virtually no restraints exist for the Dutch.

A survey by the Alcohol and other Drugs Council of Australia puts the number of amphetamine users in Queensland at 85,000, who are a far greater problem than the state's 17,000 heroin addicts and "drug workers were buckling under (their) weight" [85]. (Queensland's population is about 5 million). In the entire state the total number of beds is 275, and burnout by health workers is "a really big problem in this area. It's full on".

The conferences continue. In December 2001 one was "Drugs: National Problem, Local Solutions", at which Ethan Nadelmann, described as an international drug policy expert, advised of the following:

> We know there has never been a fully drug-free society in the history of human civilisation.
>
> We have always been looking to alter our consciousness, whether it is through drugs, exercise, sex, prayer or gambling.
>
> How many more people will die while you're in office with a policy you have deemed moral which ignores science and public health? [86]

The last sentence referred to Prime Minister John Howard, a conservative whose Christian values more than once determined his judgment on public policy, but more on this in the next chapter. Responding to these comments we have Australians For a Drug Free Society NSW convener Deborah O'Neill who labeled the above a "philosophical argument which lacked scientific data". Note again the selective differentiation, appropriating the concept of 'science' for her purposes although her own approach is exactly the opposite—a tactic found in defenders of myths, the gist of illusion trees.

In January 2002 Queensland's police minister confirms his determination on an "all-out war on drugs" and wants to copy a New South Wales trial in which police use sniffer dogs to randomly search public places [87].

Such a plan must seem highly altruistic towards the moral custodians outside uniform, given that the vast majority of attacks on police officers in Brisbane is prompted by alcohol [88] and the dogs are unlikely to leap at the bottle racks in bars. According to the police themselves the problem in the city "was escalating with some officers being assaulted on several occasions". Queensland Police Union general secretary Merv Bainbridge criticised some of the judiciary for being too soft on drunken hooligans:

> The (union) is sick and tired of our members being assaulted, and sometimes very seriously, by thugs who walk out of court without a custodial sentence.

According to the 10th Australian Illicit Drug Report the amphetamine market in Queensland has become a $400 million a year industry, heroin is down, cocaine use is "possibly the most complex illicit drug market" (probably a back-handed compliment to the sophistication of its users), and cannabis "still prevalent and widely used, with cultivation and distribution boosting the profits of organised crime groups" [89].

Remember Mark Le Grand, former head of the Criminal Justice Commission, who spoke out against trialing injecting rooms in Brisbane back in 2000? It is now March 2002, and he writes an article on the fight against drugs [90]: Access Economics puts the illicit drug trade in Australia at $7 billion per year; more than 100,000 people had used heroin in the preceding twelve months; the proportion of people aged 14 and older who had tried heroin increased by fifty percent between 1995 and 1998; and estimates put the total proceeds from illegal money laundering in Australia between $1 billion and $4.5 billion each year. He complains about the lack of political will and surmises that one reason could be the growing investment by a lot of people in the status quo—a real decrease in the problem would remove the basis for their income, their power, their very existence. A typical resource space scenario.

He has a point. Consider the following chart (Fig. 16-7) included in the House of Representatives Standing Committee on Family and Community Affairs report on substance abuse in Australian communities published in September 2001, which is meant to convey the functional framework of the nation's strategy on drugs.

The incursion by the drug trade into the broader economic system is nothing new of course; we have come across it already. The effect is relative to the size of the country: take Guyana, a transit post for a substantial amount of cocaine transported from the south into the US. During the northern summer of 1993 one kilogram of cocaine was intercepted by the local police who swooped on a small plane as it touched down on one of the country's wide rivers. For the next ten days or so the street price of the US dollar against the Guyanese currency shot up by a few points, from around G$124 to around G$128 per US$1.

The attempts by parents to get their kids on to ADHD prescriptions must have paid off. By May 2002 the Queensland Crime and Misconduct Commission warns about anecdotal evidence that attention deficit disorder drugs are traded by adults and children alike for between $2 and $5; after all, that particular market enjoyed a growth rate of 490 percent during the 90s, and that was just its legal side [92].

In February 2003 the Queensland Police Minister thought drugs being available in hotels was "mindblowing" [93]. He went on to say,

> To sit here today and see people trading in death, and that's what it is ... across the counter just like it's another drink, is extremely worrying.

Presumably he meant the kind not marketed by the licensed venues of his state.

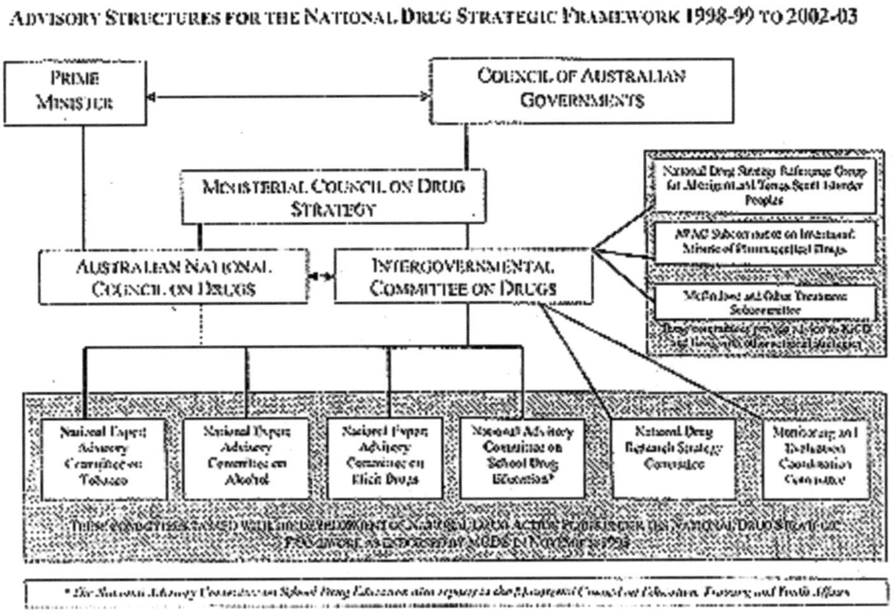

Fig. 16-7: What happens in government when too many people get high (obscure on many levels) [91].

In Thailand Prime Minister Thaksin Shinawatra also had declared war on drugs and vowed not to give up, whatever the cost [94]. As well he might, since during February more than "1500 people have been gunned down in the streets", there is a blacklist of 50,000 suspected drug producers and traders, and extra-judicial killings have become the norm. The International Narcotics Control Board has declared Thailand the worlds biggest user of metamphetamines.

And while not in the same global league, Queensland is Australia's capital of illegal amphetamine production, according to a report by this country's Crime Commission [95]. On average police discovered an underground lab every 2.5 days over the last year (2002), cannabis accounts for 72 percent of all drug offences over the same period, and cocaine use is up as heroin is down. Etc etc etc ...

The preceding is only a small selection of what is on offer, never mind the full international context. To some people this ongoing iteration of reports, recommendations, higher and higher-reaching statistics concerning use, financial clout, and efforts by law enforcement agencies year after mind-numbing year may seem tedious, but they demonstrate the sheer inevitability of effects which have their basis in a functional framework that represents the essential dynamics of complex systems, only at different scales.

Both sides are entrenched in their positions. 'Mainstream' society looks upon the 'other' with disdain, fear, and anger, and engenders just about the same reaction in its target. Demographics whose members are less capable of weaving their way amongst the obstacles of life follow their desires (however dimly perceived in their entirety) and sooner or later get caught; and so by definition the majority of prison inmates have something to do with drugs. Others with more nous recognise the issues more clearly, compensate in their behaviour and get away with it, thus not only escaping the statistics but also contributing to their society through ways which are being appreciated in intelligent circles (just analyse art and creative performances anywhere). Societal islands form, such as underground networks to ensure supply and the insulated ambiences in enforcement agencies no-one feels like assisting. The media carry a steady stream of data and events designed to sell their stories and no matter what the topic may be, a negative, sensational angle surely helps. Children are exhorted to read and watch and so naturally none of this escapes them—notwithstanding the compacted perception of one's surroundings that young age or immaturity in general contains. The very people who are supposed to be the leaders and guardians of public standards are also those who make fools of themselves by issuing ridiculous statements nobody with any street cred believes. Society is shaking itself into sectarian clusters of open and hidden interests through the insidious combination of self-preservation and public cohesion.

And throughout all this the articulated sentiments of everyone flow and ebb, leaving their flotsam to be picked up by studious selection. Academics who are more at home with straight figures and probability curves on defined rasters, they are invoked to furnish an objective background to what has not even been defined as a proper question. Comparisons are made between a precisely random sample of the population and some group who has been momentarily categorised according to what officialdom believes to be representative of its perception. The former provides the statistical baseline, the latter the figures in yet another table. Even in less confronting demographics the result is barely convincing and has to be labeled with riders.

Take the 2002 study on cannabis and psychosis in the Netherlands for example (see at reference [57] above). There is no doubt its mathematics are sound, but the initial sample of 7076 persons had to be taken from a population of just under 16 million (the estimate for 1996 [96]). To ensure a valid reduction to something akin to 0.04 percent of the total a rigorous selection had to be employed which covered municipalities, private households, and the individuals living in each household. So of course the baseline consists of truly randomly selected people, but its very randomness mitigates against the need to have a measure through which to capture the demographic nature of cannabis users. Since the aim was to understand the presence of psychosis, and the mental configuration of such patients is a function of a mind's grappling with its environment, the mathematically correct sample pool is not necessarily the best way to decipher the complex relationship between social norms, mind-altering substances, demographic mobility of users, physiological dispositions or otherwise, and a condition which had been categorised in terms of a regimen with a culture all its own.

And yet such studies and therefore these kind of evaluations are required by a system which functions under the auspices of statistical tables and correlative accounting methods. It is not that they do not produce some figures, reproducible and detectable as they should be—the problem is the initial definition and its conceptual dependencies. Unless an algorithm delivers a figure which can be fed back into the system from which it had been gleaned in the first place such that it still makes sense at that point; in other words, unless such a PAT chain remains congruent throughout the process, neither the data nor its source nor its recipients will be served. No wonder then that after literally generations there are still administrative entities for whom it is feasible to conduct yet another study, another survey. As none had been capable of addressing the issue once and for all the theme needs a constant rearticulation ad absurdum. The flow chart in Fig. 16-7 illustrates the flowering of the bureaucratic landscape on such fertile soil.

The rot sets in at both ends. The gigantic illusion tree resembles a hurricane, and just like its meteorological counterpart it wreaks havoc at ground level while beyond the energy-enhancing updraught there revolves a stable consolidated system of ordered formations, putting the entire land under its shadow. The family unit, so adored by the pillars of society, turns into a manipulative tool for enforced morals. Parents are driven to spy on their children, the children are enlisted by authorities to spy on shop keepers who might sell them tobacco [97], Environmental Health Officers stalk children which in turn alerts parents and police, and mothers and fathers wonder how to explain the logic of alcohol advertising on TV with drunken hooliganism on the streets and such sentences as "All illicit drugs carry risks and are dangerous. There is no safe level of use" [98], when it is quite possible that all the kids have experienced is loving bliss amongst their friends after a joint or a tab. Not to mention all the adults who must spend a considerable amount of their time and acumen making their lives livable while at the same time satisfying the rules of the day. In the meantime governments are being seen as ineffective weights around the public's neck but manage to preserve their culture amongst the corridors of self-confirming entities. In this age of globalisation their officers have more in common with their colleagues from around the world then with the constituents they are meant to work for, generating the atmosphere of an isolated will to survive, prompting ever more remoteness. There comes a stage when a critical line has been reached, a threshold beyond which the ability to reconnect with the larger system has been lost. It is a point that cannot be recognised; not by the public because the operatives are hidden from them, nor by the members themselves since they have forgone the necessary reference. The only signs are small hints here and there, differing in context maybe but similar in their functional appearance: it comes when basically unconnected members of the public are being recruited to perform for the state's executive. It is a step towards the ultimate resolution of incongruent abstraction levels brought forcefully together—a collapse within the complexity levels of a resource space. The final stage? When the public is arraigned on one side of the street and the troops on the other.

The existence of incongruent PAT chains and resulting illusion trees showed their problematic influence on several occasions during Part I. Later, on a more detailed level the mismatch of abstraction complexes could likewise be understood. This time the much larger scale of societal dynamics shows the same occurrences in principle, although the contents are now different. Since the interdependencies amongst resource spaces rely on the degree of connectivity and the sheer mass inherent in the system, we can expect the effects to become more intense as those two factors rise too. The stated attitude of officials towards drugs is only one element in the process of fractionation, and the diversion of immense funds from more productive projects only one destabilising force (in the United States the figure so far is about US$300 billion [99]).

The social fabric which carries the feedback loops of individuals and groups defines not only the role of what is acceptable, but the treatment of the deviant. As such it reflects its own nature, at the functional as well as the contextual level. The interdependency in a resource space is not merely a convenient descriptor to catch a possibly complex circumstance; it is a functional entity in its own right. An emergent system is a product of its constituents and the removal or modification of any one of them affects the whole—in fact, it is now *another* whole.

When Michel Foucault observes the changes in punishment during 18th century France he sees them in the context of a general development.

> ..the criminals of the seventeenth century were 'harassed men, ill-fed, quick to act, quick to anger, seasonal criminals'; those of the eighteenth, 'crafty, cunning, sly, calculating' criminals on the fringes of society. [100]
>
> It was as if there had been a gradual lowering of level—'a defusion of the tensions that dominate human relations, ... a better control of violent impulses'—and as if the illegal practices had themselves slackened their hold on the body and turned to other targets. [101]

The entire social environment was changing; a lessening of economic pressures, a rise in the standard of living, demographic expansion, and a need for security brought about by an increase in wealth.

We can see the rise in abstraction levels. From bodily confrontations of the 'me-you' kind representing nothing more than an object here and one there; to the widening towards additional appendages serving as defining elements of the 'other' and simultaneously requiring a more sophisticated dealing with consequences (a dead body can be kicked into a hole—a stolen item needs to be processed). The response by society moves in tandem. For Foucault the evolving penitentiary during the 19th century and the delinquent are almost two sides of the same coin, one defining the other, "in a sense twin brothers" [102]. He does not even see them as one influencing the other, whether the scientific analysis of the delinquent causing modifications to the prison system, or the "internal elaboration of penitentiary methods" somehow casting a new light upon the persona of the inmate. Rather,

> They appeared together, the one extending from the other, as a technological ensemble that forms and fragments the object to which it applies its instruments.

The ability to define creates the categories that underpin the references. For the Inquisition in the 1500s the background of religious mania included humanity in general by definition and so "neighbours denounced neighbours, friends denounced friends, and members of the same family denounced each other" [103]. In the 1800s the emergence of the working class provided a new demographic to be extrapolated into supportive and destructive elements, the latter being the criminals and meant to be understood as such. Since abstracts are usually more important than the objects they stand for, a crossing of these boundaries attracted a more stringent response than delinquent behaviour within the confines:

> Workers' action was regularly accused of being animated, if not manipulated, by mere criminals. Verdicts were often more severe against workers than against thieves. [104]

In the above the diversification of abstracts occurred within a society that contained demographic strata but also a unifying, over-arching culture. The distinctions therefore referred to variations on the theme, the main course of which everyone could relate to (provided they had the desire to do so). Should the control be in the hands of an occupying power however, no such commonality applies and its constraining—enhanced because it was self-serving—influence disappears. It is not a matter of content; as usual the dynamics follow their course in principle regardless of any inherent values.

When the British governor-general Lord Bentinck denounced the Indian customs of sati (the burning of widows on the pyres of their dead husbands) and thugi (a ritual murder to honour the goddess Kali) he exercised an interpretation of values as they pertained to his 19th century England. Since he did so in Calcutta it aroused opposition amongst the orthodox majority [105]. Having the might of the British Empire behind him he succeeded in eradicating the ritual murderers, although some women were still burned "well into the twentieth century".

From our perspective the action towards sati constitutes a conceptualisation from the top which can only be effective if its substance can be transposed to the levels where it is meant to be applied. It clearly could not, as history has shown. The traditions practised in rural districts of India represented a mentality that could not absorb any ideas which required it to elevate to more complex considerations. Neither could the mindset of the thugis; but being a group that was ready to be identified as a criminal minority their presence per se was removed. Sati survived because there are people who do not change their minds; thugi did not because its followers are dead.

A cohesion within the relevant abstraction complexes helps to smooth over the discrepancies caused by momentary circumstances. 'Cohesion' in this case is not some esoteric concept but the very concrete existence of affinities which are ready to be invoked should the need arise, whereas no affinities make for isolation and thus negation. The systematic articulation of characteristics which could have been part and parcel of a common dynamic turns affinities into discrepancies and the result is two sides instead of one. For that reason the traveller into precarious regions will always be safer if accompanied by a local, preferably someone with status (which is merely another word for connections). Regardless of the purpose of his visit, the human link with the surroundings is more important than the visitor's persona.

Or consider the following: suppose you are travelling somewhere in South America in a diamond producing region that is full of fossickers coming and going. You take a speed boat back to the main city because the trip takes only four hours rather than the full day the ferry needs. The vessel is basically a rowboat with an outboard engine and there's the driver and five others, including yourself. You've spent the preceding hours walking amongst the timbered huts full of diamond merchants and general suppliers but have no stones on you—but only you know that. About an hour into the trip, with the jungle a few metres on the right and far away across the watery expanse on the left, and nothing else around but the heat and a little boat moving in it, the man next to you suddenly starts looking for his wallet. He upturns his pockets, shakes his bags, curses and swears, and the spectacle gets more intense as the moments go by. Imagine the situation; you the only foreigner (obvious as anything), just returning from a place that's got practically nothing on offer but raw diamonds, with who knows what in your gear, and five locals of, let's say, indeterminable demeanour. What do you do? As always, preparation is essential. Before you even set foot in the boat a careful conversation with the skipper would have told you something about him, his boat, and quite possibly whether the other passengers are just that or maybe not. Out there on the river this would have paid off. As soon as the fellow starts behaving funny, re-establish your connection with the driver for all it's worth. Now he knows you know him, the others know he knows and you know they know. If the agitated passenger really did think to turn the blame on you he could at least expect some consideration from the others; if they turned out to be on his side after all—well, bad luck (in this case the wallet turned up just as suddenly, and I'm here to write about it).

In essence it is the connectivity in society which saves us all. Without it the recognition of any civilising standards fall by the wayside. It would be folly to imagine that a police force, however well-trained, causes crime to remain manageable. While there are always the violent outsiders who have to be dealt with, by and large a society is peaceful because the populace is good enough to will it so. The dismantling of the links between demographic strata (the abstraction levels in our resource space) create the isolated clusters which from now on follow their own internal references and none other. The result is not a society but a conglomerate of demographic sects which act like the city states of old—answerable only to themselves, defensive against the outside, and becoming the incubators of specific cultures sharply defined.

And so we come back to education. The provision of data when the developing mind needs them must surely be one of the most fundamental pieces of societal infrastructure the human species has accomplished. Unfortunately it is not always recognised as such. As instances of deprived children have shown the laying down of abstraction layers inside a single system is essential for that system's later performance. Highly abstracted notions cannot be processed properly if the structures at the lower levels are not capable of connecting with them. We have encountered the compacting of complexes before; an idea per se does not disappear, rather its meaning is mutated into a shallow version of its former self and the outcome is a PAT chain that carries simplistic representations. (On that note it is an interesting exercise to posit the developmental stages of the growing mind with their pitfalls against their larger-scale counterparts in a tribe, nation, or culture)

Nor are the functional abstracts removed from the danger. In an international numeracy survey adults from several nations were tested for their maths skills [106]. Some samples out of the total twelve:

Multiply 6 by 21.
What is the area of a room 11 metres by 18 metres?
How many apples does each person get if a box of 72 is shared by six people?

Britain came last with twenty percent getting all questions correct, for Australians it was a third, and Japan topped the group with forty-three percent successful (in between were France with 40%, Holland with 38%, Denmark 36%, and Sweden 34%). The ages ranged from 16 to 60. Amongst the British "more than 25 percent could not calculate 10 percent of 7900". Given the kind of information I have used in this very chapter, how would those 25+ percent digest the meaning of it all, and what would their resultant view of the world be?

A study conducted in Brisbane's city council vaccination clinics found that while 99 percent of children's carers had with them the booklets outlining at what stages vaccinations should be given, only 8 percent had actually read them [107].

In Queensland, primary and secondary school guidance counsellors increasingly find themselves helping the teachers because of mounting pressure in the classroom [108]. There they confront what is termed "challenging behaviour", diet-related problems, severe depression, drug abuse and threatened suicide. Education Queensland recorded that more than 100 of their staff had been injured in attacks by students between July 1999 and March 2001 [109]. The majority of the attacks, we are informed, occurred while the teacher tried to restrain a child with special needs or while trying to calm down agitated students. Some got punched in the eye or bitten on the arm. (Although this should be put into some perspective: in 1999 Queensland had a total of 43,369 teachers and 594,584 students in its entire education system [110])

The demographics of violence are being analysed and twenty percent of primary schoolchildren in areas of poverty or unemployment have mental health problems which are deemed clinically significant [111]. The combination of low income, unemployment and single parent families leads to a tendency for temper tantrums, destructive acts, and eventually to criminal behaviour and suicide, with Queensland having "one of the highest suicide rates in the world".

The readiness for violent behaviour is as much a reflection of internal turmoil as it indicates a faulty environment. Brutishness, that is low-abstractive complexes acted out, can be observed when a mother chases her young boys like cattle off a suburban street; or when parents act out their inadequacies while watching their children play sports; or the father who visited the school principal in his office and punched him after he was called in to discuss the aggressive behaviour of his son [112].

The sectarian deconstruction of demographics is also apparent when "ethnic teenagers" resort to knifes and machetes to protect themselves against the attacks by others [113]. Note the insularity of standards: here is a group of people, presumably members of one society, who think it appropriate (rightly *or* wrongly) to organise their daily affairs in terms of going armed with weapons more relevant to guerilla or civil warfare. Their perception has been cultivated through experience, mentality, group cohesion, and now defines them just as a police uniform defines a police officer and a certificate defines a tradesperson.

The top layer responds with measures accessible to that stratum. Queensland's Teacher Registration Board asked for police to protect teachers in schools because their members "just could not cope with bad behaviour" [114]. They seemed not to have been successful in this regard but did have three hundred behaviour management specialists on their side, with University of Queensland behaviour management researcher Robyn Gillies being quoted as calling for more of them. Schools did not need police she said but "needed to develop systematic approaches such as mentoring and professional development", and "principals should be trained to support staff". On the other hand a guidance officer, Isaac Williams, had designed a program featuring such things as sailing on Moreton Bay to teach students to "take more responsibility for their behaviour". According to him, a "culture of anti-intellectualism made it difficult for teachers to stimulate students' aspirations".

Like the Advisory Structures for the National Drug Strategic Framework the proliferation of highly abstracted entities abounds. At the heights where academics, policy makers, and sophisticated conceptualisations live, the spectacle of a machete-wielding youth or a hysterical pupil (I prefer to reserve the word 'student' for those attending universities) gets transformed into, well, *frameworks*. Up there an angry visage is deconstructed into the elements of a resource space which lives by and through its very own affinity relationships. Whether in functional or contextual terms, those links in the PAT chains attend to *their* connectivities and follow *their* adjustment protocols.

Much has been made of the polarisation of society, usually referred to as a widening gap between the rich and the poor. A closer analysis however reveals such descriptors to be somewhat misleading. It is true that higher incomes increase further, and therefore purely in relative terms the actual sizes of pay cheques are accelerating away from anything below them (ie, a 10% increase of 100,000 is always worth more than even 20% of 30,000). But it is also true that by and large the amount of assistance available to the needy has never been more extensive than now (a glance across the Brisbane telephone book entry for Centrelink, Australia's government welfare agency, reveals no fewer than forty-three categories available to the enquirer [115]). Notwithstanding the cost of living and sometimes quite serious exceptional circumstances, in this country at least there are ways and means to navigate past the rough spots of life. What *is* becoming more challenging are the tactical measures the individual has to take in order to receive those benefits. Even for students the mosaic of university or college courses, their respective weightings, their points value with reference to the government's allowance schemes, changing offerings from year to year and, not entirely unimportant, their eventual significance in the form of a degree and what this means for further advance, can be a taunting knot to unravel (and that's assuming the person behind the help desk is always as informed as they should be). For someone of fewer resources and having a less clearly defined direction the chances of a missed access would be real.

For the more serious issues (such as crime) the distancing of administrative strata from the demographics they are supposed to be dealing with presents a phenomenon of long-term proportions. Incongruent PAT chains hiding behind their logic boundary may have been a quaint scenario in a Rousseau or a Nietzsche, but in the global village of today the ante has been raised considerably. In the mind of an individual the logic boundary exists in the form of TS clusters which are linked by diverse affinities—there is a potential vertical link but either side follows a different regimen. In society the threshold consists of groups of people who are connected by a shared concern (even if that should be the law which one sends after the other), but there is no mutual understanding and whatever is done gets misinterpreted or at least perceived in a parochial context. For example, in a research report by the Australian Institute of Criminology the authorities are criticised for representing middle class values which are meaningless for the youth gangs they intend to target [116]. Reforms are impeded because of too much "'social worky' language and other initiatives that did not strike a chord" with young people. One has to agree—it seems there are more administrative flow charts to be found than programs such as the one which puts young people on sailing boats.

Both regions of society's resource spaces have the ability to exert their influence; one downwards, the other upwards. The administrators and executors modify the functional landscape to suit their agendas and act upon the latency and applicability; the citizens below react to the policies by changing the product. Thus society has to come to terms with the quality of the perceived abstract introduced by the former, and the quality of the object abstract as furnished by the latter. Once such a discrepancy holds, the image appearing on the screen of a management researcher, social analyst or judge is no longer representative of the object which their collective lens had captured. Their reactions present accordingly.

The impact the perceived abstraction of something has bedazzled the ancients, haunted the Inquisitors, and pampers many of modern-day intellectuals. With only a few abstractive layers to go by, a person is a walking, talking, acting human being who also happens to exhibit certain characteristics that may or may not make a difference. Remove the 'human' from the equation, transform the remainder in a conceptual morph and one ends up with a derived entity that only serves the interpreters behind the camera.

In an article criticising the culture of blame Michael Duffy writes about a man who was dragged into litigation through the mechanisms of the medical and legal systems which, one would have assumed, exist to help the citizen [117]. A "relatively minor injury" ended up being worried over lest it followed its natural course and healed. The status of 'victim' began to take over as general reference with its classification of 'sufferer' on one side and 'destroyer' across the line. Duffy quotes Robert Hughes from his book *The Culture of Complaint*, and it is worth repeating:

> Since our new-found sensitivity decrees that only the victim shall be the hero, the white American male starts bawling for victim status too. It's as though all human encounter were one big sore spot, inflamed with opportunities to unwittingly give, and truculently receive, offence.

Another example from the same article. A 21-year-old trainee teacher took a metal for-sale sign and tobogganed down a snow covered mountain slope. He had an accident which made him a paraplegic, sued the New South Wales government and a ski-slope operator. The courts gave him $7 million.

Later that year a man won almost $4 million from the Council when he hit his head on a sandbank while body surfing on the Gold Coast [118]. If this case demonstrates the distance in an abstract sense, the next is also doing so literally. A 14-year-old school boy played truant, got into a rock fight somewhere else, was injured and subsequently awarded $221,079 against the school he was away from in the first pace.

Such a system does not care about the individual who is moved through its machinery. On the contrary, what gets processed is an instance of a perception that has been moulded according to a cloying, cosseting culture. And, like any domain that endeavours to preserve itself, it perpetuates its character wherever it can—with one essential difference. It is not a culture which constructs to come into being; for it to exist it seeks to build down, abrogate, and diminish.

In the US a couple of deaf parents actively sought to procure sperm to ensure their second child is also deaf (they already have one deaf child) [119]. While cultural affinity is nothing new, an initiative to prevent an improvement that is supported by the wider system reaches into the elementary engines of societal dynamics.

On a similar level indigenous people seek to re-establish their traditions and expect to be accommodated. In Australia the concept of reconciliation is intensely pursued by Aborigines and mostly supported by the rest. Yet none of the latter is able to give a clear interpretation of what is expected, by governments or individuals. Assuming people understand what the word means (from 'to render no longer opposed' to 'make compatible or consistent'), what exactly is one to agree with, what of our mannerisms or procedures is to be made compatible with what from tribal customs? Nor has the indigenous community clearly spelled out its agenda, to be understood as one definitive document (colour prejudice seems too vague for this purpose and in any case is taken care of by legislation already).

In our perspective, the elevation of a diffuse concept to a focus for some initiative is usually the sign of an illusion tree. That is to say, while the considerations leading up to it may well be valid, by the time their abstractions have been moved along the PAT chain they are now affinitive with other domains which have very little to do with the original contents. We have seen such incidents in Rousseau's work, and in the *Communist Manifesto*. Against the general background of forming clusters of specific interests and self-focusing initiatives the term 'reconciliation' has to be seen as a symbol that is used for whatever the occasion requires.

As such it enters the general resource space of society and turns into the particular currency of some special interest group without being given the common standard the entire system needs in order to make the concept per se available everywhere else. The commonality of standard, without which any cohesion between the various resource spaces of subsystems is not possible, goes out the window. Having deconstructed the means for cross reference, the elements have not disappeared but continue to exist in isolation from each other as separate entities, even though they are, or should be, representative of the same theme. What then becomes perceived as trivial or important is a matter of locale, not facts.

For example, in Australia there exists an administrative entity called the Australian National Training Authority (ANTA) which provides nationally recognised training packages for food industries such as butchery, bakery, cookery, food processing and so on. Part of these packages deal with the standards to be employed in safe food handling. In addition there are Food Safety Standards, a nationwide legal framework of food handling requirements [120]. At the same time television contributes to the topic of food preparation by showing cooking programmes designed to entertain. Two such recent series on Australian television, *Forever Summer* [121] and *the Best* [122] featured handling procedures which would definitely not be included in any accredited guide books. In both, meat, salads, and fruits are liberally touched, squeezed, spread, and garnished with bare hands, and in the second sample one of the cooks' dirty fingernails can clearly be seen (not that long nails, however well manicured, are an advance in the first either).

Without going into the entertainment value of these pieces (which is not my agenda anyway) they do introduce a certain behaviour pattern into the common culture of their audience. This may seem to be trivial, but in a wider context it is not. Consider the current spread of what has been termed 'severe acute respiratory syndrome' (SARS). By 30 April 2003 there were 354 deaths reported worldwide, with 148 in mainland China and 150 in Hong Kong (the two most severely affected regions) [123]. The reported probable infections for the same areas were 3303 and 1572 respectively. After public-health interventions sprung into action the infection rate had dropped to below 20 per day in Hong Kong [124].

To put this into perspective, the influenza epidemic spanning the globe during the years 1918-19 is estimated as having as many as thirty million people put to death, and all this in the course of about six months (the total death toll of four years of World War I reached eight and a half million) [125]. Checking the list of epidemics around the world during the 20th century we find 75 in Africa, 8 in Australia, 33 in Europe, 25 in Canada and the US, 90 in Asia (incl. Russia), 14 in Latin America, and 16 in the Middle East [126].

By the 7 May (in other words, seven days later) SARS had caused 66 more deaths in China and 43 more in Hong Kong. Nevertheless, due to an emergent state of panic riots broke out in China, the most violent one being a protest by "several thousand villagers" in front of a government building which housed suspected SARS patients [127]. Articles in newspapers offer estimates of up to $30 billion worth of losses in the region and for this country the Australian Tourism Export Council values the combined impact of the Iraq war and SARS at $2 billion [128]. Yet as the World Health Organisation has pointed out, about 200 million people had travelled around the region since the first outbreaks and only five had contracted the virus.

Obviously, a disease that can cause death and so far has no cure needs to be dealt with. Yet it becomes evident that by far the most important factor in the spread or rather non-spread of the virus is the general hygiene of people's environments. The most affected demographics do not represent what in other places has come to be accepted as good hygiene. Ubiquitous spitting on the streets no matter what the contents of one's mouth, an adoration of 'image' which dresses the front of restaurants like little palaces whereas their rears often resemble grottos with stalactites of sooty grease, and an atmosphere of superstition which prevents corpses put on vessels for instance but sees nothing wrong with them being dumped into bays or waterways, including foetuses—these are examples of mannerisms which form part of an identity and so possess inertia (to a tourist only the first may be evident). The necessary measures to be taken therefore are by no means mysterious and would suggest that a calm and analytical response is adequate. Instead the density of media coverage, the global reach of perception, and the momentum of organisational entities that have become involved had elevated the issue into emotional realms that in fact prove counter-productive.

The connection between a TV programme and Chinese riots may seem tenuous, but in this case the entertainment industry did contribute a watered down standard of food handling to a society, the effects of poor hygiene can be traumatic in any case and as SARS has shown, even a relatively minor outbreak has the power to seriously disrupt the economic systems of entire countries. Therefore the observance of fundamental standards is not trivial after all.

In Australia the general checks and balances are more rigorous because European society had been forced to charter its course through cultural and/or ideological waters before. For example, up until the 1800s the Thames was the source of the thick, foggish air spreading through London and giving this city the stage ambience so beloved by present-day admirers of gothic tales. In those days however it was anything but romantic. Typhoid fever was only one of constant problems and epidemics were a permanent hazard. Under certain weather conditions "the sewage of nearly three million people had been brought to seethe and ferment under a burning sun, in one vast open cloaca lying in their midst" [129]. How many people today even know the meaning of the word 'miasma'?

Identity containers drive their host environments; they can also be imported under the mantle of 'understanding'. Yet what is meant by 'understanding' does not refer to an analytical assessment of procedures but a vague notion of acceptance, where 'acceptance' has now assumed the opposite of the equally indistinct 'intolerance' it is meant to replace. We have discussed the role of the 'not' in the creation of affinity complexes in the previous chapter. It is being used to identify the 'other', not through a knowledge of one's own character but by pointing to what 'we are not'. Its more serious effects can be observed any time a society with a more complex resource space comes into contact with another of a lesser expanse (notwithstanding of course that many such judgments can be based on perception). The clash between colonisers on one side and indigenous populations on the other have demonstrated the dynamics over and over again. In the West the increase in its resource space made it possible to accommodate the considerations needed to transcend one's immediate social horizon and so to build down the moral and emotional barriers between 'them' and 'us'. However, if the idea of 'tolerance' is not understood as a functional abstract (containing as it does the principles of applied actions) but as an object abstract (with the word itself becoming the salient feature without having a grounded base), its practice results in ambiguous implementations which are unable to address the actual detail. It's another example of an illusion tree creating higher-level abstractions brought to bear upon the stratum the structure grew from originally.

Thus a critical examination of the issue does not take place. In Australia every year a number of people succumb to meningococcal disease, an inflammation of the brain's meninges which can cause death, especially in younger children. The website of Meningococcal Australia Inc warns not to share food, dips, icecreams, or drinks and bottles, amongst other things. It also tells parents to prevent certain practices by their children, such as sucking shared pencils or toys, since the bacteria are passed on by saliva [130]. At the same time Australia's indigenous people paint by mixing earth-based colours with saliva in their palms which then gets applied to a surface. Over recent years many primary and secondary schools have adopted the practice as a symbolic sign of reconciliation, and events where children make their own pictures that way are held regularly. Despite many reports, either in the media or through the offices of the schools themselves, to my knowledge there has not been any criticism of the custom in terms of the health hazards involved. Given the ever present threat from something like meningococcal disease the connection should be made sooner or later, or so one would think. In private conversations with individuals (one of whom happened to be a registered nurse) the issue of hygiene could be acknowledged, but as soon as the conceptual complex 'Aboriginal customs - reconciliation - education of the young' enters the picture there is a marked distinction between common standards on one side and ideological bias on the other.

Or consider a Federal Health Minister in 1998 attending an initiation ceremony in far north Arnhem Land in which Aboriginal boys around the age of seven were circumcised (the Minister for Aboriginal Affairs was also there) [131]. We will deal with the practice of mutilation and sexuality in the service of one's gods in the next chapter, but for now let us focus on that scenario. The Minister, himself a medical doctor and, so the title suggests (does it not), an elected member of parliament of a Western nation at the end of the 20th century and charged with upholding currently accepted standards of health and medicine, such a man then had consented to be a "proud witness" to an event in which young boys have their foreskins cut off in order to become 'men'. The procedure was carried out without anaesthetic but did indulge in the use of a surgical knife, the only concession to the colonisers in what otherwise had not changed over the past 40,000 years (we are not told if in the absence of the good doctor the more traditional sharp stone would have been used). Leaving aside the point of a modern-day politician (some even go so far as to use the epithet 'leader') taking pride in participating in something for which the very idea of progress has been totally alien for 400 centuries, I can only speculate on precisely what kind of inner sustenance he derived since my own emotional landscape obviously does not contain the same features as his:

As Nathan's screaming rang out across the land, the stormy sky opened up and teeming rain pounded the red earth on which the ceremony was conducted.

The sweltering heat and the dense humidity added to the tension. It was a dramatic climax to one of the most emotional moments ever experienced by either minister.

Nathan's father was ATSIC chairman Gatjil Djerrkura, and no doubt his office particularly compelled him to see his son under the knife. The phenomenon of parents beset with the inner glow of something they most cherish and so sacrificing their offspring to underscore their officious status is nothing new (it certainly survived those 40,000 years everywhere). Mr Djerrkura said it was important to show that "his culture was alive and well" and,

It is a thing that we can't, that Nathan, can't escape from. It's part of the culture and tradition and he's got no choice.

Our ongoing descriptions of resource spaces, higher-level abstracts, affinity complexes and such may sometimes seem dry and removed from reality, but they are articulations which stand for something very real indeed. The expansion of these within the European mindset meant that every now and then some human of flesh and blood found himself in the torture chamber, faced a line of judges with hate in their eyes, or paced the streets fearful of every glance, every encounter. And why—because they dared to challenge what everyone else accepted as law, divine, or cosmic absoluteness. Those who live off Western standards owe their comfort to these persons, and without them there would be no parliament, no separation of powers, and no ministers (health or otherwise) going on emotionally enriching junkets into humanity's past.

Seven months later that year, in a jurisdiction out of reach from Australia's Federal Minister for Health, Dr Margaret Somerville of McGill University in Montreal received a human rights award at the fifth international symposium on sexual mutilations at Oxford University [132]. As "one of Canada's top medical ethicists" she declared that "circumcising baby boys is criminal assault" (in the next chapter we will see how that should be equally applied to older male children). She is joined some time later by Bond University psychology professor Gregory Boyle in Queensland, who wants circumcision of baby boys to be outlawed, just as female circumcision is "now illegal in several states" [133].

If we can assume that the separation of parts from a live body meets with a critical attitude by most civilised people, the paucity of such reactions when it comes to the same acts performed under the auspices of tradition suggests that the relevant connections in the mind don't get made. What keeps the domains detached are their mutually non-affinitive elements, and it is only through the formation of incongruent PAT chains that this type of diversity can exist. On the wider scale of society the same principles apply. Instead of TS complexes we find groups and organisations which pursue one goal and not another, and the more intense their interest the more exclusive their focus tends to be. Given the dynamics of affinity generation, this is to be expected. The more significant an issue is held to be, the greater the contrast to anything else; that is to say, the more self-referenced a series of abstracts are, the lesser their potential ability to associate with other structures.

In what kind of context the event is performed depends on the constituent elements current at the time; in what manner it goes through its paces is a function of the degree of complexity the resource space can muster. Not so long ago at a jail riot in Guatemala City gang members entered a phase of violence during which fellow inmates were hacked to death, some were decapitated and others had their faces destroyed [134]. Note also the conceptual relevance for an audience: would residents in Paris, Sydney, or London, upon hearing the phrase 'jail riot', automatically visualise spittle-flecked visages, claw-like fingers hacking away at another face, and heads being ripped away from their bloody ligaments? By the same token, how many concerned Westerners associate 'the need of the poor' with cutlasses slicing open the bellies of mothers and arms torn out of their sockets by enraged villagers? And would Western Aid documents be written in the same fluid style if their authors had to hold pens slippery with blood under the light of burning huts?

The distancing from antagonists and the simultaneous alignment with preferences receive additional impetus by virtue of scale. What could have been an encounter between two local factions evolves until it becomes a massive divide spanning the globe. The advent of articulated fury by anti-globalisation protesters at just about any meeting of international representatives has entered the annual calendar of industrialised nations [135]. Through websites and email the emotions of demographic clusters are organised, their responses directed, and their targets chosen. As always, the affinity relationships are mutual: protesters seeking confirmation of their pain find it at those contacts; individuals looking for a reason to be angry do so likewise. Together they form a phalanx.

In terms of more local phenomena we find the same principle at work. Pressure groups concentrating on one suburb (but utilising the conceptual principles gleaned from elsewhere) agitate for an initiative which they see as instrumental for their well-being. For instance, a leftist paper in West End, Brisbane, focused on the planned (and eventually implemented) removal of a couple of hectares of bush land in the inner city area [136]. Whatever the environmental reasons given, their followers wanted the area free from housing etc and the place dedicated to wildlife. In an earlier issue of the same paper the Brisbane City Council is criticised for, as the writer sees it, reneging on promises to maintain the level of infrastructure such as West End's police station [137]. No doubt there are details in these arguments which provide enough material for years of debate (as indeed they do), but what is absent are the over-arching considerations pertaining to metropolitan areas. One of them happens to be the provision of public services measured against the density of ratepayers. Brisbane is a rather poignant example because of its traditional small-plot, low-density nature, and the increasing demands on taxpayer-funded infrastructure per unit of area. Local pressure groups may well feel for their own corner of the earth, but their quality of life is directly tied to the effectiveness of the entire administrative entity of which they are a part. Thus in our functional terms their activism represents the application of a limited social horizon to a wider system, whose own elements are now being forced to deal with such input.

A similar example is the ideological turn many environmentalists have taken. While the initial sources for their concern do have their basis in observation and analysis, the process of demographic clustering has led to the creation of pressure groups drawing *their* themes from a much wider and often political (and politicised) spectrum. Take the comments by Margaret Karembu of the Department of Environmental Sciences at Kenyatta University in Nairobi. At a conference in London she berated bodies like Greenpeace for spreading emotional propaganda which had the effect of scaring Kenyan farmers off any products simply because they are being associated with the generated spectre of genetic engineering [138]. Despite having the opportunity of planting bananas that are cheaper, bring more yield, and are healthier, they do not get used. "We don't get data, we get opinions," she is quoted as saying, "Greenpeace has a loud voice, but most of what they say is not factual, and they don't provide alternatives. We can't make policy based on opinions."

Greenpeace, like many other organisations born out of the personal activism of individuals, has the use of facilities which are global in reach, made effective through technology, and tempered through marketing know-how. They also happen to function outside the legal and administrative framework of accountability and as such represent transnationals above and beyond any government. Through the romanticising of the wild the idealists who are entering the web of political environmentalism are nurtured on a diet of 'nature' as some benign being who looks after its children like a mother after her offspring (they never spent a night alone in the jungle—their beloved rain forest—and listened to the soft sigh an animal makes when a big cat breaks its neck).

The existence of identity complexes leads to the transposition of concepts from one's own domain to those on the outside. Just as the adjustment process induced changes in the connected matrices of OtoomCM, so do the sentiments held and expressed within a certain demographic act as references for the evaluation of the 'other'. That a thorough understanding of another person's inner thought processes is virtually impossible has been mentioned already. Yet the interdependency of cultural feedback provides at least some common ground, and the social delimiters of how close to one's centre an outsider can probe are a matter of constant readjustments. The fact that they *can* be subject to modifications is due to their situatedness within the shared resource space of the system. On the other hand, the more isolated the domains become, the harder it will be for any cross-reference to be made. Add the self-preserving properties of identity containers and the windows of opportunity will mostly be closed altogether.

US foreign policy is a case in point. Its assessment processes, driven by the combined weight of its military, political, and commercial machinery, hardly facilitate the subtle sensing and diligent observations necessary for in-depth awareness. Data input occurs, but there is no willingness, in fact capability to interpret the information in any other terms than those prescribed by its own values. Whereas the prudent traveller treads lightly and observes the manners of his host, for the arrogant visitor the land is merely an extension to his own back porch. By the same token, prudence does not mean subservience, nor does arrogance imply control. Although it may seem to Americans that their type of politics, their social agendas, and their moralism is the best in the world (as they keep reiterating to the rest of us), given their money-based electoral system (from the presidency down to the local sheriff), emphasis on violence (from the death penalty to private gun ownership), and their evangelical obsession with combating sin (from prostitution laws to seeding the globe with bible thumpers), other more civilised people may beg to differ.

As we just witnessed, one of the countries that came to the righteous attention of the US was Iraq. Although there was a time when that nation played a more useful role for Western interests, its changing texture of political significance within the Middle Eastern context moved it from one end of the American spectrum to the other. And yet, as a demographic consisting of religiously influenced, emotionally charged, and hierarchically governed citizens it stayed true to its nature throughout. The distant view of Western countries, but particularly the US, Britain and Australia, that a Middle Eastern Islamic people can be yanked out of millennia of history through a bombardment of their cities and poured into the shiny new mould of a liberal-humanist democracy (assuming for the moment that the US really does represent such a thing) accompanied by much rejoicing on the ground, is laughable were it not so tragic.

Apart from the historical perspective which should give any analyst of the Middle East pause for thought, the inherent instability caused by tribalism, religious fervour, and emotional disposition gives some hint as to the social dynamics one can expect. It is possible that most tourists manage to glide through Cairo or Dubai without much of a problem, although being in Port Sudan or Salàlah may not be quite so straightforward. In any case, their image of these lands will have been generated behind the buffers of tour guides, hotel attendants, and officials who have an interest in trouble-free movements. Step outside the cage however and reality starts to intrude. US policy makers would have benefited from visiting a traditional bazaar without a guide and without buying a single item to experience the feel of an expectant but ultimately disappointed crowd (and not understanding a word of their language). They would have added to their knowledge of local customs by bothering to learn the Arabic numerals *before* ordering a meal in some shop and then comparing the spoken with the written cost. They would have learned valuable lessons in Middle Eastern negotiating habits by organising some transport across town and observing how often the entire business had to be redrawn. Or, given the US' declared penchant for human rights, they could have tested the resolve of local officials towards themselves and maybe even a female companion at the point of a slight indiscretion towards something like a holiday. A mere month of such encounters would have been more instructive than employing armies of political analysts.

On the other hand, it is possible that the information had been there all along. For instance, it could have been known that one measure of maturity is the period of time a group of people can be suddenly left without supervision and *not* erupt into destructive behaviour (in our schools teachers are paid to utilise such observations all the time). The state of mind manifest in agitated jumping up and down on the same spot at a moment of either joy or disappointment would also give some clues about the person's inherent rationality (especially when it concerns an adult). And quite possibly there would have been some observers who, upon realising that in the entire realm of Arab culture there is not a single region without at least some form of despotism (and never has been), could envisage the possibility that just maybe there is a certain mutual relationship between mentality and governance.

Indeed, one could assume that. On the other hand the human mind is also capable of seeing a Madonna in a cave, a Buddha in a tree and Jesus in a pizza. So, A$105 billion [139] and more than 9000 bombs and over 700 cruise missiles [140] later, cholera has appeared in some Iraqi cities, there is looting in the streets, returning clerics are hacked to death by devout followers of another [141], and a Shi'ite majority clamours for an 'Islamic democracy' as if it had never heard of oxymorons [142].

On that note it is interesting what Samuel Huntington had to say about the Americans' encounter with the Serbs and the Croats in the 1990s [143]:

> Throughout the war the American policy remained stubbornly committed to a multiethnic Bosnia despite the fact that the Bosnian Serbs and the Croatian Serbs overwhelmingly rejected it.
>
> ..American idealism, moralism, humanitarian instincts, naiveté, and ignorance concerning the Balkans thus led them to be pro-Bosnian and anti-Serb.

In the end, he writes, they alienated the allies, furthered the war, and "helped to create in the Balkans a Muslim state heavily influenced by Iran".

Such adventures (and from other quarters this is how it must seem at times) undermines not only the nous of the West, ineffective as it is rendered by the looseness of its conceptualisations, it contributes to the erroneous image held by non-Western nations which many times gets re-imported back to us. When Huntington describes the emerging self-confidence in East Asia he correctly identifies their sentiments based on discipline, collectiveness, and the determined will towards progress [144]. What he does not mention and what gets forgotten during those moments is the technical/intellectual framework without which such progress would not have been possible—a source provided by European thinking. While it is true therefore that thirty years of diligently copied texts produced a result, it is also true that it took the West three hundred years to *write* them.

The formation of anti-Western attitudes within the West, helped along by demographic conglomerates opposed to technology and industry while enthusing about 'natural' life styles of tribalistic deprivation, has contributed to an ambience of passivity. In an article about student achievers in the Australian education system, Michael Duffy compares Chinese, Vietnamese and Asian students in general with the rest [145]. Although comprehensive statistics are hard to come by (interesting in itself) it can be said that approximately "50 per cent of entry-level dentistry students are Vietnamese, 25 per cent medical students are Asian, and the majority of pupils in selective schools in Sydney and Melbourne are Asian". At the same time, in 1999 only 3.45 percent of the population were made up of Chinese and Indo-Chinese. When Duffy talked to Chinese Australians they regarded our national image "either as irrelevant or as a decaying carcass, with its emphasis on losers, defeat, oppression and victims":

> They cannot see how a nation mired in the whining of Pauline Hanson, the moans of anti-business academics and artists, and the anti-work whingeing of many Aboriginal leaders, can stimulate the children to go out and create and succeed.

We have already encountered the leaning towards victimhood, a stance even defended by the judicial system which sees it quite in order to apply the notion of 'guilt' to entities that should never have had any guardianship over individuals in the first place. But if the nation's highest offices acknowledge the dependency of a private citizen's welfare on various institutions to such a degree, the message enjoys a particularly effective articulation. The gain in the status of 'victim' is only one outcome; the gradual adaptation of society towards an apparatus which *is* such a guardian is the other.

On a smaller scale of affinity relationships we have seen how specific elements of a domain become the links to other complexes because there existed a mutual correspondence between these elements; there was no need for the rest to agree directly. Once the connection had been made data can be used for both of the former areas since in broader terms the affinity combines them. The same principle applies to society at large. If there is a notion that individuals are not to be trusted and institutions fall short of their responsibilities, it sets in motion concepts which are affinitive with whatever negativity can be ascertained towards persons and authorities in general. Both abstracts become congruent with other perspectives that prove relevant.

The ideological form of environmentalism not only concerns itself with technology per se; its conceptual roots go much deeper, right down to the divide between primitive and developed societies. Although lower life expectancies in rural areas compared to the cities have been confirmed repeatedly [146], the exact causes are less well defined. From our point of view we can identify rural and urban settings in terms of their respective resource spaces and therefore recognise the differing rates of supply of just about anything that is necessary for human existence—the overall result then determines the quality of life. For the committed environmentalist however the statistics are irrelevant; what matters is the perception of 'nature' being synonymous with 'wild' and 'healthy' and apart from humans. If 'health' can only exist with 'wild' and 'wild' is 'nature', and if 'human' means not(nature + health + wild) the inference will be that any human activity cannot possibly coexist with the idea of 'nature..'. Some go so far as to posit humanity as a direct opposite and since 'nature' equates with 'earth' the necessary conclusion demands for humans to be an enemy of the planet; in some statements the word 'plague' is even used.

The derived nexus between humans as helpless victims in the face of the uncaring system surfaces often indirectly. In a review of David Suzuki's book *Good news for a Change: Hope for a Troubled Planet* mention was made of certain errors in the text. For example, despite the book's presentation of pesticides causing cancer a researcher who followed this up found that what almost always did cause the disease was a lifestyle choice such as bad food, no exercise, smoking, drinking, and so on [147]. A similar bias was applied to genetically modified crops. The possibility that GM crops have harmed butterflies or any other creature has not been confirmed but the book rails against them anyway. Note the conceptualisation of the players: the established facts of individuals contributing to their own demise is dismissed in favour of a factor that can be ascribed to industry, an abstract complex in opposition to the 'natural' which now encompasses 'people' as its hapless children. Same with the crops—here is anti-earth technology destroying the planet's innocent butterfly children—not because it can be confirmed or otherwise, but because the ideational modules fit.

From the image of humans as a plague on earth to the perception of having too many of them is almost a foregone conclusion. As such it has become a major plank in the environmentalists' argument about a sustainable planet and how to achieve it. The fears mainly centre on the carrying capacity of some land and the population density in various countries. However, as Ron Brunton has demonstrated in his text *The End of the Overpopulation Crisis?* the figures do not justify the assertions [148]. For one thing, countries with the highest density rates are not those with the lowest standards of living, and the carrying capacity of a region is an unquantifiable concept because it relies on too many variables; which is why estimates of sustainable populations have varied from "one billion to more than 1,000 billion" for the whole planet [149]. We can see why: since 'carrying capacity' is another way of saying 'resource space' and therefore definable in terms of the elements in our e-chart, a discrete figure is impossible to come by.

In a sense, this is just the point. As the data for Europe for instance show, a high density population can indeed coexist with high agricultural yields, and the problems in under-developed countries are not so much the numbers of people but their inherent capacity to live. Since the resource space of a demographic consists not only of the products but their latency and applicability, the ultimate standard is based on more than simple figures. For the green ideologue the population size must be the only delimiter since it allows him to exclude the means of moving beyond the low-abstract constraints of primitive demographics. For animals such constraints are real, but humans have abstracting minds and so physical limits can be overcome. From the moment humans realise their higher potential however they are leaving the conceptual domain of 'nature' and are now the enemy, the 'plague of the earth'. Thus when it comes to the issue of sustainable development the potential of technology is left out of the equation and the debate ranges around simplistic perceptions of multitudes.

There is more to the antagonism towards humans than the waving of theoretical discussion papers. For decades now the message of environmental concerns has been largely hijacked by the emotional ideology of primitivism. No wonder then that people's minds have gradually imported its principles. From Griffith University in Brisbane we hear about a research assistant from the US who helped local staff to capture a magpie which had been attacking children [150]. As they set a decoy "a neighbour approached, grabbed the young researcher by the throat and threw him to the ground. 'You leave my bird alone, you bastards. He has never hurt anyone.'"

If one can dismiss the fate of a few children, how about a large part of a city's population? The north-eastern seaboard of Australia is home to the Torresian Crow, and for several years now Brisbane has experienced an explosion of these birds. As scavengers they eat whatever they can find around the backyards of the larger metropolitan area. They are so noisy that a single crow doing its "aark aark" within ten metres of an open window makes it quite difficult to conduct a conversation over the phone, never mind when there are two or four or ten of them. In summer they start just after four in the morning and continue for the rest of the day, although a flock may move on eventually—to be replaced by another one. At regular intervals residents complain to the media and over the years there appeared many articles by columnists and reporters describing the problems the crows cause. In other words, the issue is not insignificant. Loss of sleep and its effects have not gone unnoticed elsewhere: fatigue has been identified as an important component of accidents, so much so that it is held as the major cause of road fatalities [151].

Nevertheless, the crows have been classified as an indigenous species and are protected by law. It is illegal to kill them and fines are up to $225,000 or two years' jail [152] (I am not making this up). A pamphlet issued by the Queensland National Parks and Wildlife Service acknowledges their nuisance factor in a characteristic fashion: "Occasionally, large numbers of very vocal crows substantially affect people in special circumstances such as those in hospitals and nursing homes" [153]. While the issue here is patronisingly downplayed, if any persons *are* being disturbed they are not seen as normal members of society but merely as those shunted away in sick quarters and hiding in special facilities for the infirm!

In the Appendix I have included a copy of a letter written to me by a director of the Parks and Wildlife Service addressing various points I have raised (see Appendix - Torresian Crow). Its contents are instructive for the manner in which the preservation of the birds are rationalised. The effect is an obfuscation of the logic behind the measures. What follows is a brief summary with reference to the relevant points in the letter.

Decibel levels: although formal measurements of sound levels can be made, in an urban setting the subjective sound corridors can change very quickly, sometimes within a matter of metres. Buildings, wall configurations, trees and bushes, walk or drive ways, they all modify the effective sound level at any given location. A set distance is therefore meaningless. Further, to compare sound sources such as birds with vehicles disregards their fundamentally different nature; cars move on roads, heavy vehicles do have their designated routes (and therefore *are* subject to council laws because of their nuisance value), and none of them move amongst the trees at window height.

Bird numbers: it certainly seems that numbers have increased dramatically (the public reactions are testimony to that). However, if the link between numbers and feeding by people can be acknowledged, an education campaign to change the public's behaviour is not unthinkable (it has been done in the case of the ibis for instance).

Disease carriers and contact with people: it is not explained what "direct" contact means (eg, a parrot sitting on my shoulder would constitute a 'direct' contact, but surely so would a flock of birds fluttering centimetres from my face). In any case, the assertion is made that research about possible disease transmission is considered unnecessary because there is no, let's say, interaction. Yet in the next paragraph the role of crows as doing away with decaying matter is perfectly understood, as well as its occurrence in places which are frequented by children (the places are even labeled 'dangerous' because of the presence of such matter). So, if crows eat things which are dangerous because of their bacterial content, and such things lie around places which are visited by humans, then by inference crows are coming into contact with humans (it's called a syllogism). Furthermore, the letter's argument rests on the benefit of having decaying matter removed from public places. If we accept that crow numbers had increased substantially but have possibly reached a plateau, and if we also recognise that the theoretical maximum age of crows is 35 years, it can be assumed that their actual median age is quite lower. Their numbers having stabilised there must be a time from which onwards the number of birds dying equals the number of birds being added, so that the total weight of decaying matter consists of dead crows to a considerable degree (given that in an urban environment natural predators are less significant although specific cases may eventually become more important). Thus the initial assertion ‘it’s a good thing crows get rid of carcasses’ changes to ‘crows can get rid of carcasses because a lot of these carcasses are crows’—somewhat less inviting. It is also true that the above relies on the functional characteristics of the parameters, not their actual value. As such they do not lend themselves to factual proofs, and this goes for both sides of the argument. In other words, in the absence of proper figures, percentages, and hence trends, the original statement has no value.

Community support: it is true that people are supporting environmental concerns at an increasing rate. Unfortunately it is also true that many if not most do so out of ignorance about the destructive potential of nature, weaned as they have been on the concept of ‘natural = good’. If therefore community resistance is a problem, the responsibility lies with governments to educate the public in a rational, comprehensive manner (for example, would the same considerations be applied to rats? Maybe soon—there are already greenies who put out water containers to support mosquitoes, the carriers of diseases such as malaria, Ross River fever and dengue fever).

Since the original points in the letter have been made by environmental researchers, their presumed intellectual nous raises the question of intent. How far does their social horizon stretch, and through what detail does it interact with other issues of a more human-related complexion?

The conceptualisation of a benign mother nature being applied to life in general gels with the notion of passive victimhood mentioned earlier. What we get is a societal character making itself dependent on vague notions of naturalness within the preconfigured ambience of need. Under these circumstances a substance will be found desirable if it falls within the semantic (but not factual) envelope of ‘non-technology - non-corporate - unchosen and unhealthy modern life - required nature-derived balance’. Alternative medicines fit the bill. Every year Australians spend an estimated $2.3 billion on products of unproven efficacy [154]. Not only are consumers handing over money for something without factual value, in some cases the effects are actually detrimental to one’s health. The Food Standards Agency in the UK stands behind a health warning against using vitamin supplements indiscriminately [155]. For instance, vitamin C creates problems with high uric acid levels, B6 has been linked to nerve damage and memory loss, and beta-carotene supplements increase the risk of lung cancer. As for the flora in general, any good herb book will advise its readers about the positive *and* negative influences of plant extracts and throughout history certain people have honed their skills to kill with herbs.

Wandering through life without the inclination to take charge, particularly at the personal level, evaporates the inherent ability of the individual to control the situation. The vested interest industry-driven agenda to proscribe the details of one’s life not only deprives people from acting responsibly in potentially precarious situations such as driving a car or taking drugs, it permeates even seemingly trivial indulgences—but they add up. Ridiculously absolutist statements declaring that ‘no illegal drug is safe’ (see earlier in this chapter) or ‘every k over is a killer’ don’t exactly help (the last refers to overstepping the speed limit on roads and is printed by Queensland Transport on its receipts for driver’s licences).

The general deterioration of self-confidence is making inroads into the statistical tables of nations. The imaginary unloading of problems by filling the void with food has managed to raise the fat levels in the West. A study by the International Diabetes Institute finds that the number of obese Australians has increased by 250 percent since 1980; at the moment sixty percent weigh too much [156]. Overweight people are prone to heart disease and diabetes, but even without such ramifications the annual cost to the government is somewhere around

$13 billion. Diabetes-related health costs alone are in the vicinity of $1.2 billion every year for Australia, but the condition is also causing significant difficulties in the Asia-Pacific region at large [157]. Not so long ago on a national radio talk-back programme such concerns were discussed between specialists in the field and members of the public. None from the panel of experts even hinted at the fact that first and foremost it is the individual who makes the personal decision to gorge themselves with food but all of them put the blame with society at large, that vague colossus out there somewhere [158].

One reason why people are exhorted to act the way they do is through example. Without the ability (indeed, readiness) to copy other persons' behaviour we would not learn and any kind of socialisation would not take place. On the other hand, when the mind goes into copy mode too often critical thinking becomes suppressed, and this has consequences. Over the years many studies have been made to discover the exact influence of films and TV, but one way or another they tended to suffer from an imprecise articulation of 'influence'—something is happening, but what?

Research by Herbert Krugman, an academic and manager of corporate public opinion research at General Electric, showed that TV viewing switched brain activity from the left to the right hemisphere, thereby releasing endorphins which are naturally occurring opiates (ie, in our brains) that can be just as addictive as the naturally occurring opium (ie, in the poppy plant) frowned upon by authorities [159]. In addition the higher cognitive functions of the brain shut down, reducing mental activity from its usual level of critical thinking. What remains depends on the general state of one's TS formations. Thus to the question, "Does watching television influence your mind" the answer is "Yes, in terms of your personal make-up". Still, if the average American's brain gets washed with TV-induced endorphins for over four hours every day of their lives, something has to give sooner or later.

In Otoom's context, the difference between active and passive information gathering can be easily demonstrated. Go to an unfamiliar part of the city and work your own way from point A to point B; make it a three kilometre distance. After that exercise the route will mostly be known to you. Now do something similar but use a taxi and tell the driver where you want to go while you just sit there. How well will you have learned the way now? The result should not come as a surprise: in active participation the TS complexes are more numerous than in its passive counterpart because the branching of detailed decision-making needs the associative resource space; to simply watch something only requires what comes from the outside.

As mentioned, every now and then humans need to get into copy mode. It becomes a problem when that state is made into the standard by which the intellectual level overall is being judged and the standard gets defined through questionable contributions. Four examples should suffice to illustrate their spread and nature.

On a national radio programme David Morgan, Westpac's CEO, stated that "Perception is reality" [160]. This from a not exactly uninfluential member of society, heading a major bank.

During a radio programme about books and their writers an anthologist talked about science for non-scientists and, referring to Einstein's book *Relativity*, said that he "can't understand the first paragraph" [161]. None of the others on that show bothered to challenge him. For those who might want to test their own intellect I have presented that paragraph at the end (see Appendix - Einstein's 'mystery'?).

While discussing the British Queen's relevance to Australia in the context of how rational a monarchy is these days, a listener phones in and mentions she had just been to a party celebrating the Queen's birthday. The host of the talk-back programme expresses disbelief at such an act, but when informed it was about the Queen of Thailand he exclaims,

"Marvellous!" [163] (This, if it needs to be pointed out, is not about politics; rather the illogical articulation of an issue, even the reinforcement of illogicality on the public air waves)

A marketing manager who has close contacts with the media informs me that newspapers are generally written at the intellectual level of 10-year olds.

Another conceptual clustering occurs when the notion of 'equality' is included with the provision of services of whatever kind. If the general idea of 'equality before the law' (certainly welcome) gets paired with the meaning of 'sameness' due to the negligent use of the respective semantic structures, the result is an applicability which is forced to encompass possibly incompatible abstraction layers. Take for example the provision of punishment for individuals from differing demographics. A five-year old cannot be approached with the same complexity a twenty-year old can, and vice versa. In other words, to prevent a child from playing with power points requires a different tactic than teaching an adult about the dangers of electricity; and a quick survey around the world should convince anyone that there are marked nuances amongst adults as well. Therefore, a response to anti-social behaviour which neglects the inherent differences in mental resource spaces will either turn out brutish or ineffectual. Or take the Brisbane City Council buses which have been equipped with cloth-covered seats. In other jurisdictions the seats may be wooden—less comfortable but more hygienic as they accommodate a variant public. On a bus in Brisbane one gets softer coverings, but quite a few seats carry the marks of mismanaged menstruations—a clash of standards under the mantle of egalitarianism (the same goes for public libraries).

In chapter 15 we dealt with emotions, and considering the neurobiological basis for their existence the idea suggested itself that in terms of cognitive functional entities their presence at any given time and place is finite. That is to say, if we can identify a particular emotive module as part of a TS, that EM will have been drawn from a finite pool which prevents its participation somewhere else. The more obvious effect can be seen when an individual experiences a degree of exhaustion in conjunction with some emotional event. The question can then be asked, if such limits exist in a single mind, does the wider realm of society offer similar observations?

As it is already difficult to ascertain the precise location and influence of EM-related components in the details of a PAT chain, to expect a similar possibility with society at large would be somewhat optimistic. And yet, the overall effect is there and the sheer presence of emotive bias is not invisible either. Here then is an assortment of applicabilities within the resource space of a population, expressing the existence and influence of EMs by the outcome—something on this side but not on the other, as if the two compensated for the difference.

- Youth is given priority over age in most everyday affairs, to the extent that in some industries anyone over forty need not apply for a position. At the same time there is a degree of acceptance of indigenous cultures and the role of their elders which is at odds with the knowledge base they can draw on.
- Similarly, as the manifestations of primitive art and myths are respected, the emphasis on the skills every individual has to offer has never been greater.
- Members of the establishment are very careful not to step outside the bounds of their regulated behaviour, but find excitement indulging themselves with the works of individuals who are their exact opposite.
- While most forms of criminal behaviour catches the attention of experts who attempt to discover ways of shifting the blame away from the individual, there are some acts

left which bear the full brunt of condemnation from just about anyone which includes hounding by the media.

- Child-parents relationships are viewed with rising suspicion regarding the potential for abuse; yet a steady reconfiguration of economic circumstances ensures that children remain dependent on their parents longer than ever.
- Current affairs issues are becoming more and more complex and represent ever higher volumes of data, but the time allocated to them is shrinking and the data suffer over-simplification.
- Democracy is ascertained and demanded at every opportunity in the public and commercial arena, but in both the policies are increasingly driven by special interest groups which are not elected and not representative of the wider population.
- Female fashion seeks ever new ways to bare the flesh, simultaneously setting more complex social and legal traps for the unwary.
- A celebration of individualism, but when things go wrong it is the society which is at fault.
- Again the emphasis on individualism, but marketing psychology finds sophisticated means to generalise from the observation of a few.
- Still on individualism—differentiated behaviour is looked for but has then conferred upon it the mantle of some category turning it into yet another -ism.
- On a similar note, stereotypes are rejected but similarities are welcomed.
- Accepting the principles of multiculturalism and spurning intolerance and presumption, while importing demographics which discriminate to the point of murder.
- More and more emphasis on privacy but embracing quite the opposite when it comes to reality shows.
- Before homosexuality was accepted schools were segregated and gay pupils could use the lack of opportunity as a form of protection. Now, being gay is no longer a problem officially but classes are mixed and it is harder to blend in within an atmosphere of gender assertion.
- Western societies are getting less and less belligerent, yet the appeal of gratuitous violence between individuals is on the rise.

None of the above can be said to have been planned, in the sense that an overall agenda had been designed and enacted. Rather, those examples are the effects of initiatives that people have entered into without being aware of the ramifications. The initiatives can be compared to the trains of thought inside minds, and were it not for human consciousness none of them could ever be expressed openly. In a society the actions themselves are known, but their wider affinities and deeper resonances are not, except to certain individuals who have made it their business to describe them. For that reason such efforts are not always appreciated.

No matter what the eventual outcome, the beginnings are found in the dynamics of abstracts. From the identity containers to the peripheral entities of momentary alignments, their characteristics are shaped by the fundamental forces of human nature. After the biological realm, the envelope for plants and animals, our central nervous system has produced the mind, and here too we find major and lesser players. There are three which had determined our destiny from the earliest beginnings, had always been with us but are now, at this stage in our evolution, more significant than ever. Why this should be so we will investigate in the next chapter.

17. Heralds of doom

Religion - Woman - Sex

Sunday scenes:

People are filing into church, dressed for the occasion. Inside they seek out a place on the benches and wait for the minister to begin. During the service they will listen to words, sing some songs, consult their text book, and glance at their children and others. Afterwards they leave, linger outside for a while to greet the priest, make contact with other parishioners, then continue with the day.

The above could have been recorded on paper or on film. It represents the evident behaviour of sentient beings who have had the time and opportunity to develop cognitive structures which at certain times and in certain parts are brought to the attention of their consciousness and so acted upon. Their availability for conscious awareness points to an underlying network of notions, linked together for the occasion and drawing on relationships that require an interdependent congruence to be actualised. The content can be vast. It can extend across concepts of the self, family, nationhood; of values, priorities, hierarchies; and of needs, desires, and demands. Such a range does not assemble easily with a few texts and speeches; rather it must be steadfastly constructed, layer after layer, with any discrepancies attended to as they arise. As such it comes to stand as a carefully built edifice, worthy of protection. One may call it an identity, a culture; but since it contains elements of the non-real the language of these beings has another name for it: religion.

Leaving the surface of ostensible manifestations we move towards their structure. Now we can observe the scene once more, and with the different perspective other things come into focus.

There are humans who have employed the feedback process of language and mannerisms to identify a set of others by their affinity with themselves. Depending on the power relationship at the time of contact, such affinities are either tested for their existence or ensured through disciplinary measures; the former is used with adults, the latter with children, although overlaps can occur. The affinity consists of the perceived similarity between expressed ideas, whether the articulation is made through words, acts, or appearances such as their dress code. One of the means to confirm the validity of membership is the regular attendance at designated meeting places, where the architecture of the building, the details of the ritual, and the status of their presence constitute the functional elements of the group. Each of these three types is a construct in its own right and derives its significance from the history and the current purpose of the occasion. Therefore the momentary expression of interest is neither arbitrary nor is it a chance event, but a deliberately set marker in a continuum that delineates the participants' status as members, citizens, and human beings in general. To put this another way: a church service is an affirmation of what was, what is, and what is to come.

As far as Western culture is concerned, the main basis for its current mythology comes from the Christian Bible, a mosaic of texts assembled over the centuries. It carries historical material in the form of personal views directed at either the past or their present, and descriptions of imagery by people who felt the need to communicate their visions to the world at large. Since it spans a considerable length of time (in terms of its creation as well as its story material) the book as it exists today cannot be considered a collection of items put together under some objective regimen (such as a scientific book on butterflies for example). Instead, its broad thematic purpose ensured that only those pieces found themselves as appendages which fitted its tone, and only those details made their way into its pages which concurred with the dedicated sentiments initialised from the very beginnings. Notwithstanding the opinions of its users, the Bible is therefore not a book of reference but one of direction.

The result is in accordance with the dynamics of emergent systems. Hence it needs to be seen as a step-by-step process where current data are the determinants of potential additions. Although adherents to its message tend to perceive the contents as a final whole whose parts have been arraigned according to the plan as asserted by them, in evolutionary terms the progression moved from fundamental characteristics to elaborations on the theme. These developments do not follow beliefs or political machinations but are a reflection of a general mechanism found wherever dynamic, self-regulating, pattern-seeking systems hold. As in any resource space the beliefs etc of the participants certainly influence its parts but only within the bounds of established rule sets. In order to understand its character we need to begin at the origins of ideational constructs. They have defined its framework ever since.

As reference I am using an edition titled *Good News Bible*, published by the American Bible Society as a British Edition and handed out to primary school children within the New South Wales education system (at least during the 1980s) [1].

It is not the purpose of this chapter to provide a comprehensive historical account of biblical texts. Suffice to say that we find two types of information, a description of incidents derived from historical events, and accounts by those who felt the need to articulate and so publicise a personal perception of their faith. Since our view relates to the format of Otoom, a few words on the significance of these types.

In previous chapters we have examined the dynamics of memory creation. For information to be recalled it had to be stored in affinitive form, where 'affinitive' means the potential to associate with already existing structures. Once stored it becomes part of these TSs, and will not be recalled unless there exists a train of thought (ie, the PAT chains) which enable its con-TSs to make contact with existing complexes. During this process there is the possibility that the recalled information will have been altered because of the degree of congruence exhibited by the participating PATs; thus there is a relationship between the degree of congruence and the degree of modification. Ultimately the truthfulness of this information can only be ascertained through active feedback from various sources, where 'feedback' refers not only to communication with other individuals but also to internal communication between diverse TSs belonging to the single mind itself (given the possibly unequal qualities of either it would be presumptuous to prefer one to the other in a categorical fashion).

When it comes to the recounting of historical events therefore, the validity of this or that statement is a function of somebody's memory, the event's situatedness within the cultural atmosphere of the people, and the current need of the person to recall the incident. If information is laid down in written form any subsequent usage will be performed under the auspices of its contemporary value system, which means that disagreeable elements can exist but could be simply removed or replaced by antagonists. Should the information be handed down in verbal form only (such as in cultures who do not have a written language) there tends to be a commensurate insistence on word-perfect transmission because it is only through such rigidity that the detail of the message does not get lost (anyone who has ever played the game of passing on a sentence through several people will know *how* different the original can be from the last)—hence antagonists are nipped in the bud. Thus there is a certain trade-off: verbal history is more metaphorical but stable, whereas written history is more factual but may not survive.

Because the Bible deals with the distant past as well as more reachable epochs, the former is expressed via contemporary perceptions influenced by the writers' mystical components, and the latter can be expected to be truthful but may not be as comprehensive as possible. For the reader who wants to test the validity of either account the influence of personal bias will apply in both cases, although their origins and contextual placements differ. In other words, in an object-oriented sense the Bible student will differentiate between statements of pure faith and those describing situations at a certain time in a certain place, but in the functional sense the cognitive aspects of illusion trees and the logic boundary will have to be considered regardless. As the use of the Bible is an ongoing enterprise, from the writing of its sections millennia ago to modern-day Sunday classes, for us the development of cognitive structures is as relevant to an ancient author considering the story so far as it is to today's students assimilating the information and building their own personal imagery from it.

In chapter 4 we examined some of the concepts under Manichaeism which itself drew on features that have their origin in ancient Persian and Mesopotamian civilisations. The concepts of 'good' and 'evil', conceptualised in non-material and material forms, date back to the Neolithic (the New Stone Age) approximately 6,000 years ago (historians would modify this figure according to the specific area they consider). The cognitive clustering of 'good' with unearthly concepts in tandem with the notion of 'evil' being surrounded with objects pertaining to the physical and from there to 'dirt' and 'filth' can be observed early on. Not everywhere in the world has that categorisation taken place but, as it happens, Middle Eastern religiosity was beset with an antagonism towards the physical aspects of human existence from its virtual beginnings and this has been propagated down the ages.

Such a sharp divide of the world as seen by a people had to produce equally distinctive borders of moral interpretation. The mind had to deal with an identity container which on one side was meant to provide an intellectual home to its owner yet at the same time accommodate the perceived certitude of existential misery—all within the selfsame body. As we have explored earlier in Part I this led to the formation of the mind-body dualism and the logic boundary, plus the necessary functionalities to deal with them on a daily basis. In regulatory terms this means a set of emphatic directions for one's behaviour to ensure continued adherence to such principles and therefore the survival of the perception, carried as these were by individual mindsets as well as the wider societal/political arrangements.

It is not surprising then that one of the corner stones of religious piety is the unquestioning of its doctrine. In the Bible this becomes the explicit promise of death to the first man the very day he eats from the "tree of knowledge" [2]. The setting is the "garden of Eden", a conceptualised 'original place' which naturally contains Middle Eastern rivers, thereby indicating the demographic centricity of the story. Its connectedness goes further than geography however; since the mythical underpinning of one's societal identity is a matter of confirming the entire spectrum of the persona, behavioural characteristics become articulated through the mythical medium.

For example, Adam and Eve (the original human personae) had two sons, first Cain and then Abel [3]. The older son became a farmer, the younger a shepherd. One day, so the story goes, Abel gave an offering to his god which consisted of the "best parts" of a first-born lamb, while Cain did the same with some of his harvest. The generationally determined killing was accepted by the god, the plants were not. Cain was "furious", was told by the deity that his brother did the right thing but he did not, and so enticed his brother into the fields where he killed him. There follow recriminations, some bartering, and eventually Cain leaves for another region.

Let us consider the cognitive elements of that episode. As all this is meant to be a religious statement about the first moments of human existence, its inherent quality ought to be taken as a descriptor of the god-human relationship. Thus we have a god who speaks to humans, thereby placing any messages firmly into the realm of explicitness and not metaphor. The god requires offerings, that is demonstrations of his servants' goodwill towards him. The correct offering is parts of a dead body, expressly killed for the purpose which is indicated by the type of victim: not a carcass used for nourishment in general but a first-born, thereby linking the sequence of births to the potential fate. So important is the element of killing that some vegetarian substitute is not only insufficient but in fact removes the god's grace entirely. If the substitute is offered anyway the punishment follows although some haggling won't go astray. (That passage alone, if read and understood by Western officials, would go a long way towards preparing them for what can happen in the Middle East; then again, how many professed Christians have actually *read* their Bible?)

That rigid distinction between 'good' and 'bad' and so between 'good people' and 'bad people' comes through many times. In another section of text that underlines the inherent evil of humanity all the people of the world are judged to be worthy of being killed, except Noah, his wife, his sons and their wives. The god tells him to build a boat, fill it with every plant and animal so that the species may survive and then causes a flood to drown the rest of humans on this planet. After the waters have subsided the chosen survivor's first act was to burn an animal of each kind as another offering to his god. Note the conceptual context exemplified by such a tale: in a spiritual dictatorship (ie, the one God) the servile masses (ie, the humans living under his command) are being judged unworthy and so every one of them is condemned to death; the only exception being a single family who has demonstrated its subservience and so is spared. Being the theme of the reigning spirituality, the story is deemed fit for the education of children as a moral to be learned and revered. Something similar, being removed from the grace of contemporary power, would be judged quite differently (suppose the masses are called Jews and the place has a name like Treblinka).

Given the historical background of these texts the emotional ambience of the people living at that time needs to be taken into consideration. Lower abstractive abilities in the mind make for lesser shades of grey, and the death penalty was not subject to long debates as to its ultimate morality as it is today (the word 'today' is used in the context of a Western-style liberal-humanist democracy largely removed from religious overtones). Nevertheless, we should not loose sight of the fact that I am citing a book that is presented to students without age restrictions and without any riders attached to explain possible cultural or historical differences. And, above all, it is called the *Good News Bible*.

The judicial framework in which killing is placed in order to change wrongs into rights is worth noting. Still in connection with Noah a few laws are laid down, one of which is the provision for punishment in case of murder. "If anyone takes human life, he will be punished" the text has the god telling Noah and some sentences later it is made clear that the penalty is death [4]. Such a setting has profound implications. If the act of taking someone's life without any further qualifications necessary invites the god-given right, if not duty, to take life once more, the possibilities of never-ending abuse are considerable. Firstly, the definition of 'killing' in this case is too vague to allow a rigorous forensic interpretation at that time; secondly, an invitation seemingly from the god himself would put the onus on a believer to apply such a punishment not so much because the deed warrants it but as an exercise in sheer piety; thirdly, if the interpretation of the crime is vague and the reasons for punishment are a matter of religiosity the chances for a miscarriage of justice are high with the added danger that the act is irreversible; and finally, if the entire process is fraught with that many ambiguities, there will be different versions depending on the social domains involved and so any punishment will not settle the matter but will instead propel it along and may even inflame the situation further. In short, such an inspired sense of justice is a recipe for continuous disaster.

Ritual sacrifice may be frowned upon in a modern European-type society (where the Vatican is held firmly in check), but its status here is more the exception than the historical rule. The act of giving something which is held in high esteem becomes more influential as the value of the object increases. Arguably one cannot reach much higher than one's child, and if males are worth more than females to offer a much-loved son must have a considerable effect. In one chapter Abraham is asked by his god to sacrifice his only son, to which the man willingly agrees. In the end he is told to desist—at the moment of the knife held high to be plunged into the boy's body [5]. This is meant to demonstrate how a faithful human servant ought to behave.

As a further example of what kind of divine justice people should adhere to, take the following quotes: "Put to death any woman who practises magic. Put to death any man who has sexual relations with an animal. Condemn to death anyone who offers sacrifices to any god except me, the Lord." [6] We will deal with the Middle East's obsession with sex later on, but generally speaking the sentiments are in line with a mind full of hate, revenge and, it has to be said, a lust to destroy. In a later chapter we learn of a man who was brought before a leader called Moses because he had gathered firewood on the Sabbath, a day on which such labour is evidently forbidden. Moses and his community did not immediately know what to do with him, so they consulted the god. His advice? "The man must be put to death; the whole community is to stone him to death outside the camp". "So the whole community," we read, "took him outside the camp and stoned him to death, as the Lord had commanded" [7].

The sharp distinction between adherent and antagonist creates particularly formed identity containers. Since the purpose of one's core structure is to define one's self, and thereby to define one's surrounds in conjunction with the self, the nature of such a domain is of fundamental importance. A society takes its input from the actions of its members, and the members draw from the wider generality the inspiration for their own agendas. When the response to irreverent behaviour is meted out to an entire community, the message is meant to have more import than using the example of a single individual. In addition to the type of crime the manner in which justice is applied ought to be taken into account as well. In the story about Moses he comes back from a mountain to find his people dancing around a golden calf, an incorrect icon [8]. This is his response: he put himself at the gate of their camp and called out for everyone who was on the "Lord's side" to come over to him; when the group had gathered he told them to go through the camp from "this gate to the other" and murder all their brothers, friends, and neighbours. They obeyed "and killed about three thousand men that day". Not even Pol Pot had such underlings at his service.

In a similar incident others who had run foul of Moses' idea of piety were punished by divine intervention, with themselves and their families being swallowed up by the ground, "so they went down alive to the world of the dead," and "the earth closed over them, and they vanished" [9].

Some have remarked that the Old Testament, from which the above has been taken, represents a dark, vengeful god, consistent with the ambience of the time; but that with the arrival of Jesus the atmosphere became lighter and more humane. It is certainly true that the biblical texts have been composed according to the writers' own inner composition and the message reflects a set of attitudes which need to be placed within the relevant time frame. The world of antiquity, indeed the world up until quite recently, was a dangerous and forbidding land, where a false move had cruel consequences and where success for the few meant power and influence unimaginable to today's citizens. The existence of compacted abstraction complexes, filled with the searing awe in the face of an imagined god and placed in conjunction with pitched emotions that pushed their owners helplessly towards an abyss of utter horror, became their master like a fever against which there was no cure. When Flaubert writes about the priests of Moloch, a fervent group of adherents in ancient Carthage around 250BCE, their actions appear the product of a diseased mind. And so they were, but their antics were evoked by the same mental structures we use today, the only difference now being a more varied and balanced content. Back in Carthage the priests used to assemble a blazing fire between the legs of a huge hollow bronze statue representing the god Baal, turning the metal red-hot. Children were driven towards the officials who snatched each one and placed them upon the moveable arms of the statue, to feed the deity [10].

> It looked as though the walls crammed with people were crumbling under the screams of horror and mystic pleasure. Then the faithful came into the avenues, dragging along their clinging children; they beat them to make them let go and handed them over to the men in red. The musicians sometimes stopped in exhaustion; then the cries of the mothers could be heard and the sizzling of fat falling on the embers. The henbane drinkers crawled on all fours round the colossus roaring like tigers, the Yidonim prophesised, the Devout sang with their split lips; the grills had been broken down, all wanted their part in the sacrifice; and the fathers whose children had died earlier threw into the fire their effigies, their toys, their carefully preserved bones.

It is easy to dismiss such brutal fervour because within our own cognitive landscape it lives as an alien presence, sheltered behind the asserted notion of 'the other'. Yet the Christian heritage, supplying the moral and cultural basis for the European mindset, uses the selfsame functionalities to paint its own scenes: the butchering of a son, the burying alive of an entire community, the slaughter of three thousand men, women, and children at the behest of one. Nor did the arrival of Jesus Christ, the name being a type descriptor derived from 'salvation' (the Hebrew *Jehoshua*) and 'anointed' (the Greek *Christos*), change the moralistic commitment of his followers. In the New Testament the writers proclaim their own tendencies for righteousness on behalf of Jesus, to whom the manifestations of divine guidance are attributed. The author Matthew for instance describes a scene in which Jesus explains the significance of their god. He, Jesus, tells the assembled disciples that should anyone cause a child to loose their faith "it would be better for that person to have a large millstone tied around his neck and be drowned in the deep sea" [11]. In the same vein, his followers are exhorted to cut off a hand or a foot and "throw it away" if the limb should make them loose their faith, or take out an eye if it should do likewise.

Whether such advice is being followed to the letter or to whatever degree depends not so much on the abstracted reasoning by the individual but on the organisation of their TSs in terms of the extent of the logic boundary and the associativity of any ameliorating complexes. Such organisation is not some arbitrary phenomenon but a function of the person's experiential stages which provided the input.

Up to now we have described the identity container as a domain seemingly distinct from the rest of an individual's TSs. As a form of conceptualisation to highlight the potential difference between one's actual identity and any ideas not directly associated with the self, such a view served us well. In precise, technical terms however, it is not quite correct. During the initial stages of the mind's existence elementary structures are laid down upon which later complexes are built. The word 'elementary' however should not be mistaken for a substitute of 'simple'. At this point the TSs are non-complex because the network had not had the opportunity to develop, and therefore any representation cannot be differentiated into subdomains. If a unitary value could be put on the spatial extent of a thought structure, such values would not be substantially different in the case of a young child or a much older adult; what would be different are the sheer number of associations any given unit could have access to. (For us to ascertain such a unit, ie a repeatable measure of what constitutes 'thought', one would have to be able to deconstruct the spatial extent of its manifestation into discrete entities. Complex systems do not lend themselves to that exercise, because the identifiable phenomenon produced by a complex system is not a summation of its many smaller elements but the composite of the elements evoked through all of them acting together—the whole is greater than the sum of its parts. To put this another way, in a non-complex system I could have a bag of ten apples, and taking away one apple after another I would be left with what could be called 0.9 bag, 0.8 bag, 0.7 bag and so on until 0.1 bag; in a complex system, doing the same the result would be something like 0.9 bag, 0.8 bag, 0.7 bag, *no* bag)

Since each newly created TS is an abstract representation of some input, its ultimate usefulness to the individual evolves as its meaning becomes modified through constant feedback with the outside and through similar dynamics conducted amongst its connections. That is to say, a 'rose' is in its initial cognitive manifestation not a certain species of flora but a red object featuring a shape and colour that is understood in its most immediate sense, without the semantic appendages a rose would have if thought of by an adult or especially a botanist for instance. Thus originally it is not even a flower nor a plant—it's just some 'red thing'. Only gradually additional TSs evolve which imbue that elementary representation with further depth, and so it can turn into a symbol of love, death, or of some political party. Note the interdependency amongst such domains: at any given stage of complexity an association will happen (*if* it happens) in accordance with the affinity potential of the structure. As such it becomes a matter of probability relationships as we have mentioned in chapter 15. Recall Liane Gabora's description of autocatalytic closure and the chance an idea will be connected to another. The greater the number of details and therefore abstracts (large or small), the greater the possibility some other notion will evoke a resonance. In general terms we can express the phenomenon through the e-chart where the number of products, their realised potential and the resultant applicability precondition the resource space for further advance. In this case the products are the abstracts, the realised potential is its meaning to their owner, and the applicability refers to the realised affinity potential the entire complex is capable of (see the Appendix for a technical explanation of its most fundamental form in the computer model).

That cognitive evolvement leads to the following. Incoming sensations are turned into representations of a complexity commensurate with already existing thought complexes. The higher the complexity, the higher the degree of affinity and therefore the greater their potential for further associations. Because the TSs serve as the mind's identity and subsequent additions are linked in terms of their associativity with current domains, the identity of the person will be a construct with relatively elementary abstractions at its core and more sophisticated (ie connected) abstraction complexes towards its periphery. As the system grows more centres come into being but they all are relatable to varying degree to what we have termed the identity container all along. The core will always remain at the centre although later events can modify the significance of its information value. Hence the saying, "You can take the person out of the city (or country or whatever) but you can't take the city (etc) out of the person".

We can say then that as a culture determines the ultimate availability of information and so its members' perceptual depth, its expressed values and precepts become a function of the affinity potential of its cognitive elements. A high potential engenders diversity, a low potential makes for paucity.

Suppose now the initial input is some kind of conjecture. Once a child is being told that there is a god and his name is Jehovah it will accept the information and represent it however simply in its mind. If the child is told the name is Allah the abstraction will be accordingly; and so on. Further input will elaborate and the mind's identity container will form around the initial, low-abstract, low-affinitive TSs making up the core. There will be other input coming from the multi-faceted environment and the representations will format themselves in terms of the feedback induced by family and strangers. Thus are mannerisms and accepted and unacceptable types of behaviour born.

The feedback process occurs between person-to-person and between person-to-surrounds. In other words, representations can be confirmed through the articulation made by someone, or they can be in the form of some observation. If the observations concern some natural event (eg, the falling of a rock, the traffic on a road, a hot iron) TS complexes are set up which increase the relevant knowledge base. The same goes for information sourced from another person. If that person should express religious sentiments these are added to the knowledge base affinitive with the mythology and these domains grow too.

What happens when during the course of these developments the two regions clash? For example, someone exposed to the Bible is led to believe that the earth was created in six days, but later attends school and learns about geological epochs. Firstly, if there is nothing to clash the two domains will exist side-by-side and hardly interfere with each other except for higher-abstracted contemplations of the 'what if' variety. But should there be a discrepancy as our example suggests, the outcome depends on the contextual weight of either domain. An existence mainly constructed (or construed) around biblical assertions will have spawned a great number of associated complexes representing a significant slice of life's eventualities and they have to compete with the new information. Maybe the new data consisted only of a passing sentence, or maybe they formed part of the person's career. As shown before, it becomes a matter of affinity relationships, overall complexity, and the richness of the data themselves—a scenario described by the e-chart.

Combining the aforesaid with earlier observations about the contextual spread of content across TS complexes (ie, a diversified content will offer more interpretations and alternatives than a more compact one—those 'shades of grey'), we can see how an idea's closeness to the identity core of the individual determines the representative value in the hierarchy of priorities articulated by that person. The nearer to the low-abstract, original formations the content is situated, the more intense and unforgiving the actions will turn out to be. In general this can be witnessed any time a situation becomes dangerous to one's life. Something less significant can be brushed off, talked about, responded to congenially. The more precarious it gets, the less recourse there is to wider considerations and so eventually the reaction can be as extreme as murder; in other words, the degree of destructiveness exhibited is in opposite relation to the abstractive reach of some content and in direct relation to its proximity with the core identity.

This leads us back to the issue of resource space (whether with regards to a single mind or a society), and the type of response manifested in the face of some provocation. For example, refugees who do not achieve their aims through the authorities and inflict harm on themselves and others demonstrate a conjunction of emotional bias coupled with conceptual intensity which grows into a particular problem when such actions are performed within the context of an unequal environment. In Australia there had been several riots in detention centres caused by refugees who either faced lengthy delays in their processing or had their claims rejected. In one series of such incidents the total damage came close to $10 million; quarters were burned down, fire crews were attacked and some threatened suicide and self mutilation (in some instances that was actually carried out, such as the sewing together of lips) [12]. Since Australian society does not run along the lines of emotional upheaval there is the tendency to equate such violent outbursts with the kind of serious cause had a similar action been taken by a Western-style resident. Hence private citizens and refugee support groups adopt a position of empathy and activate the legal system under the auspices of a more rational ambience. The costs involved, from societal discomfort to the financial burden on the nation, are not in tandem with the underlying cause but are a form of ransom instigated by the martyrdom of certain demographics for whom such behaviour patterns are a common bargaining tool.

The beating of one's own head as an expression of grief, the wailing as a public display of hurt, serves to underline one's status as victim and enhances the perceived weight of injury done to the person. The range of intensity runs from mere gestures right up to purposeful suicide, and the Middle East and indeed now the rest of the world are forced to be witness to this type of behaviour. Given the interdependency of affinity relationships we can go so far as to say that with such options at one's disposal, a particular situation will not only be explored as to the exits it offers, but also as to the possibility of further intensifying the conflict so that more extreme reactions can be carried out, or that others will be more likely be enticed to act likewise (this can happen on the conscious as well as the subconscious level). In terms of progression lock within a resource space, from a certain stage onwards the positive will be spurned in favour of violence because the actualisation of abstracts and the feedback they evoke are now firmly within the representative domains of destruction and victimhood.

Returning to mythology, the crucifixion in the Christian context can be seen as such. Performed at a time of low abstracts, social and political turmoil, emotional demographics and intense competition from various mystics, in the end the implemented strategy had the desired effect. And, in another sign of identity containers triggering denominative associations, what under different circumstances would have been classified as emotional blackmail had become elevated to a central icon of an entire belief system.

The cognitive elements of that event—the brutal fate of a victim, his relationship with the deity, the identity of the perpetrators, and the underlying notion of implied obligation and revenge—have moulded the character of the Christian religion ever since. As the faith became established in Europe as an organised religion, its followers had the Jews in their sight. Historical events helped shape the appropriate configuration in terms of the church's reach, (political, social, and financial), and the necessary secular power arrangements across the lands. The Inquisition did not begin as an instrument to persecute and burn heretics; first and foremost it served as an administrative tool to settle the communal problem of Jews who had converted to Christianity and ran foul of neighbourly suspicion and piety [13]. Nevertheless, a resource space that contained a dictatorial spirituality and the relevant vengeance that goes with it soon found adequate excuses to apply its version of the world to all and sundry. For a demonstration what these rather dry words mean in practice, refer back to chapter 9 and its section on Nietzsche where an eyewitness account of a burning is cited.

It often seems that the word 'heresy' has lost its meaning in a time of humanist democracies, where the church and state have been set apart, a separation of powers ensures adequate protection by and from the law, and the likes of Karl Popper can actually celebrate the questioning of whatever an expert had pronounced. So much so in fact that any recurrence of a traditional response to it is not even recognised because hardly anyone knows anymore what those people in the past actually did (this is the converse to a similar phenomenon which can be seen in the accusation using a word like 'fascist', applied to just about anyone who does not agree with ego-centric protesters—a real, fascist, government representative doesn't just block your path in the glare of TV cameras, he quietly takes you to a strong room and gets to work on your limbs).

If the religious view represents such a comprehensive explanation of life and the universe that most PAT chains have become associative with some of their aspects, we can expect a high probability of some challenge being perceived as a dire threat to the very existence of one's identity. In this case the question of mitigating circumstances, indeed of the overall worth of the perpetrator, become largely irrelevant; just as a life threatening attack would not prompt the victim to muse about the criminal's domestic situation. Take the fate of Michael Servetus, a medic who was arrested by the Viennese Inquisition in 1553 and who managed to escape to Geneva where he thought he was safe. At that time the city was under the religious and moral jurisdiction of Calvin, and Servetus' questioning of the Holy Trinity made him a target of Calvin's righteousness. The man was tried and burnt to death [14]. "His execution is said to have involved a slow roasting on a spit that took two hours to kill him". As it happened, Servetus was just about to discover the circulation of the blood through the body, seventy-five years before William Harvey, its historical discoverer.

Or Giordano Bruno, a monk who early in his life was suspected of heresy and spent the rest of his travails under the church's suspicion. About his philosophy Michael White writes:

> Bruno believed in an infinite universe, a universe far greater than the cramped, rather ridiculous, parochial little place imagined by the Church fathers and theologians. He considered Copernicus' heliocentrism to be too simplistic. His was a far more modern vision, one in which the sun was viewed as nothing more than one of many stars in an infinite firmament. In this philosophy, the people of the earth, the entire human race should be considered as just another group of living things in a universe in which all was interconnected, each element interdependent and interrelated. [15]

He also interested himself in the art of mnemonics, the techniques through which memory can be cultivated and enhanced. In May 1592 he was arrested by the Venetian Inquisition, incarcerated by its Roman counterpart in February 1593, and after seven years of imprisonment and repeated torture burnt at the stake in Rome [16].

One of the advisors to the pope had been Cardinal Robert Bellarmine whose inclination was to have not only all the heretics burnt but also all Protestants and similar dissenters. His view was definitely anti-science and "fifteen years after Bruno was in his grave, the reverend Cardinal would instigate the arrest and trial of Galileo". "As a reward," so White tells us, "the Church would canonise Bellarmine in 1930" [17].

The mental strictures which compel their owner to apply the god's wrath to another human survive to this day. In Morocco Islamic vigilantes haunt "bad Muslims" who are seen as neglecting their faith. In July 2002 the body of a teenage girl was found in Casablanca, "violently stabbed to death". This was followed within days by the body of another girl, "both her arms hacked off by the killers, and a week after that, the decapitated remains of a third girl were discovered" [18]. Police suspect the work of extremists, a group going by the name of "divine judgment for infidels". While some of the terrorists' statements refer to political situations such as Israel and Palestine or the quest for independence in the Indonesian province of Aceh, their conceptual framework derives its fervour from their minds' intimate embrace with an idealistic purity that brooks no diversion nor compassion. Within such thought patterns there is no room for abstract differentiation from the theme, a state of being that is the hallmark of the fanatic.

The intellectual separateness goes both ways: as the religionist does not comprehend the wider view, so is the more liberal person struggling to come to terms with potential manifestations of compact TSs. The demolition of the World Trade Centre in September 2001, the bombing of a Bali night club in October 2002, the repeated attacks on Westerners in Islamic regions by extremist warlords using their victims for ransom, all these acts and others are only now beginning to be understood in terms of their sheer depth of hatred towards anything that does not accord with a compacted perspective of desired morals. Such inner reality is just as 'real' to the extremist as is the other to a laid-back Westerner. The negligent use of the word 'understanding', brought about by the West's own imposed adherence to religious ideals that fear any intellectual proximity towards an undesirable content, turns any attempt to approach a foreign concept into a difficult task. Yet contemporary events, in the political as well as other arenas, compel an increasing number of people to overcome their own reluctance and come to grips with existences that so far have never been fully confronted.

The ability to get close to an object, a person, or an idea which one perceives as repulsive is an art that needs to be learned. As the dynamics of affinity complexes show, there is a 'natural' tendency to flow with the intuitively offered associations, where 'intuitively' refers to the main contingent of sub-TSs which organise and administer the material which eventually makes its way to the cortical regions (recall our discussion about consciousness in chapter 15). Since any ideological content is stored in the form of representative strata and the abstractive potential of these is a function of their introduction through time, the conceptual weight and extent of relevant sub-TSs will be re-representative of their content, as we have already mentioned. But not only that: as such content and therefore such strata belong to the identity domains of the mind, their output will not only be in perceived congruence with the type of religion at hand but they will also act *as* the identity of that person. Thus persons acting under the regimen of their god actually see themselves as part of the theology, and what they do and say represents, as far as they are concerned, the body of that religion. On a more trivial level we have the employee working for a company for a considerable length of time and, no matter on what rung of the enterprise they may be, consider themselves as instrumental to the company's well-being—in a sense they *are* the company. The situation has more import when a cleric endeavours to change the world because for him it has run foul of the god's code. The response from such people is done by them, but perceived to come from a higher authority, be that a CEO or the god himself. Johnny Hart, the creator of the comic strip B.C., has summed it up succinctly (Fig. 17-1).

Note once again the element of interdependency; although the denominative TSs forming the identity container can be incongruent and the result of illusion trees, their associative potential is realised through affinities with material fed back from the outside (actually, the resonance can also come from within the mind—the ruminations of believers who seek to make sense of it all are testimony to that). Since the dynamic turns on *affinity*, the respective environment of the individual is important to the feedback process. While adherents may dispute this (after all, their focus is on the perceived generality of their belief), the conceptual atmosphere breathed by the believer determines their spiritual character and the outcome can be as varied as one's surroundings. For instance, a certain demographic of Protestant Christians arrived at the conclusion that the observance of Sunday is the work of the devil, the Catholic pope his virtual disciple, we are witnessing the beginnings of a cosmic, yet earth-based, struggle for the supremacy of it all, and one of the most, if not *the* most important issue is the acceptance of Saturday as the day of rest instead of the seemingly dangerous timing represented by the Sunday; so their text goes which runs under the title *"Earth's Final Warning"* [20].

Fig. 17-1: Who is doing what when the god speaks. [19]

Conservative politics has its own shades. During the recent quite severe drought the Gold Coast City Council sent letters to the two hundred churches in its area asking their congregations to pray for rain [21]. By the way, the weather on Australia's eastern seaboard more or less continued until the beginning of this year when the Southern Oscillation Index began to move upwards (this indicates wetter weather).

On the other side of Brisbane, the Sunshine Coast's Caloundra Christian College seemed to have made more direct contact with whatever it is that religious hot-housers make contact with. With the headline *Holy war over 'possessed' pupils* (note the pragmatic single quotes) the state's Courier Mail nevertheless devoted a 26×34 cm space to the report (including a photograph of one addressee) which described how "up to 30 students" claimed they had been "touched by God" [22]. This experience made them shake, jerk and cry, but the principal and the school's duty pastor assured themselves that there was a "plan" at work and all was well. One 11-year old girl whose "uncontrollable shaking" caused her to "fall to the floor" has been quoted as saying "I could tell it wasn't the devil". The father of that pupil, a math teacher at the same school, "knew what was happening" because the same had occurred with his older daughter who is thirteen. The 11-year old apparently had started the whole event with the other 30-odd children taking it from there. Leaving aside the issue of a family whose internal mindset triggers attacks of religious hysteria in its children, the incident demonstrates the intellectual (or otherwise) culture of Queensland's non-metropolitan areas. While in the article some academics have suggested not to be too concerned about all this lest the phenomenon got worse (certainly a sensible advice), there has not been one voice in that text which pointed to the fact that here is an Australian jurisdiction in which elements of its education system generate manifestations of medieval proportions.

Every now and then there emerge some dissenting voices who point to the inherent brutality of a book that lays down the fundamental framework of the Christian belief system. In 2000 two lawyers wrote to the German Family Minister with the request to "officially class the Bible among books considered dangerous for children because of its violent content" [23]. "It preaches genocide, racism, enmity towards Jews, gruesome executions for adulterers and homosexuals, the murder of one's own children and many other perversities," they wrote.

On a similar note (and still within the context of education) the last years have seen a gradual uncovering of the type of approach so typical of religious presumption. Former inmates of orphanages that were run by religious institutions have come forward to publicise the treatment dished out in the context of moral righteousness. Punishments such as regular beatings with cricket bats, bashings against a concrete wall, pushing one's head into a toilet bowl containing faeces, the attachment of electric wires to genitals etc etc were all committed by people who felt they had god and therefore the law on their side [24]. Yet any accusations made at that time (for example in the year 1978) and received by the Children's Service Department in Queensland were not acted upon [25]. While I do not have the necessary information in order to identify the precise reason for inaction, the framework of ideological security on one side and the social acceptance of its principled underpinnings on the other contribute to a culture of comfortable acquiescence for everyone concerned (with the exception of the victims of course, but they fall under the category of 'deviant' and therefore don't count in that system).

A former student of the Sunshine Coast Christian Outreach College has sued the school for damage inflicted while practicing its policy of corporal punishment [26]. The social demographic of Queensland's Sunshine Coast falls within a region often referred to as the 'Bible belt', where something as fundamental as religious intensity gives rise to dynamics involving the political, social and judicial arena. How can it be otherwise: denominative TSs centred upon a people's identity interfere with daily affairs in a manner that can only be described as comprehensive. So it happened that it took until the 21st century for an institution to be taken to court, something which under most other circumstances would have been possible generations ago.

In another orphanage girls were beaten, held under water, and allegedly branded in the back with a red-hot poker to "exorcise the Devil" [27]. In 1999 the Queensland Government received the findings by an inquiry into those and other allegations of physical, sexual, and psychological abuse in church- and state-run institutions for children, known after its chief officer as the Forde inquiry. Twelve months later a report by the Forde Implementation Committee criticised the churches for not coming up with sufficient funds to address the fates of the victims [28]. The Catholics assured the rest that their support lay with counselling services and independent complaint processes, they just wouldn't disclose the details for reasons of confidentiality. The Anglican church was unavailable for the article, but its authors note that the previous year the church's Archbishop Peter Hollingworth "while receiving a $200,000 government cheque for the $25 million cathedral building project—warned the church might not be able to afford to contribute". In August 2001 ex-residents of some of the institutions complained that the recommendations by the Forde inquiry were not fully implemented [29]. One comment:

> "...very few successful criminal prosecutions of abusers had occurred, no civil actions for compensation against state or church organisation had succeeded and former residents faced constant battles for access to their records."

By April 2002 the Queensland Government had contributed $2 million to a fund designed to help former inmates rebuild their lives. "The largest amount so far donated by a religious body was $15,000 from the Uniting Church last year" [30]. As for the Catholic church Archbishop John Bathersby is quoted as saying,

> "There is a general consensus that we would like to contribute something.
> In this ecumenical age, it was felt it would be good to do something together."

His Anglican counterpart did not go into any detail; the church was "still considering the matter".

Unless there exists a conceptual structure that encompasses society such that its various instrumentalities are affected by its influence, the process of uncovering, analysing, and responding to generic violence would not be as protracted as the one illustrated above. Not only is there the background, its very nature must be taken into account when identifying *what* it is that proves so problematic.

Compared to the speed with which public institutions respond to criminal acts identified as such by the system of their culture, the difference is obvious despite the fact that the acts themselves are in exactly the same league. In New South Wales physical punishment for children by the education system had not been outlawed until 1995 [31]; Christian schools were against the change. On a radio programme Peter Barnes, a Presbyterian Minister, referred to a section of the Bible called the Book of Hebrews where in chapter 12 the application of physical force is sanctioned (it is described there as an act of love [32]). Accordingly then there is "nothing wrong with applying on hands, legs" and abolishing the custom would be "the thin end of the wedge". At least he did not go so far as to suggest the burning of witches, as an Anglican vicar in Britain did (the year was 1994) [33]. Although "immediately rebuked by his bishop's office", this sentiment towards dissidents has only lost its threat because over the centuries the church had been disenfranchised of its power. Go back four hundred years and such utterances meant a very real terror indeed. In any case, when it comes to the religion's inherent attitudes they have neither been altered nor even questioned in a fundamental way.

The logic boundary as articulated throughout these pages referred mostly to the cognitive complexes in the mind; in terms of the functionalities involved however they are no different from similar dynamics observable in wider society—only the scale has changed. The ideological domain on one side and the structures put together by the exposure to the here and now form a dualistic system which is bound to run into the discrepancies generated by either. Which side wins in the end becomes a matter of distributed weight across the connections.

It is now almost 150 years since Charles Darwin wrote *On the Origin of Species* (see chapter 9), and still the biblical version of our earth's history maintains its hold over individuals and institutions alike. In 1992 the Australian Institute of Biology conducted a survey testing the influence of creationism amongst first-year biology students in Australian universities [34]. The poll was topped by students at the University of Technology in Sydney with 20 percent declaring their belief in creationism; the next figure was 17 percent at the University of Sydney. "Of 4,255 students from 17 universities who responded to the survey, one in eight choose the creationist option," "forty-one per cent believed in evolution over millions of years guided by God, and 43 percent in evolution without God". Since evolutionary biology *is* being taught in secondary schools, the result shows the type of background knowledge offered to humans within certain demographics. Given also that the creation of the world is only one aspect of the Christian perspective, the overall ambience existing within a mind which is under such an influence points to a general set of complexes that do not gel with the reality as ascertained by more congruent PAT chains. The question then becomes, to what extent does the content on the ideological side of the logic boundary hold in people's interaction with the world?

Debora MacKenzie, in an essay on 'creation science' as it is also called in the US, explains the political/social background to the rise of fundamentalism in that society [35]. A combination of science-bound challenges to people's core identities and an acute discomfort with rationality produced an antagonism that went as high as some state legislatures. Alabama for example added a disclaimer calling texts describing evolution as "controversial', a similar statute had to be overturned by a Louisiana local court in 1997, and the Kansas State Board of Education "rejected the school curriculum standards written by its scientific advisors" (it is no coincidence that those names evoke memories of other examples related to narrow-mindedness and hate—just ask the blacks there). In Australia Ian Plimer, a professor of geology, dared to take a Sydney creationist to court by claiming that his views were based on false information which he used to sell his books, tapes and videos [36]. Although the court agreed that Roberts (the producer of these books etc) did make misleading scientific claims he had not breached Australia's fair trading laws and Plimer lost the case, appealed, and lost again. For his efforts he received "many aggressive telephone calls and letters," some "abusive and full of bile and invective". "The creationists," he writes, "tried—and failed—to have me removed from my Chair at the University of Melbourne". Because of the threats he had to keep his address secret, remove his name from the electoral roll and even use a foreign passport. In terms of assistance from others he says,

> I have had moral and financial support through all this from colleagues and the press, and from groups abroad such as the Geological Society of London and the German Geological Society. But I have had scant support from professional bodies in Australia and at the trial I felt I was virtually fighting the case alone.

As Plimer points out in his book about the doings of creationists, *Telling Lies for God*, the message espoused by them sits comfortably with people who find it difficult to make sense of a complex world [37]. For an analyst of thought structures the problem goes deeper: if some individuals do have a need for simplistic answers, why seek them in the Bible? The conceptual spread of a particular viewpoint through society, altering perceptual elements in order to fit them into a framework of a spiritual ideology, and thereby creating cognitive structures which from then on re-represent reality in a very specific format, they all lay down the foundations for affinity-seeking patterns. When it comes to reacting to the outside, such minds will behave within the scope their imagery has allocated to them.

Plimer's lone stance against religious conviction should be placed against the background of—in this case—Australian society and therefore its main players who have worked their way through the ranks against that very tapestry. One of Queensland's former Premiers, Joh Bjelke-Petersen (he eventually received a knighthood from the British queen in another display of societal affinity relationships), is a Christian and Sunday school teacher who never tired of declaring the Bible's relevance to everyday life. And yet, "the Bjelke-Petersen years, from 1969 to 1987, saw arguably the most corrupt government in Australia since the Rum Corps days in colonial New South Wales" [38]. "Four of Bjelke-Petersen's ministerial colleagues went to jail for various offences, surely a record for Australian politics," Peter Charlton writes on the occasion of that man's 90th birthday celebration, which was "marked by much breathless adoration".

When Barry Jones, a former Federal Minister for Science, explained the reasons behind the "Joh phenomenon" he might as well have articulated a description of people with ideological thought structures [39]. Here are the points as he wrote them in the form of subheadings:

> Those who look for simple solutions to complex problems.
> Those who are apprehensive about the future.
> The inarticulate.
> Christian fundamentalists.
> The lazy and ill-informed.
> The "white shoe brigade" [this refers to a label given to developers with no concern for any other issues besides the promise of money to be made].
> Extremists in small business.
> The Alf Garnett vote. This sector is hostile to higher education and sceptical to middle class issues..
> People who don't like politics.
> The smaller government push.

The spread of particular TSs from the centre outwards create denominative complexes with decreasing intensity (ie, bias) as we move towards the periphery. While ideology-related formations can—and often do—inhabit the wider system, in this case the entire resource space is predisposed in terms of the core's content. If an incongruence occurs, as it sooner or later does, the outcome is not only a matter of discrepancy with grounded PAT chains. Its other effect concerns the applicability of those domains which happen to be involved with the clash. An affinity relationship is another label for an interdependency that has become embedded within the connectivities of a particular region of abstracts. Since they exist on both sides of the logic boundary, a change in the applicability of a region will have repercussions for the neighbourhood. If the surrounds are largely congruent a change can be integrated within the wider framework of the resource space, provided the potential is there. And, since these potential affinities are congruence-based, the probability of re-forming associations is relatively high (assuming the entire space is capable of such interference). However, if the complexes are derived from ultimately incongruent links then, in terms of the overall system the structures are more or less arbitrary and imposed modifications are less easily handled.

From the perspective of cognitive dynamics this is the underlying reason why in a liberal democracy church and state ought to be separated. The two domains, the church with its ideological strictures on one side and the state with its multi-tiered instrumentalities on the other, are two competing systems each with its own functional drivers. Although it cannot be claimed that politics is free of its own convictions, they nevertheless are part of an organisational whole which imposes checks and balances along the way. How well they work in the end is another matter, but they are still a far cry from the closed rule set of religious doctrine. In practice this becomes a difference in societal economics. Should there be an upheaval on the church's side, the nation remains a functioning entity because in a liberal democracy the church is not needed for the nation to function; if the problem lies with some part of the nation's political apparatus, the nation as a whole can still function because it only ever *is* a part (extreme circumstances excepted of course).

As a sign of how close to the religious fold Australia's present federal government is, the appointment by the Anglican Archbishop Peter Hollingworth to the position of Governor-General in 2001 speaks volumes (according to Australia's Constitution, the Governor-General represents the British queen in the country, and amongst other things can exercise this function subject to the Constitution, is able to assign session times to the Parliament and may discontinue them, can dissolve the House of Representatives and at times the Senate, recommends the appropriation of revenue, and could intercede in certain decisions by the High Court of the country if they should go against the interests of the queen—a not inconsiderable office therefore [40]). The symbolism seems to be explicit: he will continue to wear a clerical shirt on Sundays, his official government-provided residence will get an extension built for the use as a chapel, but he does concede that "there are some who believe in the ... separation between church and state" [41].

The proof for the intrinsic incompatibility between the two systems was not long in coming when the church got embroiled in sex abuse claims, the government tried to defend its appointment, and the office itself got caught in the middle.

The overlay of one's identity with incongruent content does not diminish the functions of identity containers; we have stressed the difference between functional and contextual elements all along and so no matter what the data its carrying structures will perform their purposes in any case. Since one's identity is that what gets projected as the 'self', its nature forms the articulations through which we recognise the persona. If the dogma demands the proliferation of its tenets because without them humans are perceived as not sufficiently valid, we can expect the articulation of an adherent to reflect this. What is termed 'evangelism' is therefore nothing but the projection of these containers, and because the doctrine allows for both 'good' *and* 'bad' behaviour, whatever gets perceived about someone else will find a relevant denominative congruence with the observer. Note the difference to other types of conceptualisations: if there is no affinity with an idea the potential connection does not get made—end of story. For instance, an applicant for the position of chef will not be considered if they do not know anything about cooking, but no further action will be taken. The religious context on the other hand does have something else to say, namely the alternative categorisation of 'bad' in a socially comprehensive manner. Thus evangelism does not merely entail the attempt to convert to one's own belief system; failing that the target is seen as a negative influence in the scheme of things and actions will be taken in accordance with the resource space of the instigator. Should the evolutionary standard contain the—effective—separation of church and state, the result will be merely less material in one's mailbox. If the church does have the power, the recipient of such 'love' will have to contend with anything from the destruction of their family to torture to a full-scale invasion of their habitat.

In an article on the religious forces behind the latest Iraq war Alison Cotes begins with the question, "which particular verse of the Bible will [George W. Bush] be reading?", and she quotes Matthew 5:9 ("Blessed are the peace makers") and Matthew 10:32 ("I come not to bring peace but a sword") [42]. She then points to the leaders of the most committed nations and their religious side: Britain's Tony Blair a Christian Socialist and a lifelong member of the Church of England, Australia's John Howard a low-church Sydney Anglican, and the US's George W. Bush a born-again Christian "since the age of 40". We further learn that Bush holds daily Bible study sessions besides his annual prayer meetings, confirms praying often during the day, and has rebuked a new employee to the White House for not attending Bible study "on the first day of his job". His take on the role of himself and the US can be gleaned from statements Cotes cites in the article:

> "We can be confident in the ways of Providence ... Behind all of life and all of history, there's a dedication and purpose, set by the hand of a just and faithful God."
>
> "..we go forward with confidence, because this call of history has come to the right country—may He guide us now."

> "The liberty we prize is not America's gift to the world, it is God's gift to humanity."
>
> "I welcome faith to solve the nation's deepest problems."

Australia's Prime Minister, who has demonstrated his affinity with that rhetoric, can link to a domestic version. In that same lead up to the war the Dean of the Anglican Diocese stated in his inaugural sermon in Sydney:

> "Different religions cannot all be right. Some, or all of them, are wrong. And if wrong, they are the monstrous lies and deceits of Satan, devised to destroy the life of the believers, to capture them into the cosmic rebellion against God, and to destroy the freedom they should have in Christ."

As Cotes says in her last sentence, "The God of John Howard, Tony Blair and George W. Bush is no pacifist, so be afraid, be very afraid".

The effects of evangelism can be studied in history books where the large-scale examples are written. They can be seen in their finer detail amongst religious congregations around the world who work themselves into a fever to rid the world of evil. They also play a part when believers offer their children as sacrifice in their struggle, a habit that can be traced back to the Old Testament section of the Christian bible which in turn reflects a spiritual Middle Eastern ambience in general. Or it can be a small street scene that is just one of so many fleeting moments, never to be played again and so lost to society which is awash with propaganda.

There is a preacher, one of many in the state of Queensland, who sometimes stands on a street corner in the Brisbane suburb of West End. He loudly proclaims the reign of Jesus, rails against homosexuals, and represents at once the best and the worst in the principle of free speech. One morning, on the 7 March 2002, another zealot poured petrol over the man and tried to set him alight. Someone from a nearby shop prevented a serious accident and the preacher got away with minor damage. Five days later, on the 12th, he was back again—same corner, same voice, same poison. This time however he had two others with him, a petite woman and a toddler barely able to walk. So, did the trio offering themselves to perhaps another attack draw from their ability to make rational decisions, or was that scene the articulation of a mind convinced of its righteousness and quite willing to confirm someone else's violence in a most dramatic way?

In Part I we have observed the attempts by thinkers to delineate from their everyday existence towards the conceptualisation of the divine. It sprang from the mind's ability to idealise a perception and consider it as an element of reality. For us the phenomenon of abstraction (of which the aforesaid is a manifestation) is significant in itself: since there is a system which abstracts, the dynamic creates the notion of the god-concept which in turn modifies imported data in terms of the existing affinity relationships. Having an abstraction complex that contextualises the relevant content in the form of a mythological format, the need to confirm its significance in relation to the outside will predispose its owner towards the formulation of some proof. We can say that from the perspective of cognitive dynamics, the steps taken in order to construct a proof that god exists is not so much an explanation of the god's existence but illustrates the ability of the mind to abstract. Hence attempts to prove the existence of god is an indication of a person's belief, and not a test for the existence of a deity. It does not matter how the proof proceeds, the mere fact that there is one indicates the existence of a believer. Thus, a proof of the existence of god is really proof that somebody exists who believes in god (keeping in mind that the whole argument turns on the presence of incongruent TSs).

Another aspect of idealisation is the concept of perfection, that is the assumption that there is an all-knowing, all-powerful god. But if the god is all-knowing, he cannot be all-powerful because he would recognise the inherent limitations wherever they occur in a system. And if he is all-powerful in terms of some system, he cannot be all-knowing because he demonstrably does not have access to a wider reference outside that system.

Even the attempt to conceptualise something like 'perfection' is bound to fail. If 'perfection' means 'completeness', and given that a con-TS can never be complete compared to the entire structure (which also contains the sub-TSs), then how can the notion of 'perfection' be articulated in the first place? It cannot be done in an explicit way, but it is very likely that the recipient of an articulated TS (the conscious attempt therefore) will have connected it with their own associations. Since the resultant total TS in the receiving mind is now dissimilar to the sender's total pattern, it follows that the articulation of something which is taken to be 'perfect' consists of the overlaps shared by those TSs or, to put it more formally, of the intersections of a finite number of structures. Since they are finite, they cannot be perfectly complete.

'Ideals' present another problem as well. Suppose we define an 'ideal world' as a world where faults cannot occur; a non-ideal world is one where faults can and do occur. First consider a system, any system, in a non-ideal world. A system receives data, processes them, and provides an output; a fault occurs when the output is flawed. How can we have any assurance that the system works correctly in a sufficient number of cases? We can test the system, and if the output is correct say, in ten cases, we can reasonably assume that it will be correct in the 11th, 12th etc case as well. The more often it works, the greater the chance that the output will be as expected. Therefore the best we can hope for is to establish a degree of probability in terms of the output being correct. To ensure 100% reliability (remember, this is a non-ideal world) we can push the system beyond its limits and thereby establish a performance envelope which delineates the extent to which the system can be pushed and still maintain correctness. So in other words, we can have 100% reliability but there are inherent performance limits. Or, we can have no set limits but a less than 100% reliability.

Now suppose there was an ideal world. Returning to our two options from above, if we chose to push the system beyond its limits then there would be no mistakes either, because it is an ideal world. But that would be a contradiction since in this case we could not have pushed it beyond any limits in the first place. On the other hand, if we preferred no limits but less than 100% reliability, that also would be a contradiction because in an ideal world the reliability could not be less than total.

Since there is a contradiction in either case, it is impossible that an ideal world can exist. Note: a system is not a system unless it can be defined as such, and part of the definition involves a performance envelope. That envelope must be finite (hence the limitations—see above). If it is not, the system would be omnipotent and that in itself is impossible because we run into the kind of problems encountered with the notion of an all-powerful entity.

While the above can be seen as rather academic, it does point to the inherent discrepancy between such structures and more grounded domains, and in a very fundamental way at that. For the relevant cognitive frameworks to exist, there must be affinity relationships which enable such content to link with similar constructs, and the clustering effect on both sides produces the logic boundary we have had ample opportunity to discuss. Since life in general does not care about the single mind's capacity to deal with incoming information, the import of extraneous data is bound to interfere with complexes which do not provide the necessary congruence to digest these data. Provided the congruent PAT chains have sufficient resource space to contain the ramifications of this or that content (in terms of the products, realised potentials and applicabilities of the e-chart), and provided the other side can maintain its existence without having to intrude into more pragmatic realms because of its sheer volume and hence significance in the mind overall, the result will be relatively unproblematic. But the difficulty can only ever be relative to the contingencies of the environment at large. For a society that exists at a level of complexity of the ancients for example, the finer points of evolution are neither here nor there; but when it comes to modern biology, gene research or the evaluation of bio-systems a knowledge base that relies on some mythological interpretation will not only be useless, it can be downright dangerous.

The more fundamental the incompatibility between grounded and ungrounded structures is, the more pervasive its influence on the system as a whole. Researchers have found that induced stress leads to the release of the steroid hormone cortisol in the brain [43]. In the case of chronic stress the continuous presence of this hormone results in the withering of dendrites and their subsequent death. In humans prolonged exposure to trauma has been found to cause anxiety, memory disturbances and intrusive thoughts, and imaging the brain of affected individuals revealed degenerative changes occurring there. The link between depressive illness and a stressful life has also been established [44]. Although the clinical investigations referred to here have concentrated on childhood abuse, wartime experiences or being the victim of violence in general, the similarities exist.

One of the more critical cognitive states occurring during an exposure to violence is the frantic attempt by the mind to reconcile the act with one's identity, a search after the reason for being attacked that increases in urgency the less capable one is of coming up with an initiative of one's own. The desperate question "What have I *done*?" is a manifestation of the self trying to come to grips with a perception that just does not make sense. With a violent act the conflict is immediate and extreme; in principle however the mismatch between the potential of the core complex and that of other TSs is the cause for stress in any case. The mind being an interdependent system, whether the irrationality exists on the outside and challenges the inner domains, or whether the lack of logic is part of the self and faces the more rational outside, the effects are the same.

It is therefore understandable that in studies exploring the link between goal seekers and depression the pursuit of abstract goals heightened the tendency towards that condition, in contrast to concrete goals [45]. Concrete goals, that is to say the re-representation of grounded content, offer less possibilities for incongruent chains than their abstract versions (and if they *are* abstract, the successful realisation by individuals of however long-term visions against the background of exactitude and diligence is no secret).

In another study the confirmed cause of emotional distress by irrational beliefs has been the basis for an investigation into the means of measuring them [46]. The authors note that "it has been increasingly recognised that irrational cognitions play an important role in distressed relationships" (the research was conducted within the context of marriage). Other points made were a higher total irrational belief and greater need for approval amongst women than amongst men, that people who possess irrational beliefs "tend to have a poorer self-concept", and people with significant religious leanings "tend to be more concerned with guilt, blame for mistakes, or deviations from strict values or norms" [47].

All this indicates the influence of denominative, incongruent containers. Since the ultimate effect depends on the respective weight of all the structures, a religious influence sometime in life is not a guarantee for bloodthirsty monsters. It does however set the conditions for affinity relationships beyond the constraints of reason, and in conjunction with the need to assert one's identity they come to the fore within the context of the entire system. Naturally, the same relationships can be observed at the larger scale of society. An often produced argument for the adherence to a religion is the good it exhorts its followers to do. What is not mentioned at that point is the basic, community-related, altruistic nature of humans which compels them to act agreeably no matter what belief system they subscribe to, or indeed when there is none at all. Furthermore, true tolerance is not proven by the pacifism towards those who aren't seen as a threat but at precisely the moment when the person is found to be disagreeable. For that reason the effective mark of a liberal democracy is the treatment it reserves for dissidents, not its adherents no matter how colourful they may be (by the same token, a lover of parachuting does not display courage when jumping out of an aircraft, but those do who are horrified of heights). On that note it is interesting to observe the conceptual evolution of 'religious freedom'. It began with the members of a religion declaring themselves to be infallible, hence exercising zero tolerance. It has moved to the mutually agreed recognition that every belief has its idiosyncrasies and should therefore be tolerated. Is it time for a 'version 3', a position that derives its status from the understood realm of reason, leaving behind ungrounded ingredients—that is to say, a freedom *from* religion?

In Europe and therefore in the West the most influential religion had been the Christian derivative of Middle Eastern mythology. For two thousand years it has pressed its subjects into a mould that touched every demographic, every period. Our protection today against its more virulent articulations is due to the steady efforts by intellects who saw reason as their illumination, not a searing image of some deity. The road they travelled is littered with their ashes and their bones. And yet, just as a thought requires a willing mind and a word a receptive ear, the carriers of wilful interpretations are still with us. While not gender-based as such, there are certain dispositions which invited one bio-demographic to be called 'God's silent army'. What these are we are going to examine now.

Despite operating within the conceptual region of Otoom, we have had ample reason and opportunity to observe the dynamics exhibited at a more fundamental level in the course of our bottom-up approach. The differentiation of the sexes should be no exception—firstly, the physical system must be in place for anything at a higher level of complexity to manifest, and it is the physical mechanisms which supply the moment-by-moment background to cognitive functions. While subsequent contingencies may apply their own bias to the unfolding of higher dynamics, these are nevertheless situated within the functional hierarchy. At this stage in our history gender-based cognition plays a significant part across the entire societal landscape, and its functionalities, from the fundamental to the complex, project their character in a similar fashion.

Androgens and estrogens are two groups of steroid sex hormones which are both found in males and females. The former occur in higher concentrations in males, the latter in females [48]. Because these steroids are fatty they have direct access to the inside of cells and so can affect various areas of the brain. Moreover, the hypothalamus (a particular sub-system of the brain) influences the output of the ovaries and/or the testes (connected to the production of sex hormones) via the pituitary gland, and since the feedback mechanism is bi-directional wider brain functions influence hormone levels and vice versa [49]. For example, levels of testosterone (an androgen) increase in male spectators if their team wins, but decrease in those from the other side, although in females the hormone production adheres to a regular 28-day cycle (but when it kicks in during puberty, thoughts turn to babies in a rather elemental way).

As Steven Pinker has summarised in his section *Men and Women*, the basic fact of one egg being fertilised during a process of competition amongst many sperm cells sets up an environment in which the investment by the female is relatively larger measured against her product than the investment by the male, again measured against his product [50].

Since one sedentary egg (an output from an interdependent wider system), is paired with only one of many mobile sperm cells (an output from another type of interdependent wider system), the process necessary for a successful conclusion can be expected to reverberate through the wider dynamic. A limited set of eggs is produced by a female, to be allocated their specific opportunity for fertilisation in her life. A constantly produced mass of sperm cells are supplied by any male, to be exposed to a chance of meeting one egg. Things start from there.

The customs of largely traditional demographics feature the males vying for the acquiescence of the woman, and placed within the context of their particular geography and economics the procedure can take many twists and turns. Watch any folkloric festival (especially when several styles are on offering) and the music and dance reflect the specific nuances that had emerged during the evolution of their culture. And yet from a system's perspective they are all variations on the theme: egg cell needs sperm cell, gets one, produces offspring. Same at the festival: girls move in from one side of the stage, boys from the other, some music and dance interlude, and they leave in pairs (for what it's worth, the only entertaining element would be the different colours of the dresses—everything else is basically the same).

And so to Otoom. If the physical part of the entire organism saddles the female variant with the necessity to produce the egg, carry the foetus, and go through the motions of nurturing the infant, then the central nervous system will have to play an appropriate role. On the other hand, if the body is meant to handle the demands of competition, that is of seeking, identifying and evaluating threat, defeat and victory and all the associated logistics, it is the brain which is left with this particular kind of processing. This does not mean that females will never be faced with threats, nor that a male will never exercise a passive role; it does mean however that within the sum-total of environmental, biological, social, and finally cognitive contexts millions of years of evolution evokes trends. They may not always directly determine, but they will certainly guide indirectly.

The purpose of 'mother' has other ramifications. Since the ultimate goal is to nurture the child so that yet another member of the species can continue, its main focus is the offspring's viability. This implies the additional investment with all its consequences of caution and discrimination. But it also means the underlying willingness to carry such burden and for this to happen the cognitive dynamics have to be configured such that only extreme types of TSs can ever unseat the biologically fundamental purpose of living. In other words, the mind of a mother tends to ensure that no matter what the exterior circumstances or what the characteristics of the infant, *her child* is sufficiently good, valuable, interesting etc etc to be the focus of her attention. To put it more formally, the female mind is predisposed towards applying a considerable bias towards any content which can be identified as the person's own. That bias expresses itself in the form of emotional content, discriminatory affinities in terms of supporting its special status and rejecting its potential diminution, and as such through the particular development of the identity container along these denominative lines. The ramifications are twofold. Imported data will be assessed for their potential to be associated with identity-related complexes and treated accordingly, and the output will be configured along the preferences of domains which carry more weight. Secondly, when in the male version the persona interacts with the surroundings in an exploratory fashion, what gets amended, adjusted or modified are the TSs that have to deal with the incoming data. If it is the child which requires support first and foremost, any perceived shortcomings are related to the outside and change will be sought there.

Since the system, male or female, is an interdependent one (we should expect that in a subset of Life) and clusterings of emergent subsystems are possible on the physical as well as the mental side (again nothing new here), the described phenomena can attain a significance in their own right separate from the evolutionary link to procreation itself. Biased care can be applied even if the recipient is not a child, and competitive behaviour happens outside the context of family. Therefore the behavioural modifications through the ages were subject to a variety of influences, keeping in mind the main theme; that is to say, while the content may change, the functionality remains intact.

In a study that examined possible connections between increased stimulation seeking at age three and intelligence levels at age eleven a direct relationship was found in both boys and girls [51]. The authors hypothesise that "young children who physically explore their environment, engage socially with other children, and verbally interact with adults create for themselves an enriched, stimulating, varied, and challenging environment". In turn such an environment contributes to an "enhanced cognitive ability and better school performance". Significantly they note that high-activity in an unstructured environment are "markers for hyperactivity and inattention", while "highly active social play and goal directed exploration of the environment" contributes to superior cognitive abilities [52].

We would have to agree. A varied input evokes not only more associations with TSs, it also triggers the creation of additional complexes which become subcontainers to other affinity-seeking structures. As we have seen in our discussion on abstracts generally, the greater their number the higher the probability of developing formations. It matters therefore whether the environment is organised: mutually coherent stimuli are more easily integrated into a wider framework than incoherent ones.

The study limited itself to an evaluation of three and eleven year-olds. After puberty the cognitive equation would have to include gender-related factors. Recall the affinity-seeking properties of abstraction complexes and so the emergence of relatable domains which are bound to confirm their specific situatedness within the system. The complexes not only include participating sectors at any given time, by definition they also will have to involve associated structures which serve as the carriers and indeed triggers of memory. If the principle of denominative linking applies to the former, it will also have to apply to the latter. Further, within the context of the logic boundary we observed the emergence of regions under the divergent auspices of either congruent or incongruent TSs. The functionality of affinitive properties is the same on either side of the boundary; in this case it illustrates the separate development of patterns where each set answers to different containers.

We can say therefore, if there is a set of containers across which a particular bias is in force that bias will affect its representative content and can be expected to treat incoming data in terms of such content. Whether the data participate in PAT chain formations or whether they contribute to some articulated re-representation later, the functional outcome is the same.

If the interaction between an established domain and non-related content turns out to be incongruent the situation can be compared to the religious scenario where the mind has to reconcile its belief system with an unmatched exterior. In the previous section I mentioned research that confirms the influence of irrational ideas on stress levels, and stress leads to depression. In another setting Clare Roberts begins her article dealing with the prevention of depression in children and adolescents by noting the World Health Organisation's concern that "uni-polar major depression was the leading cause of disability worldwide in 1990". It is predicted that "depression will be the second leading cause of death and disability across the world" by 2020. Roberts cites evidence that depression is affecting "more people than ever before and at an earlier age". It is also becoming evident that adolescent girls and women are "twice as likely to suffer from depressive symptoms, as are adolescent boys or men". Why this should be so is not clear, she writes, but "the gender bias affecting girls begins around 13 to 14 years" [53]. It is interesting to note that some contributing factors are taken to be concerned with cognition, namely "pessimistic attributions" and errors associated with negative self-perception. Later on in life a better developed cognitive ability tends to ameliorate the negative influence and as such is considered worthwhile focusing on when it comes to remedial action [54]. That children of depressed mothers are particularly affected should come as no surprise then. Conversely, a supportive social environment enhances the ability to keep depression levels down [55].

When it comes to the well-being of older adults the effect of one's cognitive qualities seems to matter as well. Depression levels tend to decrease in advanced age, and the improvement is attributed to a general state of acceptance at a stage in life when the inevitability of one's condition becomes clear [56]. In the case of difficulties the authors of that article suggest that emphasis should be placed on "emotion-focused strategies" rather than "problem-focused coping" [57]. The choice is made with respect to older people's diminished capacity, but for us the preference for emotion-based remedies has an indirect value: if positive emotions relieve symptoms of depression, then surely the presence of the negative variety would at least raise the possibility of enhancing them.

These two studies (and all the other findings contributing to them) seem to confirm our view from the point of affinity relationships. Denominative TSs modify the input and place it either with congruent or incongruent patterns. A biological predisposition to identity-related bias and EMs in general increases the tendency towards incongruence (especially from the age of puberty onwards) and the mind's ability to process its content in terms of wider-referenced frameworks gets suppressed in favour of more subjective re-representations amongst the PAT chains. In different demographics these underlying circumstances have played their part in settling the male-female relationship into a state of more or less partitioned regions, with contemporary specifics making for the contextual variety of marriage customs, economic strategies, and social interactions. That such co-existence had always been a rather precarious one can be gleaned from the degrees of separateness enforced here and there. Harmonious relationships do not need fences, but potential warriors do (it is also an interesting exercise to compare the strengths of these fences with the general emotional intensity of their respective builders).

In the West the heightening of complexity in its resource space and the rising social and economic opportunities for anyone have brought more people closer together; at the same time, the emergence of feminism has accentuated the gender differences through their exercised functionalities sharing the same space. What started as a movement dedicated to the equality of the sexes in wider society has gradually turned into a power structure following the dynamics of association, clustering, and rejection. Given the particular nature of these dynamics (derived from their cognitive counterparts) the tendency exists for them to be self-referencing, identity-biased, and emotion-driven.

The images from romanticism depicting the men in a heroic pose with sword in one hand and the shield in the other, and the women in the act of sensating their textiles and reveling in the beauty of it all, may be an exaggerated attempt at classicist revival (Narcissus was the exception of course—still, fate made him into the type of currency employed by the very class he imitated). But they do hint at the kind of depictions which *can* be made to fit, regardless of the technical detail. In modern times the advertising industry takes over where the old painters left off; placing the persona in status-specific context in order to express the message.

By now the objective observer would have gathered a fundamental difference dividing the perception of the male by the woman and the perception of the woman by the male. If there was no difference, biological contingencies would suggest a serious problem with system implementation. Within a wider reference the variance can be tolerated because the picture is comprehensive, whereas in a closer one a suspicion of inequality creeps in.

For example, in 1997 National Women's Media Centre spokeswoman Janet Ramsey complained about the implementation of advertising standards and the insufficient safeguards against sexism and derogatory images [58]. In particular she referred to a recent car advertisement that "showed a naked woman, without the head in view, as a comparison to the flowing lines of the car's latest model". It is no secret that a lot of men love cars; in many ways they represent a symbol of engineering achievement and technological triumph which could be easily appreciated if only more people actually understood what it took to get there. In any case, the abstract connection between car ↔ sleek functional lines ↔ woman would escape those who do not relate to functionality in the first place. The ad therefore should first and foremost be considered as an example of reverence rather than an insult. Now consider another image from the same year, this time attracting no criticism from ethics watchers (see Fig. 17-2). Here the woman may expose certain areas of her legs, but they are directly linked to a different articulation, one which possibly builds upon the desirable kind of interaction that feminism holds in promise. An assertive kick to the crotch can be interpreted in whatever gender-specific ways one likes, it hurts in any case. Although technically speaking the raised foot is placed on the chair, the body language by the man suggests that the possibility of an aggressive move has occurred to him.

Fig. 17-2: A politically correct version of ethical advertising. [59]

On a similar note, the complaint that more traditional ads showing the woman in the kitchen were demeaning suffers from cultural subjectivity. At a time when a wife doing housework was considered respectable the imagery would obviously fall into the same category. To claim otherwise just does not make sense.

The criticism from the National Women's Media Centre came from an attitude that is uncomfortable with a perspective not one's own. It does not bother to enquire whether the values espoused outside the local reference could actually be different. In an interactive society such communication ought to be possible. That it is not being utilised indicates a focus on the identity away from the wider horizon. The car-body combination portrayed something positive, and it was condemned; accepted imagery on the other hand features negativity, yet this is not. In another ad (Fig. 17-3) 'quality' is directly associated with gender.

Fig. 17-3: 'Good' means female, 'bad' means male. [60]

Marketing needs the extra angle to make the product stand out and deprecating humour, even sarcasm, are one of a variety of means to make a point. The point looses its artistic flavour when it becomes part of a seemingly consistent array of value judgments, all looking in the same direction. When the underlying message is part of a sketch it becomes even more unambiguous. For example, take a TV commercial about a cleaning product in which the participants appear to be members of a family who find themselves in the kitchen at mealtime. The father pushes the young girl to his own messy place at the table, the wife is forced to wipe away the slops for the little darling who manages a resigned benevolent expression on her face against such uncouth environment; the mother commiserates, as if this is what one has to expect when having a male in the house. Or scenarios where a dumbfounded sweating male leans helplessly over a car until a woman saunters casually past who solves the problem with the flick of her wrist. The scene itself may be funny, but the recurring theme of the same type of association is not.

A survey covering the same period in which Janet Ramsey had sexist marketing in her sights (ie, 31 March until 1 May 1997) revealed the following: during that time there appeared six articles written by the columnists of the Courier Mail critical or derogatory of men, none of women [61]. Even the comic strips showed a marked imbalance. During the same period the tally was, *Beyond the black stump* (critical of men - 0/critical of women - 1), *Calvin & Hobbes* (9/0), *Hagar* (7/2), *Peanuts* (1/0), *The wizard of ID* (0/1), *Garfield* (2/0). In the event of some concern about injustice women had 189 organisations dedicated to their interests in Australia's capital cities whereas for men the total was 13 [62]. The breakdown is as follows: Brisbane- 23 women's organisations, 4 men's; Sydney- 45, 3; Australian Capital

Territory- 11, 1; Melbourne- 39, 3; Hobart- 9, 1; Adelaide- 23, 1; Perth- 35, 0; Darwin/Northern Territory- 4, 0.

Repeat patterns invite the idea of intent. With marketing campaigns or comic strips a link between a professional designer or cartoonist and some ideological centre could be rather tenuous. The case would be stronger when it comes to newspaper articles or indeed any text which has been produced with the direct intent to convey a particular message. It is this message which offers itself up for evaluation.

What is one to make of an opinion piece that puts the cartoon character Dot on a pedestal as a symbol of female insight, summarised in the last sentence which runs, "If it works for Dot, maybe it will work for you!" [63]. Flatulence (on behalf of the man of course) in "front of one's beloved" is described as a "togetherness taboo" which sooner or later has exhausted its buffer and now becomes, as the writer quotes a male colleague, "an endless source of entertainment, self-expression and male bonding". No doubt there are male circles in which male bonding takes such forms, but from my experience around the world they are in the minority. Which begs the question as to why exceptional characteristics are used as some kind of standard against which the female has to crusade (although a medical analysis will confirm that there are no inherent biological impediments to farting in ovaries-sprouting humans). In another scene, as Dot

> hovers over her spouse while he uses the computer she remarks with acerbity: "It's amazing how this computer has captured your imagination, considering it doesn't have breasts or play football."

Note the conceptual slant derived from a self-referenced perspective: while a preoccupation with the woman presumably is considered appropriate, giving attention to something else elicits acerbic remarks.

The juxtaposition of football-computer is especially noteworthy when considered against another article (not by the same writer, it has to be said) which laments the preoccupation with sport in our education system [64]. Drawing parallels with the atmosphere in the US, Susan Hocking writes of a father who is trying to find a school which will provide an encouraging climate for his son who is good at playing the trombone and likes debating and the arts; a school which does not put kicking a ball or rowing some boat "next to godliness". Quite so, quite so—but it does seem that when it comes to the choice between anything and one's spouse, the bets are off.

A hint that criticising the glorification of panting jocks may not have been the leitmotif after all comes in a subsequent essay (now by the same author who wrote the above) which devotes its entire columns to ridiculing men who, by their very constitution we are told, are incapable of handling the mechanics of a barbeque. Apparently, administering the gas cylinder and its heat-providing properties are beyond male skills since she asks,

> If this same group of apron-clad, tongs-wielding men should suddenly discover that the reason the heat beads went cold and the flame flickered out was that the gas bottle had run dry, what do they do then about the scientific art of sausage poking?

Needless to say, they wait for the women to come to the rescue. Others may consider relocating their social occasions away from the mental ward, but not those with a certain agenda—they would seize on such shortcomings and elevate them to a weapon-like norm with which to delineate male nature.

Even in the darkest days of patriarchy, how often would have readers opened their metropolitan daily and be presented with an insight view about the toilet habits of the writer's wife? Yet here we are, to be enlightened by a description of the kind of changes that occur in her household when her man is at home, a place where "girls usually rule, okay":

> Namely, barring those precious weekends in which we are joined by our number-one bloke, all the toilet seats have remained down. At all times.
>
> Furthermore, there have been no little tell-tale dribbly bits stretching from toilet bowl, through bathroom and on to carpet in hallway. [65]

The text is initialised by the following:

> Except for our dog—who has been neutered anyway—and my husband—who has not—ours has almost always been an all-female household.

This kind of sentence goes somewhat further than the often vague connectivity of Freud's subconscious. It is not quite as explicit as the articulation by a Channel 10 Brisbane newsreader who was still on air when she said "... but this arsehole I'm married to, he goes, oh I want to go skiing, oh okay, I wanna go to Europe, oh okay ... the minute he's slightly ..." [66].

If issues such as the position of toilet seats are obviously considered weighty enough to be mentioned in newspapers and female journalists feel that the domestic state of one's bathroom plus its surrounds is worthwhile airing in public, inferences can be made regarding their perceptive horizon. While an individual's personal priorities as to what constitutes a matter of privacy can be debatable, the contexts presented above point to the identity-focused criteria under which these expressions are being made. To press ahead without taking into account possible consequences for the unwitting participants is a sign of the kind of self-centeredness we have described earlier—unless it is argued that the damage inflicted has actually been part of the intent all along.

Not every female would publicise their private affairs in that manner. On the other hand, the freeing up of education and opportunities in general have seen professional women coming from sometimes deprived backgrounds in which their earlier lives had not been exactly ideal. It is a theme not uncommon, at least according to the accounts given by some individuals who had managed to reach positions in which they had a chance to give these interviews. During their studies they would have had ample occasion to learn about feminism and its views. How relatable to the wider world those prove to be is open to question. Restricted backgrounds encourage narrow perspectives, especially if the assertions are made in the spirit of solidarity. As we have seen before when discussing ideologies in general, the placement of an idea within the defined sphere of 'us' so that it assists in defining the exclusion of 'them' is a familiar concept. We find it again in statements like the one by Andrea Dworkin in her book *Our Blood - Prophesies and Discourses on Sexual Politics*:

> The real core of the feminist vision, its revolutionary kernel if you will, has to do with the abolition of the sex roles - that is, an absolute transformation of human sexuality and the institutions derived from it. In this work, *no part of the male sexual model can possibly apply*. (Her italics) [67]

Outside a bias-driven cognitive dynamic the exposure to a wider framework would engender an equally wider-ranging set of source material, but given the potential of exclusion-seeking affinity relationships the outcome will reflect the agenda. Whether a particular action does or does not start with the intent to exclude all the information bar the supportive parts becomes almost irrelevant once the information filters through similarly configured perceptions—a recipe for myths in the making.

Consider for instance the story of Erin Brokovich, a law clerk who took on the Pacific Gas and Electricity power company accusing it of leaking the rust inhibitor Chromium-6 into the water supply of Hinkley, a small desert town in the US. As we read in a critical review of the eponymous film, the real story is somewhat different [68]. First the film: Brokovich gets 652 people to sue the company claiming they had suffered a variety of diseases such as miscarriages, cancer, Hodgkin's disease, nose bleeds and bad backs. PG & E relents and pays the victims hundred of millions of dollars.

On the other hand... while Chromium-6 (Ch-(VI)) has been shown to cause cancer of the lung and the septum when inhaled, the US Environmental Protection Authority went through all the studies and found no data which indicated Ch-(VI) caused cancer via the "oral route of exposure". "What's more," writes Andrew Bolt, "the Journal of Occupational and Environmental Medicine reported that Hinkley's cancer rates were no different from those of the rest of California". As to the question of why then would the company settle for such a large payment certain points are noted. A lawyer who eventually took over from Brokovich got a private arbitration company to decide the case rather than dragging it through the courts. Three former judges from that company were linked to the lawyer who, after settlement, invited them for a cruise on a Cunard ship he'd chartered. As well, he learned that PG & E's lawyers were illegally investigating his bank accounts and private affairs, which may have been the reason the settlement for his clients included the "proviso they not sue PG & E for breach of privacy". As to the money for the assumed victims, "She [Brokovich] and her colleagues took 40 percent of it plus costs. They also charged the 100 children who were part of the action a third of their payouts, when Californian rules say children should pay no more than 25 percent".

Three years later and Anthea Gerrie interviews Erin Brokovich [69]. Titled *Wonder woman* her report says in the second paragraph,

> The once-impoverished single mother—whose astonishing journey from under-educated law clerk to ace investigator brought her fame and fortune, courtesy of Robert's Oscar-winning portrayal—has evolved into an A-list celebrity with her own TV show, fiercely protective entourage and a wardrobe full of Armani suits.

We learn that she is a dyslexic college dropout who has won the biggest environmental lawsuit settlement in US history; is in person "every bit as stunning and charismatic as Roberts, who had to wear falsies to emulate her character's huge and proudly flaunted embonpoint"; has a "razor-sharp brain beneath those frosted forelocks"; gets "very emotionally involved in cases"; had "two disastrous marriages"; and has "variously suffered from anxiety, anorexia and extreme allergies (including to her own home)"; and has been "blighted by allegations that she is a charlatan profiting from people's suffering". Towards the end of these revelations Gerrie writes,

> But with her great looks, fabulous acuity for sound bites and seamless progress from rags to riches, the beauty queen turned activist embodies the American dream, and Hollywood is standing firmly by her.

By the way, in Andrew Bolt's article that "fabulous acuity for sound bites" is referred to as "brash" and "foulmouthed". Gerrie's text ends with,

> Erin Brokovich, the real gal, is awesome, and should be a lesson that we're all powerful individuals who can make a difference in the world.

Several things stand out. The facts behind the film story are given only a token mention; her shortcomings are heralded not as markers for even a certain degree of caution but as a confirmation of her success, despite the unrealistic juxtapositions; her physical attributes are significant items, a form of portrayal which suddenly discards any connotations of sexism it may have had in a different context; the writing style is positively gushing; and it is yet another example of that so American attitude which says "everyone is powerful", a view that sits at odds with the obvious reality and which may not be an inconsiderable factor in the steady rise of insecurity and stress caused by a mismatch between imposed expectation and real life that even the World Health Organisation found reason to observe.

One could ascribe these exaggerations and falsehoods to the varied forces of a media circus, cynical marketing, and an enthusiasm that some people are bound to get caught up in once the right buttons are pushed—as critical as such offerings are, given the exposure in metropolitan daily newspapers. Even more serious are the ramifications by the time those initiatives reach the wider fabric of society through channels its members have a right to rely on. The transposed complex of notions which centres around a bias-inducing identity container and becomes now an active part in its own right regardless of any balancing factors has profound effects.

The statements like the ones presented above have as their common denominator an antagonism towards the male while at the same time conceptualising the 'female' and everything it stands for (*as defined* by the female) into a framework of absolutist reference. It is not an approach any scientist under different circumstances would consider valid, yet its results have made inroads into society's vocabulary. Don Peterson discusses the issue of violence against women in an article headed *Gender Wars* [70]. He writes that an "orchestrated campaign of falsification" began in 1987 when the Office of the Status of Women announced that one in three women was at risk of domestic violence. When challenged, federal Family Services Minister Senator Rosemary Crowley said that the source had been the 1980 American study *Behind Closed Doors*. In it the cited figure was not to be found. "What it did say was that one in three households would experience at least one incident of domestic violence and that in more than 50 percent of the incidents the woman was the perpetrator." The conclusion of that particular study is significant and so is repeated by Petersen:

> Family violence is probably the only situation where women are as or more violent than men ... an examination of violence between couples and violence by parents toward children reveals that women are as violent or more violent than men.
>
> While fathers who beat up their children do so on an average of once a year, mothers who beat up their children do it more than once every other month.

The reaction by Senator Crowley is worth reciting: "Why are you so worried about a little bit of wrong analysis?"

A survey conducted by the Monash Women's Health Centre amongst general practice patients claimed that 40 percent of women are sexually assaulted before the age of 16, and that 10 percent of women are severely assaulted by their husbands each year, with one-tenth of these carried out with guns or knives. As to the former, Petersen relates figures from the Australian Institute of Health and Welfare showing that the incidence represents only 0.2 percent. As to the latter, that combined percentage would mean that 1 percent of women receive gun or knife wounds—but in New South Wales for instance data from hospitals during the year 1995 indicate 554 people admitted to hospital with gunshot or knife injuries (and most of them males) out of a population of six million (Petersen's article was written in 1996). (As a quick calculation shows, the total number of these types of victims represents only 0.0092% of the entire NSW population, a far cry from the necessary 60,000 female victims in the view of women's advocates, a number which in itself would need to represent only a small proportion of all the assaults)

Similar discrepancies between feminist claims and actual data have been found elsewhere. Martin Fiebert, from the Department of Psychology at California State University, examined "95 scholarly investigations, 79 empirical studies and 16 reviews and/or analyses" which "demonstrate that women are as physically aggressive, or more aggressive, than men in their relationships with their spouses or male partners" [71]. The sources were not restricted to the United States. The aggregate sample size in those studies was over 60,000. In one sample of unmarried college students "19% of women slapped their male partner while 7% of men slapped their partners, 13% of women kicked, bit, or hit their partners with a fist while only 3.1% of men engaged in this activity" [72]. In a survey of the perception of justice, 109 men and 111 women were given a "hypothetical scenario in which either a husband or wife perpetrated domestic violence". Significantly, the participants were "more negative in their evaluation of the husband than the wife, were more sympathetic to the wife and believed that the husband deserved a harsher penalty for his behavior" [73]. An investigation in Canada revealed a decrease in male violence at higher educational levels, whereas for women the reverse was true [74]. A more recent investigation (unrelated to Fiebert's) by John Archer, professor of psychology at the University of Central Lancashire, found that in domestic relationships women are more violent than men [75]. Archer is quoted as saying, "It's a complex argument; but we do get more women aggressing against male partners than men against female partners". And, "The view is that women are acting in self-defence, but that is not true—50 percent of those who initiate aggression are women". As a corollary, a survey by the Australian Institute of Criminology found that "from 1991 to 1999 women prisoners increased from 607 to 1124, while men behind bars rose from 12,429 to 17,208" [76]. The first set represents an 85% increase, in the second it is 38%.

In the beginning of this section we described the biologically defined cognitive framework of males as directed towards the outside and its female counterpart focused towards the self-related centre. The consequence was therefore for males implementing their configuration in terms of competitiveness and as such leaning towards aggression. Looking at the above statistics it seems something is amiss. However, this is a superficial view. Dealing with the outside does not imply there is no centre, no self. What it does describe is the self's existence in relation to the exterior, so that the effects of imported perceptions are considered against the needs of what is perceived as one's own. While feminists would argue with the concept of ownership in this regard, the historical background illustrates the biological, social, and economic necessity of male activities being referenced in terms of the protective role towards the family, group, or society. Therefore aggression, as one component of male nature, tends to be channelled outwards, but empathy, care, or service is directed inwards (notwithstanding other factors which at any given time can intrude upon this dynamic). For the female psyche those auspices do not apply to the same degree. Although the aggressive element exists, it does so in relation to the much more closely defined identity. Should there be a trigger for violence we can expect a male avoiding harming his partner, whereas for a female anything beyond the perceived self (whether in terms of her own persona, or offspring, or interests) is fair game.

The wilful interpretations of research data hint at a second factor. If an emotional bias influences the perceptive mechanism and in turn affects the affinity relationships involved in forming the relevant clusters, a selection process begins which distinguishes between the varying associative potentials of incoming information. That is to say, data which support the bias are admitted, those that do not are kept out (or relegated to other domains—the outcome is the same). A parallel can be found through the research into visual cognitive dynamics where it becomes apparent that information which is outside the functional scope of the brain does not get processed [77]. At first glance this may seem trivial; but its consequence is the restriction of possible abstracts to those which have found a 'home' within its network.

Consider a statement such as "one in three households experiences at least one incident of domestic violence". On its own the assertion is largely meaningless: 'one in three households' in what sample space; 'at least' against which measure; 'one incident' out of what set of other incidents; 'domestic violence' defined under what terms? Given an inherent preference for an overall negativity the wider references are discarded in favour of the perceived semantic that is directly on offer. Hence the sentence morphs into "one in three women is at risk of domestic violence". Witnessing arguments between males and females the emergence of just that phenomenon can be observed on many occasions. The situation is insidious; while for the female the case seems obvious (recall the demonstrated reconfiguring influence of the conscious to suit an expected perception in earlier chapters), for the male the ongoing debate will only lead to more and more obfuscation, and very soon the actual reason for the altercation takes second place to his attempts at solving the misunderstanding. Since the problem is systemic, not only is a solution hardly feasible, the emotions on both sides are bound to rise—and for different reasons at that. The result is yet another day down the drain.

What I have described here is not a mystery to men who are familiar with long term relationships, no matter what and where the demographic. It may not be shared by women (for the obvious reasons), but it will also be beyond the experience of men whose only involvement with females had been during the early periods when the mutual liking for each other dominated their daily affairs. Since in the West longer lasting partnerships are increasingly rare, and men move from one untested alliance to another (or discontinue them altogether), that side of female behaviour looses its experiential significance. When it does occur the wider context has been removed and so it is prevented from being recognised for what it is.

Deprived of the resources coming from personal experience and having lost the sounding board of the wider community to skewed interpretations, the male part of the population is forced to deal with societal contingencies on an individual basis, which in itself has now turned into a vessel of obscurantist mixtures. How else can a book like *The Woman's Book of Revenge* make its way through the checks and balances of a culture? It is a guide to the art of humiliation of former husbands and lovers as seen by its author, Christine Gallagher [78]. Some of the advice we are given: "slicing the elastic waistband of all his underpants; cutting a small hole in the rear of his swimming trunks; hacking a few pages out of his favourite books; putting an Oxo cube in the shower head; placing an open can of sardines in a heating vent; or replacing his haemorrhoid cream with Vick's Vapour Rub".

The cognitive dynamics which construct their affinities around an identity-based exclusivity get transposed into the wider exterior. While one major concern of feminists has been and still is the demolition of men-only surrounds, at the moment of exercising their own power exactly the same happens but in a more indiscriminatory style. A conference, Millennium Women: Action Planning Your Work Future, invited a male speaker because he was thought to be most suitable for the conference's topic. One woman from an association representing counsellors objected to his inclusion, being "quite irate that a man was being featured as a key-note speaker and she believed the conference should be for women, women with no men" [79]. Although other women's groups had no objections, the outburst is significant for its espoused principles. Interestingly, the article does not name the particular association but since it represents counsellors it could be considered in touch with women-focused perspectives.

The phenomenon of specific perspectives being carried into the arena of human interactions has been explored by others, and with great effect. Sol Stein, in his *Solutions for Writers*, refers to the importance of having a script and sticking to it. While written within the context of authors trying to tell a story with plausible characters, it nevertheless shows the principle at work. He got the idea from Elia Kazan, a director of Pulitzer Prize-winning plays, who devised an exercise used by budding actors studying at the Playwrights Group of the Actors Studio in New York [80]. One actor is given the role of a mother whose child attends a school, and another actor is told he is the headmaster of this school. The woman is given a script in which her boy is assumed to be good-natured, eager to learn, and never giving any trouble at home. The man 'knows' there is a pupil who is disruptive in class, doesn't pay attention, and is careless with his homework. Neither knows the script of the other and both have to improvise a scene on stage in which the mother tries to persuade the headmaster to keep her son while the headmaster is trying his best to get rid of the boy. In a matter of minutes both end up arguing to the hilt.

In real life such a scene may lead to a positive outcome if both parties eventually perceive the other's position. But suppose the 'scripts' are derived from an intrinsic notion about one's personal reality, and the viewpoints are virtually unbridgeable because neither party has access to the functional configuration of the other. What are the chances now?

If differences occur in representation and subsequently in the re-representation of content, we can expect a problematic communication, as has been shown. The issue is exacerbated when it comes to education. After all, not only are we concerned with the transfer of information per se, the data need to be stored meaningfully for them to be turned into productive elements of PAT chains. For some decades there has been a steady decline in male teachers working in the primary and secondary schools in Australia to be replaced by female teachers (a situation not restricted to this country). While the academic performances of girls has risen, that of boys has fallen. So serious has the condition become that in 2002 a House of Representatives Standing Committee on Education and Training has compiled a report outlining the alarming position of boys in the education system [81]. Among other things the report notes:

> ..on average, boys do not perform as well as girls in each aspect of literacy—writing, reading, viewing, speaking and listening—and that the gender differences are greatest for the expressive modes of writing and speaking and least for the receptive modes of reading, listening and viewing. The differences between boys and girls are greater for children from lower socio-economic groups. Also, boys typically comprise about two-thirds of all students referred to reading recovery programs in Australian schools.
>
> National literacy benchmark data for Years 3 and 5 in 1999 also show that girls consistently outperform boys, but that the levels of attainment and the gender gap in different States and Territories are not uniform. There are also

> likely to be social, biological or developmental reasons why boys' levels of attainment are lower than girls. [82]
>
> Early school leaving is a matter of great concern because young people who have not completed Year 12 have much greater difficulty making a successful transition from school to post-school education and training, and employment. This is highlighted by the differential in unemployment rates for young people with differing levels of educational attainment. [83]
>
> The costs of early school leaving are substantial. The National Centre for Social and Economic Modelling has estimated that the lifetime cost to the country of each early school leaver is $74,000. Half of this amount is a direct monetary cost borne by the individual ($14,700) and government ($22,400) and the remaining half is a social cost borne by the individual, government and the community. [84]
>
> The timing and magnitude of the national trends in school retention rates for males and females vary for each state and school sector. However, the trends in each state and in each school sector are broadly consistent—female retention rates overtook male retention rates some years ago and the gaps between the male and female retention rates are now relatively wide. [85]

Although it is recognised that different subject enrolments for boys and girls make direct comparisons difficult, concern has been raised in terms of the employment prospects for boys because manual jobs are disappearing rapidly and a better personal potential is linked to higher retention rates [86]. One submission by a—female—classroom teacher is significant in terms of our perspective:

> The boys at this school…often come from non-English speaking backgrounds. What I have noticed here is that the boys have a lot of trouble trying to bring life into their reading. I think that is because they are not running the movie in their head. They can do the literal work very easily... What they do not seem to grasp—and that is right up to year 12—is the inferential work. They literally will be given five or six facts, and they cannot then say, 'From all of this information, this is what I see.' They do not pick up the nuance of the language, the colour of it and its various emotive qualities…The whole aim of everything we do—role plays, other small drama pieces, empathy exercises, even visuals such as drawings on the board—is to let them see that this phrase represents this set of emotions… If we do not show them, they will not see it. You literally have to stand up there and perform for them so that they can pick up that type of thing. That is a huge gap in their experience of literacy. [87]

Leaving aside a certain unfamiliarity with English (certainly an important factor) the imparting of information through "empathy exercises", "role plays", and the arraignment of emotions indicates a fundamental discrepancy with the framework of male cognition. Social shifts in family structures (eg, more single mothers) and employment patterns (eg, increased participation by women in the workforce) have been mentioned in the report. In a more general context the following comment is informative:

> The representations of men and women and their relationships in the media are important influences over both boys and girls. There are now more boys who lack adult male role models, or whose experience of adult men has been limited to those who are uncommunicative, uncaring or violent and abusive. These boys are less likely to identify with other, more positive, male images that do not resonate with their experience than with stereotyped media images. [88]

On the question of the feminisation of education which obviously has been successful for the girls, Professor Faith Trent, Faculty of Education, Flinders University, writes,

> The model that we used for girls' education was based on political questions. The question was different, in my view. The girls' stuff was a political question about the role of women in society, about a political way of changing the perception of women's access to education. In many ways, it was giving girls confidence to do well. To put that model on top of the boys, and think that you can do it that way and fix the boys up—this terrible desire to 'fix the boys up'—seems to be very dangerous. [89]

When it comes to the various types of intelligence the picture of boys preferring a logical, analytical approach and honing their skill in identifying and classifying objects, stands in contrast to the verbal and interpersonal environment enjoyed by girls, and it seems to be consistent [90]. On the question of single-sex classes the comments were telling. Although girls benefited enormously from such an option, the result was quite the opposite for the boys. However, it was emphasised that the teaching style in this case remained the same for the girls but had not been adapted to the needs of boys [91]. This confirms the slant towards a female-friendly environment as a result of policy initiatives. When it comes to the students' situatedness within their group, the consensus was that boys try to please their peers, whereas girls try to please their teacher [92].

In a functional sense these findings by the Standing Committee confirm the dynamics we have identified previously. In the male cognitive system the focus is on understanding the outside in terms of its inherent characteristics; in the female counterpart the relationship between the identity container and affinitive abstracts plays a greater role. To impose one or the other as a one-size-fits-all model upon the whole of society disrupts the respective functionalities of either group. While the desire to improve the presentation for girls is understandable (against the historical background of male-centred education facilities), the tandem agenda of employing the substitute as a weapon against the male—however subliminally it had been done—is now producing the inevitable results. The "terrible desire to 'fix the boys up'" is then not merely a by-product of the female tendency to bring up one's off-spring in a rather self-referential mode (which is quite necessary when the targets are babies and infants), that dynamic has been projected onto a higher scale of complexity through the imperatives of feminist ideology. In other words, a home-bound mannerism has turned into a state-run policy.

Emotional elements contribute to the dilemma. In previous chapters we have examined the context of associative abstractions harnessed to the demands of EMs. If the dynamics of an identity container evoke certain affinity relationships, in conjunction with EMs we can expect those same affinities to acquire an added importance. Not only do the resultant complexes reflect the associativity of the 'self', their emotional dimension contributes an intensity which otherwise would not exist. Since the basis for such an enhanced domain is formed by the content of the container, the articulated emotions are specifically related to the intrinsic nature of their owner. A less technical nomenclature describes such a state as ‘love’. Its structural characteristic works against the often held view of ‘love’ being the reciprocal attraction by two people but, as can be seen, it is fundamentally a one-sided affair. Provided the same associations can be created within the other mind’s abstracts, the outcome will indeed be the mutuality celebrated in so many tales. Note however the need once again to distinguish between functionality and content: since what we describe as ‘love’ is the effect of functional dynamics, the actual content of such structures is only of secondary importance to the mechanism itself. That is to say, what is the basis for one person’s affection is quite a different matter for the other. It’s an age-old tragedy—two people fall in love only to discover in the end that their natures are irreconcilable after all. And without such a thing as unrequited love, where would opera be?

There is a poignant story on just that theme. A little girl has a gold fish whom she loves with all her heart. So at bed time she takes him out of the water and puts him next to her on the pillow. When she wakes up in the morning the gold fish is dead.

Combine the narrower reach of identity and its affinities with the single-sided efficacy of ‘love’ in the female and we get initiatives that are destructive to the wider context in which they occur. Examples are a woman who falls in love with a person against the rules of their organisation; a professional whose actions are dictated by her relationship with someone despite the damage done to the members of her work environs. Since ‘hate’ could be placed in the same category only with some ‘nots’ added here and there, the scenario is similar in terms of the disregard towards the larger reference.

On that note the different perception of emotion per se in males and females becomes evident. Where in the woman the articulation of her EMs points inwards, for the male it is a matter of outside involvement. To compare the two in order to judge them against each other is quite futile. By the same logic an English-speaking person could complain that Germans never say “Please”; of course not—they say “Bitte”.

The significance of language has been touched upon many times. It is one format through which content gets transmitted, and its perceived—or rather, its perceivable—meaning determines the ultimate value of its re-represented content. Consider for example the concept of ‘narrative theory’, an idea born out of the desire to capture the complexity of individual lives in the field of psychology. In her paper Maggie Kirkman from the University of Melbourne argues for the inclusion of that approach as a helpful means to more informative descriptions [93]. While dismissing the concern about having a paradigm shift “in a non-natural science that is not governed by paradigms”, she welcomes the possibilities once an area such as social psychology is subjected to a change from “measurements and predictions to one interested in description and understanding; from frequency and causation to meaning; from statistical analysis to interpretation; from numbers to language and symbols; from universals to particulars; and from attempting to decontextualise people and events to the study of people in their natural context” [94].

She refers to a view by Mary Gergen about autobiographical narratives in which “any separation of the form and the content of life stories must be arbitrary” [95]. Taking advice from that source, she advocates the same for the “stories told by scientists” [96]. She cites John Kotre who thinks the use of the story form will “capture phenomena overlooked or only partly apprehended by science”, but he also warns that “in actual practice, conceptual precision is often lost, as it must be if we are to be responsive to the phenomena before us”. Moving back to Gergen we read,

> Her [Gergen’s] later rallying cry, “let us claim the tentative and fuzzy nature of all our linguistic formulations” suggests that she has embraced the narrative

> mode and accepted the metaphor of narrative. Both of these (the metaphor and the mode of thought) may entail slippage between data, method, and theory.

Such a broad band of acceptable types that get woven together, separated, and plaited once more, all without the overriding formality of categorisation, releases the articulator from the bounds of form because form becomes unidentifiable to begin with. The consequences can be gleaned in a paragraph later on:

> Because there are several ways of interpreting and telling the events of life, it may take some time to construct a satisfactory account (which is itself, of course, open to reinterpretation as life goes on). For example, Wiersma (1992) recounted the way in which a woman retold and reinterpreted her story several times over a few years to a psychological researcher. On each succeeding occasion, the narrator felt that she was not really the person she had previously described; she appeared to revise her story through the living of it before she was able to tell it again. Deciding who she was and discovering who she was could not be disentangled. [97]

To give up the differentiation between the "form and the content of life stories" can in Otoom's context be rewritten as 'functionality' and 'content'. Immediately we can see the problem. Enabling the random mix of an underlying principle with its manifestation through diffuse language may offer hypothetical insights which otherwise would have stayed hidden from the semantic view; but in the end the conclusion ought to entail a format that can be identified as to its players and what they are actually doing. Not only that—their articulation must allow a third party to revisit the event using the same markers. If those markers are merely arbitrary placements, to be shifted across the map under perspectives which in themselves escape proper definition, what is the final value of such a 'story'? From this ambiguity it is a short step to a mode that has as its focus not the event at all but the *act* of observing it. The exercise turns into a game—like children who play war to wile away the time.

We find a similar angle in an article by Susan Dormer and Bronwyn Davies at the James Cook University, titled *Desiring Women and the (Im)Possibility of Being* [98]. Written for the purpose of describing the "collectively available discursive strategies through which women constitute themselves and are constituted as recognisable and re-cognisable desiring subjects" [99], the authors were "interested in how desiring individuals juggle the reading of desire as coming from within themselves, and the reading of themselves as pressed into particular patterns of feminine desire" [100]. Prearranging the identifiers (such as they are) on the shifty ground of temporary references, each resulting image is not so much a representation of something but rather a measure of a subjective continuum. Since in the course of telling their story the women interviewed had to deal with the fluid 'narrative' as well as the ever-present background of memory (given the nature of the mind), it is not surprising that the perception of one's self becomes somewhat precarious.

> "So it's alright if I listen to me" one interviewee exclaims, as if somewhere in the talk a permission has been granted to acknowledge the voice she identifies as "me", the one that might desire something that does not fall within the orbit of what is socially condoned for women. This me that is experienced as having been silenced, and which needs to be given permission to be heard, or acknowledged, has been described in multiple ways by feminist writers. Lorde

> (1984) for example, described it as "deeper" and "nonrational": "As women, we have come to distrust that power which rises from our deepest and nonrational knowledge". [101]

Our concern here is not about social mores and the problems they might cause. A cognitive reference system which allows itself to be formatted, even in the context of subjective values, will at the very least provide the anchors needed for self-evaluation (again, as presumptuous as they may be in terms of their owner's knowledge base). Without a format however the self-related affinities will chase each other forever.

If the above can be accepted as a methodology which suits women, that is to say supplies the framework for the feminist approach to deal with reality at large, its introduction across a range of societal contingencies can be expected. Its clash with the scientific approach is inevitable, as is its unease with realms where formality is needed in order to arrive at conclusive evidence, such as in forensic applications.

On a level of lesser complexity the words used and how they are used create the linguistic ambience out of which we create the conceptualisation of our world. A biased tendency takes the semantic aura with it. The feminist antagonism towards what it perceives as the male 'enemy' has been taking its foothold over many decades now. Its vocabulary forms the intellectual tool set available for the society's members, female and male. Through it the imagery gets articulated, defined, and interpreted. Here is a little exercise: in any group of people, just about everyone will know what the word 'misogyny' means—it has been thrown around often enough. Its derivatives, 'misogynist' and 'misogynous' are equally familiar. So the question is, if 'misogyny' means a hatred of females, what is the equivalent word for a hatred of males? There aren't many who would know. Moreover, if the size of a word group can be linked to its usage and therefore need, it is interesting to note that the asked-for expression is only used as a noun; there are no other entries in the 1981 edition of the *Macquarie Dictionary*, although the 1997 version contains one derivative. In the 1989 edition of *The Oxford English Dictionary*, for all its multi-volume size, the reader is simply referred to 'misogyny' with no further explanation given [102a,b,c].

Lower levels of functionality do not replicate themselves literally at higher levels; but they do set a basic tone which provides a tendency. The biological and cognitive characteristics in humans follow the same principle. The inherent structures of any organism provides its system with the properties that settles it within its resource space, against the sounding board of its environment. While this is not a straightforward process, over time niches are carved out that can be identified as such. Female and/or male behaviour patterns are no exception. In the history of humanity the roles of men (as ill-used as they may have been) were always distinguishable from those of women (as ill-used as *they* may have been). For the female, the emphasis on the nurturing and caring role sits well with cognitive dynamics that apply a self-referencing bias to her charges; for the male, the outward-directed and exterior-processing faculties fit in with the task of dealing with the surrounds. While there are exceptions and overlaps in any complex systems, the projection of specifically female functionalities into the wider societal arena would have to be problematic. Could this be the reason why, despite the general fluidity of human affairs on this planet, matriarchies have always been a minority—a system type that possesses inherent flaws from a certain scale upwards?

In the West the infusion of Middle Eastern morals in the form of Christianity has left us with a legacy of mythological strictures. Feminism supplied an ideology that transforms the inner world into wilful vistas of subjective criteria. Both managed to meet in the human Eros. From one came fear, from the other hate; together they became ferocity.

Human eroticism can arguably be described as *the* most influential force in our affairs. To be sure, there is birth and death, hunger and thirst, and all the various spiritual and social desires that shape our landscape. But none of them are beset with such awe, fear, complications and sheer intensity that come from the biological imperative toward procreation and pleasure. For these two components do exist side by side; it is nature's way of ensuring the continuation of the species while simultaneously creating the opportunity for the individual to make such utmost necessity an enjoyable one. What seemed to have been 'a good idea at the time' has created an often monstrous framework within which the contingencies of daily life must find their way amongst enticements, obstacles, and even downright taboos. Any evolving complex system, whether it is the mind or society at large, builds its own paths and islands that can find themselves at odds with each other. The difference between functionality and content virtually ensures that. If one had to choose one single representative history of the human race, sex would have to be it.

First, the physical perspective. The genitals of a human are part of a system which connects the organs to the nervous system and the nervous system to the brain. The reception of and response to a stimulation of the genitals are handled by the parasympathetic part of the autonomic nervous system [103]. Its link to the brain on the other side makes two types of modes possible: an excitation derived through mechanical means, and another caused by thought processes. Since the entire system is interdependent (a feature we should have come to expect by now), the body, the genitals, and the brain are modular pieces that interact with each other to varying degree. The exact degree of participation is unascertainable because a comprehensive, precise measurement of the system's entire state is so far impossible, and in any case a proper definition would contain the cause and effect relationships of all its parts—something even more unfeasible.

However, given that combination of modes we can assert the effective performance as a function of the mechanical structure and its capability on one side and a function of the cognitive resource space on the other. Therefore we have a continuous spectrum of performance levels beginning with birth and ending with death, and at any given point in someone's life it is a question of at what capacity each mode interacts with the other to what degree.

As for the genitals, the musculature and associated nerve strands to take care of the non-cognitive mode are principally in place by the time the child is born. Although they reach a higher level of complexity towards the road to adulthood, in mechanical terms their functionality is already established. Hence it is possible for a young person at an age of, say, ten to confirm their existence on themselves as well as through the descriptions offered by charts, diagrams, etc etc. Having gone through such an exercise, in the face of cultural statements to the contrary that person may have learned their lesson and quite possibly will never again trust the assertions made by a constraining authority. That kind of mistrust could be essential in order to preserve one's sanity. On to the mind.

Our neuroscience book tells us, "The neural mechanisms underlying the sensations of sexual climax are entirely mysterious, in both sexes" [104]. So they are, but the conceptual realm of Otoom gives us at least the opportunity to attempt an interpretation, although at this stage it would have to be taken as conjecture. We know that organic life forms actively procreate and that the result is yet another edition of the life form. If we describe such a result as an identity replication (leaving aside the finer detail of technicality in biological replications) we could likewise describe the process leading to it as identity definition; for instance, to make a seed is to fundamentally define the plant. In higher developed species the process is sex, the result the offspring. Since the central nervous system participates, even directs the steps involved in the process of identity definition, the existence of functional modularity (which after all can be observed everywhere else) would suggest that its equivalent subprocess within the central nervous system replicates the overall theme. The theme is 'identity definition' (in fact, *comprehensive* identity definition), and for this to happen the entire contingent of data would have to be accessed. The data are the representative content distributed amongst the neurons, and at the moment of 'definition' they are all called upon at once. So, what does it feel like when the sum-total of one's mind gets thought? Perhaps an orgasm comes close.

Back to firmer ground. To make our expressions clearer to understand we are going to refer to the whole range of sexual behaviour, desire, thought etc as 'erotic behaviour'. Similarly, 'sex' becomes classified as 'eros'. This prevents misunderstandings based on different interpretations by ideologies which try to use specific words in order to slant the judgmental response of their audience this way or that. Since we are not in the business of moralism, for our purposes 'erotic behaviour' possesses as much integrity as 'sex'. Furthermore, to include everything, even sex, under eros, rather than including eros under sex, demonstrates our intent to widen the conceptual potential of a touch or a smile forward, rather than doing the same with intercourse backwards (it's also much more civilised).

The process of creating organic copies is so complicated and yet so fundamental that for the most part in human society it had neither been understood nor could its various ramifications be integrated in a harmonious way. No matter what the customs in this or that demographic they, being another product of emergent systems, could never fully encompass the effects of reproductive capacity and desire amongst their members. The outcome can be seen in two ways: firstly, there is no society on this planet which does not have laws pertaining to erotic conduct and secondly, the more a community's rules clashed with its biology the more stringent those rules were.

The Western mythological heritage imported from the Middle East is no exception. While rules governing erotic behaviour can be counted upon, it matters in what physical and/or social context they are placed. Signs of belonging feature in any declaration of identity, but their degree of intrusiveness and permanence tells us something about the intensity behind the gesture. Mutilating customs such as the filing of teeth or scarring are therefore indicative of the authoritative relationship between the individual and the community. In the *Good News Bible* we are told it is the removal of the boy's foreskin which establishes his status as an approved member under the religious regime. To initiate the social contract Abram is told by his god, "From now on you must circumcise every baby boy when he is eight days old, including slaves born in your homes and slaves bought from foreigners" [105]. Note the circumstances; the command is not given in the form of a revised procedure after some negative experience, but as a prerequisite to submission.

From a more modern point of view any kind of mutilation should be seen as intrusive and presumptuous, but even in the days of lesser subtlety *what* gets destroyed is still significant. Since the foreskin just happens to be part of the genitals, and that region of the body just happens to be under a particularly critical scrutiny, the idea suggests itself that the interest does not merely lie in the removal of some piece of skin but rather links to something more fundamental. I have already mentioned Dr Somerville's initiative against circumcision (see chapter 16). She called the procedure "criminal assault". A more recent statement comes from the pediatric and child health division of the Royal Australasian College of Physicians, saying amongst other things, "There is no medical indication for routine male circumcision" [106]. It echoes a similar position by the College of Pediatrics and the Australasian Association of Pediatric Surgeons. Professor Gregory Boyle of Bond University in Queensland is quoted as saying that the effects could be "long-term post traumatic stress disorder, resentment, anger, feelings of violation and inferiority and impaired sexual relations". On the other hand, a doctor from Brisbane disagrees; in his opinion circumcision minimises "the risk of urinary tract infections, infections of the penis, and phimosis, a tight non-retractable foreskin" [107].

Such criticism is interesting. Since infections are first and foremost caused by poor hygiene (apart from the biological agent of course), by the same logic we should cut off our children's ears, noses, and nails. As to phimosis, it is indeed a condition which, when fully developed, can be quite painful. What gets hidden from public consumption however is the simple fact that any living tissue is capable of stretching to a certain degree. In the case of the foreskin such stretching would entail pulling the skin backwards, and if tried again (and again...) would come pretty close to masturbation—at least that's what it feels like outside the moralistic reference. Naturally, the earlier in life this is done the less the chance of having a problem with it later (any physician worth their salt should be able to confirm this). Yet at this stage we are firmly within the reign of the anti-erotic rule set of the Judeao-Christian world view. Recall our mention of morality during our discussion of Marx and his ideology (chapter 10); there existed a device which was designed to alert the male of any erection during sleep. Since we have left philosophies behind us we have the time to look at this thing more closely in Fig. 17-4. Perhaps phimosis is the body's own defence system against sin?

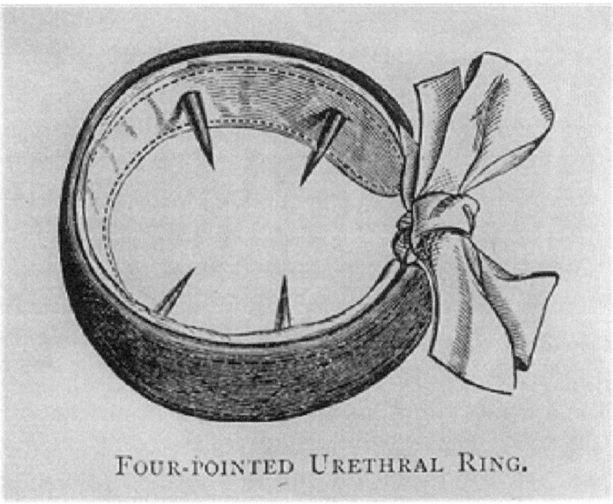

Fig. 17-4: A protection against bodily sins [108].

For religionists masturbation has been an obsession in itself. It is a major sin to be guarded against, a trauma that can only be guessed at considering the constant biological drive from the earliest age onwards. There used to be parents (and no doubt still are) who told their offspring that the practice leads to madness and mental asylums were full of children who have succumbed. Humans endowed with reason might like to imagine what it means for a child to be threatened with insanity, when the only thing capable of keeping them sane in the face of such superstition is their mind!

One often recounted biblical story about sin and its effects relates to the city of Sodom. The city and its inhabitants got destroyed by the god who did not agree with their habits. There is more to this however. Lot, a man who on occasions communicates with the deity, is visited by two angels in the form of humans (so we assume, given what happens next) [109]. The visitors have a meal with Lot and prepare for bed. At that point "the men of Sodom surrounded the house". They ask Lot to bring the visitors out to them because "The men of Sodom wanted to have sex with them". Lot's reply was,

> "Friends, I beg you, don't do such a wicked thing! Look, I have two daughters who are still virgins. Let me bring them out to you, and you can do whatever you want with them. But don't do anything to these men; they are guests in my house, and I must protect them."

The men insisted but the angels drew Lot back inside, shut the door and "struck all the men outside with blindness, so that they couldn't find the door". Soon after the angels destroyed Sodom and Gomorrah with only Lot and his two daughters escaping (the wife looked back and was turned into a pillar of salt).

It would be pointless to argue about the customs surrounding hospitality in ancient times, particularly against two millennia of moral censorship. We do know the church's antagonism towards homosexuality though, and an exemplary tale in which substituting grown men with two young girls for the sexual pleasure of a crowd is acceptable because otherwise the god's morals would have been breached, surely says something about that religion.

Having left the region Lot and his daughters lived in a cave. Because there were no other men around the two women decided to make their father drunk and have intercourse with him. Lot "was so drunk he didn't know it" but got them pregnant anyway (?). One child was the progenitor of the Moabites, the other started the Ammonites.

In the light of that incident it is instructive to observe the divine regulations on that theme as later told to Moses. It is forbidden to have intercourse with the following: one's mother or any other wife of one's father, sister or stepsister, granddaughter, half-sister, aunt, uncle's wife, daughter-in-law, brother's wife, the daughter or granddaughter of a woman with whom one has had intercourse, or the wife's sister as long as the wife is alive [110]. Daughters in general are not mentioned, although sons presumably are out anyway because that comes under the label 'homosexual'.

A demographic in which every erotic behaviour is outlawed except for one single type (ie, having intercourse with a hitherto virgin inside marriage) is enacting the same principle as one in which an accused is held guilty until proven innocent. It is no coincidence that it took the enlightenment in Europe to turn this around, nor is it irrelevant that the Roman Inquisition followed the biblical version (and does so to this day, except the office is now called "The Congregation for the Doctrine of the Faith"). If a society is intelligent enough to implement the principle of civilised law in eros it would mean that every act of erotic behaviour is allowed, except for those which transgress human rights. So far on this planet none has made the grade.

In another chapter explaining the god-given laws as related by Moses (introduced in the first section) we read about "sexual purity" [111]. The situation starts with, "Suppose a man marries a girl and later decides he doesn't want her", so he accuses her of not having been a virgin before her marriage to him. In this case "the girl's parents are to take the blood-stained wedding sheet that proves the girl was a virgin, and they are to show it in court to the town leaders". Against this proof the husband is declared wrong and so gets beaten and fined, with the money going to the girl's father; moreover, the man can never divorce the girl as long as he lives. However, if such proof is not forthcoming "then they are to take her out to the entrance of her father's house, where the men of the city are to stone her to death"—the reasoning being that she has had intercourse before marriage and done a "shameful thing". "In this way you will get rid of this evil", the passage concludes.

Consider the text away from the literal meaning of the words (as brutish as they are in any case). Since vaginal bleeding is considered a forensic evidence, its presence or otherwise can mean life or death for the female. From the outset therefore the induction of—evident—blood loss can either be in the interest of the girl if happening on the marriage night, or against her interests if happening before. If before, what to do later? From the father's point of view, the presence of blood on the sheet is desirable if his interests align with the consequences of the marriage because her husband will be tied to her regardless of his own wishes and there is the chance of money coming the father's way. If the girl's position within her own family is precarious to begin with, the opportunity for foul but quite legal play is obvious. Add a relatively small community with its accompanying interwoven and tangled relationships and you have politics as belligerent as they come. Yet no matter how widely-spun and dangerous these are, the whole mess centres on the young woman.

In the New Testament the antagonism towards human eros is continued. Paul, one of the individuals utterly convinced of the new religion's value and so acting within the context of what is now termed 'evangelism', travels through the countryside proclaiming its sentiments through his perspective of moralistic righteousness. He rails against "all the sins and evil of the people whose evil ways prevent the truth from being known" [112]. The people's "empty minds" are "filled with darkness", and his god has "given those people over to do the filthy things their hearts desire, and they do shameful things with each other". If there is any doubt about what those "shameful things" are, the next paragraph explains it:

> Even the women pervert the natural use of their sex by unnatural acts. In the same way the men give up natural sexual relations with women and burn with passion for each other. Men do shameful things with each other, and as a result they bring upon themselves the punishment they deserve for their wrongdoing.
>
> They are filled with all kinds of wickedness, evil, greed, and vice; they are full of jealousy, murder, fighting, deceit and malice.

The conceptual linking of despised erotic behaviour with criminality is exercised wherever authoritative control is desired—it occurs in virtually every human society. Through such a link the negativity of one is transposed to the other, and since everybody can see the destructive effects of murder for instance, its semantic content is spread to other behaviour forms which otherwise would remain untainted. On a small scale the same principle can be observed all the time: the mannerisms of a social outcast are rejected, just as the habits of a hero are adopted as a leitmotif—the gist of role models. Under the Nazis in the aftermath of the depression financial affairs were linked to the Jews, and hence a nation's woes were associated with monetary transactions which in turn became synonymous with Jews which in turn were held responsible for failing finances; the circle closed.

In another passage the same author refers to the scripture as an example how sinfulness should be perceived: "The people sat down to a feast which turned into an orgy of drinking and sex" [113]. Almost gleefully Paul tells his audience that as punishment "in one day twenty-three thousand of them fell dead". His own disposition (his self-flagellating attitude towards the erotic side of human behaviour is painfully obvious) becomes the emotive weapon through which he castigates his targets, the personal suffering becoming the currency by which the 'sins' of the others are weighted: "..I shall weep over many who sinned in the past and have not repented of the immoral things they have done—their lust and their sexual sins" (note the repeated use of tantalising descriptors) [114].

John, another evangelist, writes of the demise of the world which he sees as a direct result of human immorality, once again mainly defined in terms of erotic behaviour. In one of his visions he is visited by seven angels and one of them showed him the destruction of Babylon. It occurs because "the famous prostitute" is ensconced in the city and "the kings of the earth practised sexual immorality with her, and the people of the world became drunk from drinking the wine of her immorality" [115]. One can only guess at the inner turmoil of a man whose body keeps sending out its biological signals, only to be forced into submission by a mind that projects its impassioned agony onto spectres of cosmic devastation by proxy.

Of Babylon itself something more can be said than the simplistic exhortations of religious fanatics. One of its kings around 1700 BCE was Hammurapi, he of the code of law which distinguishes itself by, amongst others, the distinction made between intentional and accidental homicide; or the status reserved for women, a "more honourable one than in many countries today" [116]. It was a city whose influence in cultural and military terms had been important enough to direct the course of ancient history, and the knowledge gained by its inhabitants in the fields of mathematics, astronomy and chemistry are well acknowledged. In short, there are many things by which ancient Babylonians are known today, but facing extinction because the men were only interested in each other isn't one of them. But then again, the enmity between an advanced, diverse culture on one side and uneducated outsiders on the other is all too common.

To us the connection between a thought complex and its opposite is already familiar. If a TS encounters another which articulates an opposing difference the associative elements are sufficiently distributed to allow for specific variances. If the 'other' is nothing more than the first however except with a 'not' added to its structure, both contents are virtually the same. That is to say, an activation of one will influence its counterpart, and vice versa. In the case of fundamental TSs (our default abstracts) such an opposite does not diminish its originator, but contributes to the denominative intensity of both. Therefore, to a person convinced of their sinfulness the simplistic antagonism towards their physical self will not attenuate the emotional force of their inner state, but inflame it further. As Montaigne said, "And of our infirmities the most savage is to despise our being" [117].

When the psychoanalyst Leonard Shengold delved into the story of Kaspar Hauser, a boy who had been imprisoned for most if not all of his life and later released as a shadowy copy of a fully developed human being, he called such treatment 'soul murder'. To him it meant, "killing the joy in life and interfering with the sense of identity of another human being. It is primarily a crime committed against children" [118].

Let us translate the idea into our nomenclature. The identity container, a collection of TSs that exercise their affinity potential in the course of building the layers of experience (ie, further cognitive structures), grows into a framework of representations gleaned from the outside. Although situated within the overall system of the mind and thus sharing a complex resource space containing various niches and subdomains, it exerts a determining influence because it is the historical as well as the functional centre. Recognising the difference between functionality and content, we can say that the integrity of that centre does not stand or fall with the representative quality of its content per se, but rather with the harmony amongst its functional modules. True, the harmony becomes manifest through the realisation of associative potential between representations, but the judgment rests with the presence or otherwise of enabled linkages; *what* detail is being linked remains secondary. For the boy Kaspar Hauser human contact, the nuanced variance of open nature, the intellectual substance absorbed from education, all that was withheld from the waiting mind—like a mill churning without grain.

Associative potential does not only refer to the horizontal reach; a complex system functions as a *complex* system because it has maintained the vertical connections between the gradually built layers. Without the connections a layer could be seen as a singular spread-out structure referenced only to its own self. Its content has lost its depth, no matter at what abstraction level it is meant to reside. An example would be a metaphor interpreted in terms of its literal meaning.

Now consider the semantic content of a doctrine. If it can be placed within the conceptual framework of the mind such that its content can form congruent PAT chains with its surroundings, the necessary harmony has been gained. If the links are incongruent, discrepancies occur that can either be isolated behind the logic boundary or are forced to adopt a self-destructive symbiosis from which neither side can recover. The demands for congruence are not just made by the abstracted material from the outside, but also by its internal counterpart, the default abstracts. Because the latter exist from the very beginning (regardless of subsequent experiences), their utilisation in terms of the entire set of participating TSs can be considered crucial.

The above is reflected on the neurobiological level. For example, in the visual processing areas of the brain there exist cells which are responsible for orientation [119]. They are assumed to be the building blocks of form perception. In a brain that has had no vision since birth or shortly after, most of those cells do not respond to visual stimulation or they behave in an unpredictable manner. Zeki, explaining all this continues by saying,

> So the philosophical question of whether there are ideal forms in nature, external to the human observer, is one to which neurology is indifferent because, in practice, an observer deprived of vision during a critical period after birth cannot recognise even a small number of objects by sight, let alone have a Platonic Ideal or an Hegelian Concept of them. To speak to such a patient of the Platonic Ideal of forms amounts to making a sick joke.

In our functional sense there is no real difference between the absence of an input which is necessary for the activation of cell clusters (and hence TS formations), or an input which inevitably clashes with the associative potential of the identity container, given that its origin is due to default abstracts. The difference is contextual, a matter of scale.

The discrepancy is experienced as trauma. Evelin Lindner from the University of Oslo examined trauma in the context of humiliation [120]. In her abstract she points out that the role of humiliation in the development of trauma has long been overlooked, a state of affairs particularly poignant when considered against the issue of human rights:

> It is the unfolding of human rights as opposed to more traditional honour codes at all levels of society both national and international. This change is a major force in making the category of trauma increasingly important, and in moving such practices as 'breaking the will of the child,' that were once legitimate and even prescribed, into the category of trauma.

She observes that the word 'humiliation' itself is problematic because it has an "extremely complicated semantic field". The act of humiliating "entails a painful downward push at its core, namely looking down, putting down, lowering, degrading, debasing, abasing, demeaning, belittling, subjugating, oppressing, tainting, besmirching, tarnishing, treating with contempt or disgust, bullying, mobbing, abusing, dishonouring, disgracing". She mentions traditional societies in which humiliation is practised legally, and it is only in more advanced demographics where the realisation of its intrinsic nature could develop, "an indication that over the generations the abstractive ability of humans has increased". She quotes Dennis Smith who argues that the presence of the 'human rights' concept does not in itself annihilate humiliation; rather, it makes the concept available to the sufferer (to whom it was hitherto non-existent) so that the act is seen for what it is, the downgrading of the person: "In other words, humiliation, already hurtful in an honour society where it is routinely used as habitual means to put people down or keep them down, becomes many times more hurtful when it occurs in a human-rights society".

Lindner relates an account given to her by a member of the Hutu tribe in Burundi, where the Tutsis represent the social elite ruling over the Hutus. From the time her friend became conscious of his role as an underling his pain grew and he slipped in his school performance (although he used to be a brilliant student), as well as suffering bullying and stigmatisation. Later he became a teacher but was sacked on the accusation of having had secret meetings with students at night in the forest. Through his gradual descent into trauma Lindner shows the development of effects within an authoritarian structure which sees humiliation as a tool for control. Referring to the necessary shift towards the recognition of human rights she says in her closing remarks,

> Academia is not untouched by macro-historical change. Especially psychology, in its historic struggle to join the 'discourse of the master' namely natural science, by, at times, overdoing statistical analysis, could perhaps benefit from humbleness, and learn that natural science itself does not necessarily claim to have all the answers.

The ongoing formation of identity containers against the background of society and its cultural rules suggests to an observer who is situated within those selfsame structures that the results are 'natural'. So much so that the creation of such content is seen as an automatic product of one's status as a human being; after all, it represents one's identity, and it is the identity that in its projected format defines the culture. As Lindner wrote, it takes the abstractive ability of the mind to conceptualise an alternative. Nilanjana Dasgupta from the Department of Psychology at New School University and Anthony Greenwald from the Department of Psychology at University of Washington conducted two experiments to demonstrate the possibility of reducing the automatic preference for white over black Americans and younger over older people [121]. Their conclusion confirms the influence of either positive or negative associations with the target group in the other, despite the existence of opposite attitudes before. If the presence of culturally induced opinions can be seen as a cycle of cognitive functions, these experiments prove that breaking that cycle is productive.

A similar interdependency can be found on the neuronal level. Various regions in the brain have been identified as suppressors of pain [122]. The well-known placebo effect is an example of cognitive processes directly affecting the perception of pain. A majority of patients who were given injections of sterile saline reported relief after an operation [123]. They did not imagine it; merely the belief in the treatment causes "activation of the endogenous pain relief systems of the brain". The authors note that "depending on the concurrent level of nonpainful sensory input and the behavioural context, the same level of nociceptor activity can produce more or less pain".

The potential for discrepancy has also been explored by the anthropologist Gregory Bateson, who studied the disjointed behaviour of schizophrenics who are unable to understand the implicit meaning in a message; for them the content is to be understood literally [124]. Bateson thought the problem lay in childhood when parents are lying to their children. On one hand the adult conveys a message to the child in so many words, but on the other the child senses a different one articulated through body language and mannerism. The result is incoherent information which has to be reconciled within the child's mind somehow. From our point of view the data would be measured firstly against the knowledge base inherent in the identity container and secondly against the accumulated experience. The situation is quite similar to incongruence-inducing sentiments from religious rules; one's body and life say one thing, the god says another.

In European thinking the Judaeo-Christian limit imposed on eros has transmitted itself across the millennia. Whereas other ideology-based effects came and went due to their perceived association with the contemporary locale, in this case every new-born or, to be more precise, every new set of genitals presented another challenge to the doctrine. The challenge was taken up in the form of denunciations, torture, burnings, and barbaric tools such as the one shown in Fig. 17-4. The idea that eros is guilty unless proven innocent (via so many convoluted thought processes) led to outrage whenever a suggestion strayed from the approved path.

In the early 20th century Sydney was the home of William James Chidley, a dress, food and sex reformer who set the authorities on fire. His conviction was that for a human to be healthy a diet of fruit and nuts and minimal dress were indicated [125]. While his choice of food was more or less tolerated (even in his fifties he was six foot two tall, had sun-bronzed skin and seemed healthy enough), his dress, first a bathing suit then a Roman-style toga, attracted more attention (although basically a nudist he realised this would go too far). When Melbourne became too difficult he moved to Sydney but was soon arrested on many occasions. His dress was judged "indecent", he was fined for "hustling women and children" and even consigned to a mental asylum [126]. Whenever he was released from that institution one of the conditions was "That he should dress in men's ordinary costume" [127]. Yet his most 'outrageous' philosophy concerned the way men have sex. He believed—too fervently for his own good—that the best way for coitus between man and woman is for the man to have no erection. For this he was attacked by the population, hounded by the police, and the magistrates whished he would just go away. In his last message to the public, written in 1916, he summed up his ideas:

> "The human race must return to (1) natural coition, (2) to nudity, and (3) to a natural diet: fruits and nuts only; and each of these depend on the other two. Only thus can you obtain mutual Joy, Love, and Content. Bring children up naked, and on fruit only—clothes do more harm than is supposed.
>
> ..Mine has been an unhappy life, but it contains a moral, namely, that all my misery comes from that 'erection' in boys and men. Farewell! W. J. Chidley." [128]

A few days later he doused himself with kerosene but the fire was extinguished. He never recovered mentally and died at the asylum in December that year at age fifty-six. Havelock Ellis, the "philosopher of sex" as the biographer Vincent Brome calls him [129], had this to say about the Chidley case:

> "It is a sad thing that a new country like Australia, with so many bright prospects before it, should begin its career by setting up inquisitions." [130]

The scenario bears all the hallmarks of Christian moralism. The extremely narrowly-defined range of legal erotic articulation not only compresses every biological and eventually every cognitive functionality into its mould, the individual so constrained is now forced to exercise the remaining option within the minute degree of freedom left open. The outcome is the compulsion of what is permitted and the rejection of what is not—as the saying goes about doctrinaire regimes: if it's not prohibited, it's compulsory. For someone like Chidley to come along and offer—well, let's say, proclaim in his case—an alternative that does not involve the biblically sanctioned form of intercourse will then be seen as sacrilegious. Include his views on nudity (something to be feared by anyone not on good terms with the physical body) and his remarks fall well outside the bounds so strenuously upheld by the apostles. It is ironic, but also fitting, that this same precarious escape into an alternative also related to Chidley himself; under more relaxed circumstances an erection in boys or men would not be an issue to start with.

With divine anger alert to every 'wrong' step it is no wonder that erotic behaviour in general is anything but joyful. Sacred business is serious business and with self-harming fanatics as moral pillars of the faith the mere sign of an emotion hinting at fun would have to be destroyed at every opportunity. Linguistically the chasm dividing duty and pleasure can be seen in the circumscriptive manner in which the legal format is described, versus the head-on version of its hellish counterpart. 'Sanctity of marriage', 'fulfilling one's vows before god', or 'doing one's duty' are just about as close as the pious patient can get inside their church. But '*fornicator!*' will ring in their ears as soon as the body does its thing outside.

Not that the church had ever been fully victorious. It could not legislate against the making of ditties sung by the mob on the streets to thumb its collective nose against the authorities. Take Nell Gwynn, an English actress who lived from 1650 to 1687 and became Charles II's mistress [131]. She was generous, vivacious, open about her Protestantism and so quite a favourite with the mob, earning her the title "Protestant whore". About her it was sung,

"She was so exquisite a whore,
That in the belly of her mother
Her cunt was placed so right before,
Her father fucked them both together!"

Definitely not hymn material. *The Macquarie Dictionary*, the repository of language for Australians, has as usual the particular words listed at the beginning and end of each page appearing in the header. As it happens, by the time we reach page 721 the last entry turns out to be 'fuck'. Rather than having something confronting like that printed in bold Arial on this page and the next, the editors decided to use 'fuchsite' which comes before and 'fucoid' on page 722, which is the next after not less than eight derivatives of the offending articulation [132].

Given the animosity towards eros in general and its reach into the realm of pleasure in particular, the fear generated by drugs which in one way or another enhance the perceptive quality of the experience can be understood in those terms. An establishment which is troubled by its own inner demons generated by the body's dynamics sees itself compelled to assign its associated con-TSs to domains which carry suitably negative connotations. Since for a doctrinaire mind the 'other' is synonymous with 'evil', its consciousness conjures an imagery representing just that, and to the exclusion of everything else. Fig. 17-5 illustrates some examples.

The vast amounts of money being spent by governments on trying to gain the upper hand in the fight against such indulgences (see chapter 16) indicate a driving force which goes deeper than economic or political considerations. It is an energy that is being fed by the fundamental associativity of an identity container, configured according to religious principles.

When Evelin Lindner talks about the abstractive powers which help to understand the wider reference and therefore the abuse committed under a narrower one, she avails herself of the benefit of a distant perspective. The forming of an identity under confined surroundings does not permit any recourse to information which could counter the current experience, and anything imported will only ever be interpreted in the same light. A society which lives under the auspices of religious constraints imparts such principles under the guise of everyday activities, twenty-four hours a day. Generally speaking, by the time a child is three years old it will have absorbed the basic mannerisms of its culture, harder to witness at home but more obvious beyond one's borders where the customs are different (for the reasons just outlined). Such things as male/female roles in adult company, order of dress, and certain postures (ie, for girls to close their legs and pull down their dress) are already the identifiers at that age.

Fig. 17-5: Examples of raised expectations by the spectre of marijuana [133].

Because a local observer is part of the same environment, for such a person it is virtually impossible to recognise the functional influence—even if the effect should be a severe one. Thus as far as Christians are concerned the deeper links between drugs and eros are hidden because they exist beyond the logic boundary and are therefore inaccessible to PAT chains requiring grounded associations. They see the effects of the drugs' use for example but will not understand the connection with their own morality. The beginnings of the moral education are to be found at day one; an infant's brain is capable of pattern recognition at any level of wiring complexity and it is only the abstractive quality of its representative content that is a matter for discussion.

As we have seen in chapters 14 and 15 the primary factor in cognitive dynamics is not the content of incoming data but their pattern-forming quality. At the lowest level we either had protein complexes or sets of integer values, but the significance was not the chemicals or the numbers but the affinity relationships derived from their arraignments. At a higher scale it is again the data which are secondary; what matters are the patterns they generate across the network. Therefore an exposure to restrictive practices is not perceived as an exposure to restricted practices, but the perception complex situates itself within associative TS formations—except in the presence of TSs which are able to articulate the restrictive quality to their owner. In a closed environment this is not the case.

For example, at a reunion of former pupils at a girls' school in Brisbane some of the rules were remembered as they had existed in 1948 [134]. In those days it was forbidden to talk to boys, even to your own brother at the bus stop, and boys and girls were separated by bush known as the 'forbidden land', with even looking across that strip attracting a penalty. Arguably one of the worst excesses had been the dress code, such as "thick black stockings and bloomers" even in the heat of summer which can reach debilitating proportions in that city. 'Bloomers' I understand are trouser-like wrappings applied to the female pelvic region under the presumption that the physical parts underneath have now disappeared from public consciousness. From an outsider's point of view the constant exposure to such fear of the body, reinforced by penalties that humiliate and so apply an emotional sting to one's conception of the physical self (see above), must over the years mould the psyche into a very specific state of tension and denial, as well as turn the mind into a mutilated version of its normal healthy self. Of course the word 'normal' would be as alien to them as the Hegelian Concept to one of Zeki's vision-impaired patients.

The very real possibility of injury by heat could have overcome those earlier attitudes (space does not allow me to go into the well-documented damage done to brain cells by excess body temperature), but still the law does not look kindly on a display of the human physique. As recently as 2001 a female tourist from Norway was approached by a security officer at a Brisbane inner city beach for not wearing a top [135]. The action was the result of a complaint by other visitors. And in Melbourne five nude protesters against Victoria's laws were not arrested as they showed themselves at a beach there [136], although it is illegal to go nude.

Fig. 17-6: Nuba girls show up for an evening dance at Fungor, decorated with sesame oil and red ochre [137].

Imagine the reaction of our good citizens if exposed to a scene such as in Fig. 17-6. They are Nuba, inhabitants of an area in southern Sudan. The Sudanese government has been accused of waging a campaign of ethnic cleansing against these people [138]. The reasons given range from religion to land claims to finds of oil in the Nuba mountains. All of these may have been articulated at one time or another. However, since the state religion is Islam one does not have to be a cynic to suspect a fundamental attitude at play, considering the treatment Muslims reserve for women who do not follow their strict dress code (Islamic followers in Western countries are not the standard to be applied in this case). It is also not insignificant to observe that the killings have gone on for years, yet have largely been kept out of the media limelight. The beaming of pictures like the above into Western homes would raise all kinds of questions in the minds of well-taught Christians, the issue of nakedness itself being the least controversial.

Recall the body of a girl found in Morocco, her arms having been hacked off because she chose Western style clothes. A similar intensity can be found in the United States when it comes to the question of abortion. The National Abortion Rights Action League (NARAL) says that since 1993 "three doctors, two clinic employees, a clinic escort, and a security guard have been murdered" (the article was written in 2001) [139]. Since 1991 "seventeen attempted murders" have occurred, and since 1977 "more than 2500 reported acts of violence against abortion providers" had been listed. They include "bombings, arsons, death threats, kidnappings, and assaults, as well as more than 55,000 reported acts of disruption, including bomb threats and harassing calls". These outpourings of hate are related to the right-wing politics of this country, infused as they are with Christian morality. The systematic suppression of default abstracts dealing with erotic behaviour leads to the formation of complexes that are incapable of integrating the self with its surrounds in a productive manner; in fact they are comparable to the deficiencies confirmed in children who grew up in the wild, only less comprehensive. The enforced constraint of the erotic spectrum into biblical exclusivity affects the remainder by imbuing it with an emotionally charged negativity. The mental scenario is similar to the experience of humiliation in the communal context, where the purposeful exclusion and labeling of the target make them an outcast and induce trauma as well. The result is the rejection of the humiliated person by the perpetrators, but it also creates rejection within the recipient.

As usual, a functionality on one level of scale can be transposed into another, and so religious repression causes the schism between acceptable and unacceptable TSs. Should the structures be based on default abstracts the damage exists in a fundamental rejection of identity domains. In the case of eros, these consist of TSs related to whatever actions fall under the general heading of 'erotic pleasure', and at whatever level of complexity the age of their owner enables them to be. If the entire society labours under the same dictum, even a personally felt problem won't be seen as relating to the inherent framework because the society *is* the framework. Only an outside reference can put the perception into context.

One such error concerned the blood pressure in adults. It was believed that with advancing age it naturally increased because the arterial walls provided less volume for the heart to pump the blood through. A handbook for nurses from the year 1975 gives us a table showing the steps in brackets, where in the "young adult" the systolic pressure (measured after the heart beat) is given as "110 to 125", but "as age advances it is increased 130 to 150" [140]. Go forward to 2002 and in an information sheet for a blood pressure measuring device we are informed that for healthy adults the mean systolic pressure is less than 130 mm Hg [141]. While higher blood pressure is still possible, an improvement in the general health of adults has enabled physicians to reset their standards; in other words, the sample space has become more truthful.

The status of homosexuality has also been vehemently attacked. In a complex system the differentiation between the procreative side of its dynamics and the physical and mental contingencies dealing with the experience of pleasure suggest the subsequent clustering of various respective domains, so that TSs having to do with the continuation of the species and others related to the achievement of erotic satisfaction may overlap but are not necessarily the same. In a fluid and unbiased environment the overall state may sway towards one or the other (having emerged from an amorphous youth anyway), but once constraints are in place the clustering would be far more rigorous. How can it be otherwise—the identity container must shape itself amongst the impediments of a critical data set. In genial demographics the rather vague definition of hetero- vs homosexuality can be observed quite readily, whereas in belligerent ones the respective manifestations tend to occur in a much more extreme form. Once again it is a matter of sample space: under a Christian-induced morality it is considered normal to 'prove' one's masculinity through displays of aggression and destructiveness lest one be considered gay, and those who are gay may exacerbate their status in order to create some sort of self against all the odds. In that way the homosexual stereotype of the effeminate pansy is born, but once outside the confines even a through-and-through same-sex individual could give the 'he-man' a run for his (?) money.

Montaigne was right—the railing apostles were attacking more than just other people. In the modern age their heritage had been taken up by a new form of guardianship, psychiatry. In the 1890s Krafft-Ebing for example explained homosexuality as a "morbid phenomenon" with "mental abnormalities and real disturbances of the intellect" more common than in so-called healthy humans. Lesbianism was a "cerebral anomaly indicative of an inherited diseased condition of the central nervous system" [142]. Another, Tarnovsky, claimed homosexuals "were born as a result of the nervous disturbances of their parents, including epilepsy, brain disease, insanity, hysteria, alcoholism, syphilis, pneumonia, exhaustion or anemia". The article's author lists the following treatments dished out to individuals who had been identified as needing to be saved: castration, vivisection, cliterodectomy, lobotomy, electroshock treatment, pharmacological and hormonal injections, and aversion therapies.

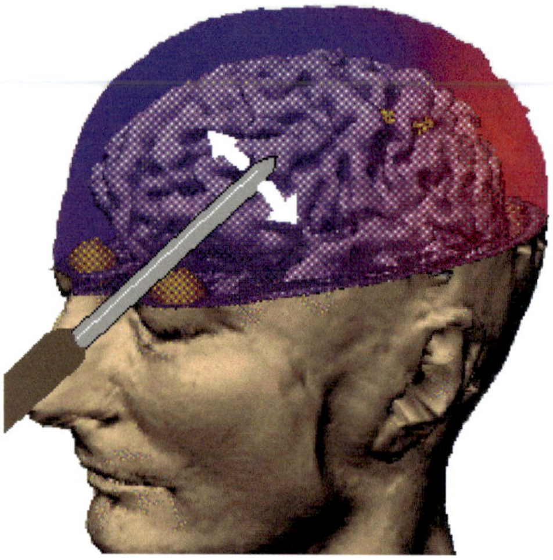

Fig. 17-7: How lobotomy works [143].

Take lobotomy ("lobe cutting"). It started with the destruction of brain substance in severely anxious and depressive patients whose symptoms were sometimes relieved. It ended up being applied to anyone who was considered extremely psychotic, and homosexuals fitted this description. Fig. 17-7 shows what the operation meant.

From the 1950s onwards the procedure got into disrepute for ethical and clinical reasons (its medical efficacy could never be proved anyway).

During a memorial lecture Australia's High Court Judge Michael Kirby described how as late as the 1960s men were subjected to "mild electroshock" treatment while viewing slides of homosexual scenes; one had approached his doctor to seek advice about his sexuality [144].

When in 1999 the Queensland government proposed to broaden its domestic violence protection laws to include same-sex couples the conservative opposition was outraged [145]. As far as these ministers were concerned the amendment "threatened to decay the moral fibre of Queenslanders". One politician declared, "Jesus Christ is the one who will strike you dead in the end", referring to those who were responsible for the reform. A year later Pope John Paul II criticised the Netherlands for legalising homosexual marriages, describing it as a "serious affront to human dignity and the family" [146].

Such deep-reaching conceptualisations of the human eros (and thereby the human psyche in general) posit the bearer against the full spectrum in a very specific way. It prompts politicians for instance to respond to their societal environment in a belligerent manner which otherwise would not have attracted the energy needed to sustain it. The drive to be watchful for signs of non-compliance with their own code, coupled with the dynamics of the public arena causes accusations to erupt that are supportive along those lines. A sudden focus on the behaviour by a public figure perceived to be a transgressor of religious standards, is an almost regular feature of ideology-driven politics. Under communist regimes it would be the suspicion of anti-socialist intent, in a right-wing state it is the opposite, and in a country like Australia we find breathless statements about sexual misdemeanour. Since the initial trigger is based on the accuser's perception, that is to say on an internal configuration of cognitive structures, outside data come secondary. The result is some headlines for a few days, comments from all directions, questions in parliament, a significant disruption of people's lives—and afterwards the usual retreat.

The moralistic ambience generates an insidious situation, and not just for the obvious reasons. If a person has been branded, from an entirely objective point of view there are two possibilities. Either there is some substance to the allegations, or there is not. If there is, the manner in which the issue has been brought to light obfuscates the facts because they have to be separated from the hysteria and uninformed mishmash of ideological presumptions. Should there be nothing to begin with, any defence would be precarious. Since the compliance with the law can be seen in terms of an individual's responsibility towards the wider society, if the law happens to be unreasonable not only are we dealing with that aspect but also with the action itself. To say "the denunciation of such and such an act is ludicrous" leaves the targeted individual's reputation tarnished anyway because under the law the act should not have been committed, but at the same time if it *had* been committed it should not have been an issue. What to do—focus on the law and there is the however faint suggestion the act had indeed occurred, but focus on the person and the juridical element remains unquestioned. We can call this situation a moral double-bind, a significant feature whenever ideological principles are at work.

Given the profound difficulties visited upon the hapless recipients and all because of some ancient manifestation of psychopathy, how would the politicians react if one applied their acronym of MP = Member of Parliament as MP = Moral Primitive?

If our history of religious interpretation has turned the moral waters murky, the advent of the feminist perspective has made them virtually opaque. Andrea Dworkin's exclusion "no part of the male sexual model can possibly apply" places the entire contingent of male sexuality on the defensive. Although written in the context of female/male exclusivity, a gender-based articulation is nevertheless a subset of wider society. Since the general expression of the human eros is already curtailed to the nth degree, to add yet another series of declared deviants prepares the field for war.

And a war it is. When the female psyche restricts itself to a more self-directed interpretation of incoming information, the data are not obliterated but are chosen in terms of their associative properties with the identity container (see the previous section). Translating this into the *moderne*, we end up with a mix of perspectives which range from outright rejection along religious prejudices to a selective optionality suitable to feminist ideology. The latter will allow erotic freedom provided it remains within the circle of approval.

Consider the furore that has erupted on some American campuses under the heading of 'consensual sexual relationships'. Some feminists "insist" such relationships (between "professors, staff, teaching assistants, etc") are so "widespread and a serious problem" that "extraordinary measures" are necessary [147]. That same article refers to a "recently completed study at the University of Massachusetts" which showed that "only 1 out of 521 students surveyed indicated she had had a relationship with a professor (the numbers were slightly higher for other employees)". The authors write,

> The new Puritans have been remarkably successful in imposing this agenda on campus, because the professoriate and university administrations have been singularly unwilling to challenge all the loose facts and tired stereotypes emanating from this group. Those female professors who have done so have been viewed as apostates. It is the male professoriate that has remained uncommonly silent. For male professors to come out as we do in this article against banning, opens us to being targeted and objectified as sexual predators.

Rejecting consensual relationships being labeled as 'harassment', the authors make a distinction between it and the possibility of a conflict of interest. The latter would also include evaluating and writing recommendations for students, or having one's children in the classes or the hiring of students to earn extra money. While condemning relationships in the sexual arena, when it comes to the other types the same voices have been "uncharacteristically silent". The accusations had been exaggerated to such a degree that a former Associate Vice Provost charged with investigating claims of sexual harassment at Chico State University was prompted to describe the campus as "composed of a greater number of unregulated rat catchers than it seems to have rats". When challenged to back up the outlandish figures the response by the feminists was "the nature of sexual harassment is such that the lack of evidence is in fact evidence".

This statement is not as problematic as it may seem. As an analogy consider the role of a confidence trickster. One can only be a confidence trickster if nobody realises their presence—it's the nature of the beast. The problem with the above however is that the definition is used as evidence, whereas it should only be treated as a starting point for further evaluation. It is its results which ought to provide the basis for judgment.

In another article feminist scholar Catherine Stimpson is quoted as saying,

> "Today the psychological and social pollution ... harassment spews forth is like air pollution. No one defends either one of them. ... [B]elow the stratosphere, in classrooms and laboratories, sexual louts refuse to disappear, imposing themselves on a significant proportion of our students..." [148]

Statements like the above apply an a priori categorisation without any due regard to the context in which they are meant to be seen, let alone provide the figures to back them up. In her criticism English Professor Joan Blythe says,

> "..it presents me, the teacher, as rapacious, predatory, dangerous even before I walk into the classroom."

Let us examine those examples for a moment. There is a cognitive complex which represents the issue of sexual harassment. There exists another which relates to the conceptualisation of the target as 'victim' (otherwise 'sexual harassment' losses its descriptive power). In tandem they become associative with a domain that requires its own imagery to project an antagonistic realm in which the outer serves as the confirmation for the existence of the inner. Since the 'inner' is synonymous with the identity container, its dynamics contextualise the periphery under the influence of its denominative properties such that its abstractive content gets channelled along an illusion tree. The result is an incongruent PAT chain devoid of its grounded abstracts, turning into a concept of its own. We have witnessed a similar phenomenon in the writings of Rousseau and Marx, where the need for a negative image caused the original perception to be stripped of its real-life variance until it became an idealised adjunct to the philosopher's ideology. At that point we are faced with a seamless concept beyond the logic boundary and hence untouchable by rational argument.

In the present case we are also dealing with an inherent antagonism towards the male. If the abstractions at that level were grounded, similar principles could be revisited through congruent PAT chains representative of some other context; as it is, the paths are closed. For example, while a concern for the 'powerless victim' may be valid in an environment of teacher-pupil relationships, its placement beyond the logic boundary ensures the biased implementation when the teacher happens to be male but shows no such thing should the authority figure be a female. The ungrounded nature of eros-related thought complexes allows presumptions to be formed under the same auspices as the general attitude towards sexuality provided in the first place. The 'guilty until proven innocent' rule has become a standard in feminised society wherever the context links with its ideological components.

A few decades of building a perception of the male as a "rapacious, predatory, dangerous" animal can be expected to affect the public. If teacher-student contact at universities invites the intensity seen before, when it comes to teacher-pupil arrangements in primary and secondary schools the atmosphere is even more charged. Suspicion and false allegations are easily raised when an imagined spectre becomes the focal point for the definition of certain people. The president of the Queensland Teachers Union called the generated negative "macho" stereotypes of male teachers "outrageous", and Robyn Zevenbergen, a lecturer at Griffith University, remarked about males who wanted to work with pre-school or primary children, "There's a lot of innuendo associated with it" [149].

Three years later and we find similar comments. Queensland University of Technology lecturer in education Erica McWilliam is sure the fear of being cast as a child molester drives males from the teaching profession [150]. In one incident a teacher was holding the hands of two little girls while singing a rhyme when he got accused of practicing some sexual ritual and the principal (a woman) then proceeded to interview the whole class. This does not help the morale. Note the cognitive elements: an ideologically-driven identity container creates a complex of notions in which the children are automatic victims, the male teacher is a perpetrator by definition, and at any given time any action could be a sign of the concept made manifest. As such the behaviour is nothing new—men in longer term relationships with women would know about the futility of arguing any further once their partner's mind was made up. Unfortunately, due to the emotional pressure during those moments the underlying processes do not get examined as they should be, otherwise the inherent impossibility of a rational assessment by the convinced mind would become apparent (and if men would only bother to look deeper, the marriage rate would drop even further).

In terms of the biologically determined mental arrangement towards one's children the scenario in schools is the same. The female sees herself as a mother protecting the young (they are her charges and they are in a state of dependency), the needs of the young are defined by the mother (the young do not know anything and the only source of knowledge is the carer), and anything or anybody intruding into this relationship is perceived as a threat (the role of nurturing is the sole task of the mother and so any intruder must always be suspicious). The cognitive dynamics are—as usual—expressed in functional terms, with the actual content coming a second. Therefore a situation in which the above is fulfilled will satisfy the instinct regardless of what a more detached assessment would observe. While for women the enacted function of carer is deeply enriching (after all, there is nothing in their minds which articulates otherwise), for the recipient the picture might be somewhat different.

For example, there are regular events held in swimming pools around Australia at which mothers take their infants and float them in the water with the intention to instill a love for swimming. In many cases the experience would be the fun envisaged, but for some toddlers the gravity-less state in a medium that offers no firm reference left right up or down would be harrowing. For the women the rather immobile facial expressions on their offspring do not tell the full story; not everybody articulates discomfort, even fear, through signs that are understood as such. In some humans the face freezes; what hides behind is not nonchalance, it is terror. However, to a mind convinced of the dictum 'mother knows best' this is neither here nor there.

On a somewhat different note but relevant nevertheless comes the study that showed 26.5 percent of women expressing "negative or neutral feelings" about childbirth, and 31.5 percent said "the birth experience was below their expectations" [151]. The researchers said, "More than one in five children has a principle caregiver who may be depressed for significant periods of their first year of life". And "Given the critical nature of the early-attachment period, this has major public health implications". Public perceptions of the 'mother' are not necessarily a true reflection of the reality.

When the same presumptuous dynamic gets extended into the wider arena the disruption escalates in scale as well. At least in marriage the damage is more or less contained—as Barbra Streisand once said, "Why does a woman work ten years to change a man's habits and then complain that he's not the man she married?" [152].

Erica McWilliam warned that "Australia had a new form of 'terrorism' which targeted white, middle-class men" [153]. "The male figure of authority", she said, "is the one person who can't claim to be the victim in any way". Consider this statement against the background of the dynamics presented above. If the relevant domain needs the concept of 'victim' as the reference against which an association with other TSs can be judged, it follows that the persona where such a description is absent becomes excluded by definition. For the entire edifice to remain valid the players need to be linked with the appropriate characteristics: women/mothers √ carers, students/children √ victims, authority figures/fathers √ perpetrators. Take any element out of the semantic circle and the structure looses the validity for the remainder. Hence feminism's strenuous efforts to maintain the setup.

McWilliam adds,

> "We are hearing now where there are fathers who are not even game to wipe the bottoms of their children because of the threat of being regarded as a harasser, and frightened of being thought abusive."

The endeavour towards the negative categorisation of males cannot be seen as part of an attempt to prevent the mistreatment of children. If that was so, a sequential approach from observation to assessment to interaction by the outsider would be in place. Instead it is the other way around. In August 1992 there appeared the following advertisement in the 'Positions Wanted' columns of the Sydney Morning Herald:

> PREVENTION PUPPETEER
> The Prevention Programme team Dympna House incest counselling and resource centre requires a woman puppeteer to work part-time. Using life-like puppets the show 'I Know That Now!' is performed in schools and to community groups. Knowledge of child sexual assault issues, *commitment to feminism* and a driver's licence essential. Counselling and performing skills preferred. Some training provided... [154] [My italics]

Again we need to remember the historical background of Judeao-Christian antagonism towards erotic behaviour in general. Since eros is mostly rejected in the first place, any instance comes automatically under the classification of 'evil'.

Quite recently a couple from the Sunshine Coast (we have come across that demographic before) withdrew their 15 year-old son from high school because they objected to the "smutty" books made available as a prescribed English literature text [155]. The example from the book given in the article involved a mother teaching her 8 year-old daughter about the clitoris and how touching it feels good. Outside the religious ambience the only question would be why someone eight years old would still be unfamiliar with her very own territory.

On the other hand, note the context. If puritan parents are upset about a mother talking to her daughter that way, how readily would the outcry have been passed on to the feminist league had the parent be the father (or any male for that matter)?

The semantic circle needed for maintaining the myth requires the victim to be firmly associated with 'innocence'. It represents another element essential for the dynamic to work. Given the layer-by-layer emergence of complex systems one can differentiate between the abstractive reach of functionalities connecting with a content that is relevant to the degree of complexity at any particular stratum. Although functionalities can also abstract into higher regions, their original sources do not disappear unless there is a breakdown of the system's resource space (see chapters 14 and 15). Hence the articulation of eros does not suddenly appear out of nowhere; it evolves from simple beginnings and adds to its sophistication as abstractive clusters are being formed. In society (at whatever scale, from the group to the culture at large) the feedback process includes the external source, and so an articulation on one level can interact with the conceptualisation (and eventually the articulation) coming from a more complex domain. As the latter is subject to its own internal organisation, the perspective from the outside does not inherently match the content found at the lower level. In the case of the feminist view the erotic expression of a child (such as it is) becomes interpreted within the framework of the ideology. The columnist Sandra McLean for example describes the eroticism of a child image at an exhibition as possibly "troubling" [156]. What is "troubling" here is not the stance of the child but the interpretation by the viewer.

Recall our discussion of the phenomenon we labeled conceptual intersection in chapter 16. The circle consisting of women √ children √ perpetrators rejects the other half of the equation. While the high-abstract complex on the side of the adult is recognised (albeit subjectively), the low-abstract complex on the child's behalf is not given its due. That is to say, for an adult the 'eroticism' may entail a far-ranging set of notions derived from their situatedness in the culture; for the child it might be nothing more than a literal gesture. Children do not exist in a different world—the world exists differently in them.

The need to categorise the fluidity of human behaviour into convenient conceptualisations leads to the definitions under the age-of-consent laws. Although they do not focus entirely on sexuality (social and economic considerations come into play as well), the moralistic component is arguably the strongest. A list of these laws worldwide shows their average to be 16 (relating to the social etc angle), but the differences would be due to the moral perception found in the respective jurisdictions [157]. The lowest figure is 12 in the Netherlands, the highest of 21 is shared by Trinidad & Tobago for male-female and Western Australia for the male-male version, followed closely by Tunisia where the minimum age for male-female sex is 20. There are a number of countries which declared male-male and female-female relationships illegal altogether (notably many American states in Western society). If in Australia for instance the 'innocence of youth' blooms until the age of 15 years and 11 months, suddenly to turn the teenager into a sexual being a few days later, how can this view be reconciled with a region such as Canada or Germany where the critical moment happens at age fourteen?

As can be expected, the strict segmentation into no/yes eroticism within days (hours? minutes?) is overlooked when the participants are exclusively female. Two years ago several feminist spirituality groups were in the news because of the unusual practises they advocated [158]. One of their publications, titled *Goddess Spirituality Handbook* and written by "wild Welsh witchwomon Ffiona Morgan who loves the Goddess and the earth with a passion" (note the aversion towards anything remotely male in the word 'witchwomon') has several lesbian stories. In one a girl of 13 has her first period. Her menstrual blood is collected in a clay bowl and "my 13 mothers stepped forward, each dipped her fingers in my blood and painted my naked body". As I said, the human eros is a wide field, but what would happen if 13 fathers were to try their artistic talents on a naked boy using his first seminal emission?

The hypocritical differentiation does not only happen in feminism of course. First and foremost the principle belongs under the auspices of ideology. In an article discussing Rodin's sculptures the Courier Mail featured a sample of his work in bronze, two naked girls performing a dance routine (one assumes), seen from the front. According to religious wisdom the sight of a naked human body is fraught with danger, a fear which carries some weight considering the artificial atmosphere church members are brought up in. A repressed mind will not digest bare skin easily, and even an uncovered arm can evoke all kinds of disturbances. For this reason conservative jurisdictions prohibit nudity outright, and the reader's letter pages of daily newspapers carry regular submissions from parents who are concerned, disgusted, or outraged by the possibility of their children coming face to face with a naked version of their own species. Entire legislations are being enacted and software companies make considerable amounts of money trying to curtail the availability of flesh without clothes. The more difficult a comparison between the artificial and the real can be made, the easier it is for the former to slip by the censors. Of course, advanced populations make use of nudist facilities where young and old mix without the presumed devastation to their minds. Every year during the holidays these destinations are frequented by the millions. Just as the geographic distances keep the two demographics apart, so does the logic boundary keep unwanted information away from its central domains. Fig. 17-8 shows a juxtaposition of Rodin's sculpture with an example from the real world. The reader might try to guess which of the two has it easier.

As we have seen, compartmentalising a behaviour type that is intrinsically fluctuating between one border and another leads to the allocating of manifestations to pre-defined sections. The allocation is performed under the regimen of the controller and does not necessarily synchronise with the intent of the actor. When fathers are too afraid "to wipe the bottoms of their children" the problem lies precisely with the wilful interpretation of the scenario by an outside authority. Given the full spectrum of erotic elements, it becomes impossible to decide where the problematic behaviour category begins and where it ends; not only are the delimiters defined by distance, the criticality of the definition pressures the participant to artificially create a margin for error just to be on the safe side. Since the original biological/cognitive dynamics are not designed with such limits in place, it becomes a matter of purposefully imposing these margins, even if they are in opposition to the natural framework.

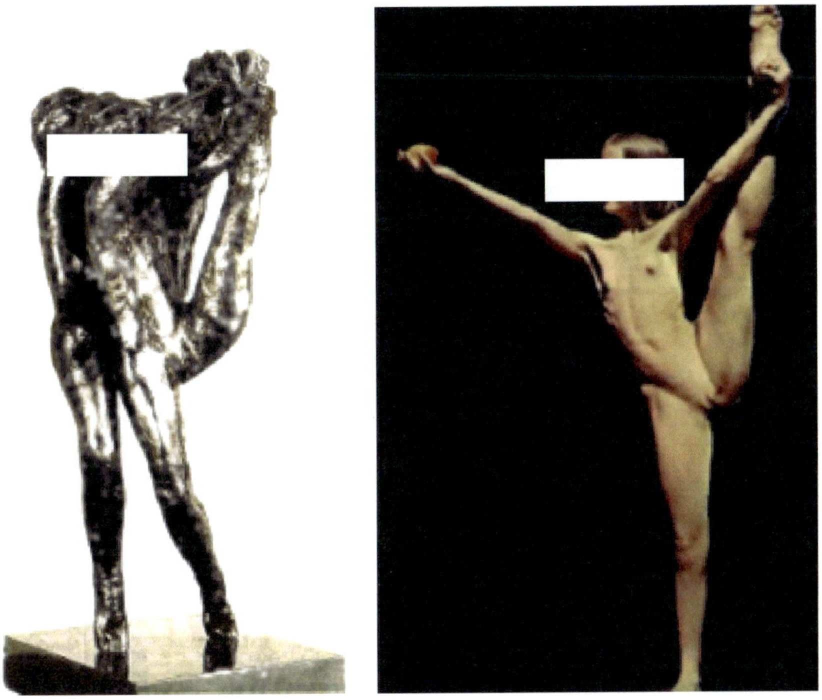

Fig. 17-8: Two images of young dancers: one by Rodin - one by nature; one of metal - one of flesh and blood; one artificial - one real. Which one is in order and which one is not? [159a,b]. (Naturally I have blanked out their faces to protect their 'modesty')

Examples are easy to find. In erotically constrained environments boys will engage in physical contact with each other, but the closer the contact the more the acts are couched in aggressive mannerisms. At that point both boys can be sure their behaviour won't be interpreted 'wrongly'. The downside is the generation of patterns that are meant to symbolise antagonism rather than the intimacy present in the subconscious source. It should come as no surprise that the same difficulty does not exist for girls (at least not to the same degree)—while homosexuality in males had been persecuted for ages, there was a time when lesbianism was not even considered. Such outside pressure sets up problematic dependencies. Suppose a person grows up under the constant need to rationalise 'erotic expression' through 'violence'. What are the chances that later in life he or she will have lost the ability to express intimacy through anything *but* aggression?

The ambiguous classification of fundamentally indefinable indices on the scale of erotic behaviour makes a mockery of the law designed to administer such moralism. If it is illegal for the parent to 'molest' their child, what does it actually mean—is touching 5cm away from the penis or vagina allowed, or is it 4 or 6 or 10? Does the word 'masturbation' constitute harassment or does 'naked' or 'intercourse'? Is it alright to enter the bathroom while the daughter has a shower, or does one have to wait until she is fully clothed or can they all sit naked on the couch together on a hot day? Unless the powers that be issue detailed charts with all the important markings in place, how can an accusation be dealt with under forensic examination? If all this seems facetious, perhaps we should concentrate on the state of the perpetrator's *and* accuser's minds at the time and subject them to an objective psychological analysis; looking at the profession's track record the option is not encouraging.

One such case of amorphous evidence involved the trial of a father whose daughter accused him of raping her when she was ten years of age [160]. The claim was made twenty tears later. By sheer coincidence his barrister spotted an entry in the medical records of the daughter which were made upon her presentation two years after the alleged incident on an unrelated matter. Since they also showed she was still a virgin the matter was eventually dismissed. The case was closed as far as the courts were concerned, but the experience proved so traumatic for the father he attempted suicide.

All this is not to suggest that aggression does not occur. The problem lies with the definition of an event for which only posteriori evidence exist. An injury is readily observed and classified but when it comes to erotic behaviour, especially in the context of an overall negativity, the interpretation relies on the subjective perception of the active and the passive participant as well as the opinion of the examiner. Even the concept of 'active' and 'passive' becomes ambiguous.

In an article dealing with the clinical approach to "children with sexually abusive behavior problems" Lucinda Rasmussen writes of "two symptoms that were seen more frequently in sexually abused children: sexualized behaviors and symptoms of posttraumatic stress disorder" [161]. Many adult sexual offenders had "started their deviant sexual histories as early as latency". Listing some descriptors of what constitutes abusive behaviour (eg, "abuse-reactive", "sexually reactive", "sexually aggressive", "children who molest", "sexually abusive children") she adds, "This plethora of terms illustrates the disparity of opinions among child sexual abuse practitioners and researchers concerning sexual behavior problems in children" [162]. While she contends that children often "use games, tricks, or other manipulative behaviors" which on their own should not be classified as abusive, behaviour "can only be considered abusive when differences in power ... or intimidation using an assigned authority role ... are combined with manipulation ... or coercion..." [163]. One treatment model to deal with behaviour that has been identified along those terms (called "The Sexual Abuse Cycle") is introduced as "Although it was not empirically derived and it has not been empirically validated, it is widely used in clinical practice with all types of juvenile sex offenders" [164].

The above is representative of an indistinct model which is designed to offer considerable opportunity for ideological slant. First note the circularity in the descriptors: 'one who molests' is 'abusive', 'abusive' is being 'sexually aggressive', and 'sexually aggressive' means 'to molest' (one can equally try any other combination). A classification of an offender entails a "deviant sexual history" in the past, defined of course in terms of the very same criteria that are supposed to be a consequence of the former; a process that does not condemn through the existence of an act but by adjusting that act's definition. And finally, there are treatments which are being applied to any child having been caught that are not based on firm knowledge, whose efficacy has not been tested, but are "widely used" nevertheless.

The notion of 'sexualised behaviour' occurs over and over again in the identification of what is perceived as problematic. It gets placed within the general expectation of articulating something but without any erotic undertones. If there *are* any then the context is deemed to be negative. Given that erotic behaviour is based on an inherent dynamic (in the physical as well as the mental arena) then its attenuation implies an outside force constraining the original state. If one sees the absence of an otherwise fundamental dynamic as a severe form of mental disposition, then this conceptualisation of the problem can be expressed as the desire to enforce a kind of behaviour which is as psychotic as that of its judges. In other words, the application of the age-of-consent regimen is a way of ensuring that no human being should be exposed to eros unless and until their own psychological state has synchronised with the kind of psychosis existent in general society.

An important factor in identifying an event as being relevant to the concern (subjective or otherwise) is the re-representation of the past. The book *Child Sexual Abuse and False Memory Syndrome* contains a selection of articles which deal with the problem of verifying recalled events and placing them in the context of a current focus [165]. In essence they add nothing new to what we have not already encountered in previous chapters. First and foremost a 'memory' is not a stored datum like a word written on a piece of paper or a series of magnetic alignments on a computer disk. It exists in the form of a stable region in terms of an attractor domain and its content is not defined through a discrete set of symbols but via the feedback mechanism amongst the evoked affinity relationships belonging to the abstracts. Therefore, if the data triggering a recall are completely affinitive with the accessed region, what is being recalled could be regarded as reliable in terms of the triggering information; if there is no complete affinity what emerges is an amalgam of more or less related and unrelated material.

Thus in the current context the following broad possibilities exist: an event A took place, and there is a recall-prompt in the context of A which produces A; an event A took place, and there is a recall-prompt in a context B producing A + B; an event B took place but not A, there is a recall-prompt in the context of A producing A + B; an event B took place but not A, there is a recall-prompt in the context of A producing A. Obviously, a scene to be recalled with lots of detail offers the potential for a large combinatorial variety of the above. Note that we are talking about a healthy system; swayed maybe, but healthy. In addition there is a medical condition called "factitious disorder" in which people "fake an illness in their children and would seek medical attention for them to gain attention or sympathy for themselves" [166]. It would be a brave doctor who offered such a diagnosis in the face of sexual abuse allegations. Even without the ideological pressure a proper analysis of something that happened in the past and has been observed through the eyes of a child would be precarious at best; within the hothouse of feminist belligerence, religious hysteria, and last but not least the dumbfounded mind of the child itself, what hope is there of achieving anything that still makes sense?

The re-representation of the past needs the material of the present. In essence one can observe the phenomenon every time a certain fad makes its way through society. The world views of its followers become imbued with its elements and one's life is interpreted anew—an impartial observer is left wondering how they could ever do without. Of course they didn't, except the 'without' was something else then. The cognitive dynamic is the same, whether the content happens to be a pop song or a belief in witches. In other words, the spreading of an idea engenders the multiplication not only in terms of received samples, it becomes actively nurtured as if its new owners had carried it in their minds all along.

Consider then the dissemination of erotic content, generally through the media or the Internet, but also more specifically through articles which are meant to portray the evil of paedophilia. From the 18 August until the 18 September 1997 The Courier Mail managed to splash one catchy headline after the other on its front page, the only interruption being Princess Diana's death which gave the readers some pause between the 1 and 11 September inclusive [167]. Otherwise for seventeen days it went from *Eavesdrop scheme to fight child molesters* to *Predator - The mind of the paedophile* to *Kids offer themselves on 'the rack' for sale* to *Brisbane video company centre of paedophile porn industry*. In October 2000 young and old could avail themselves of lurid details in the form of a reported court transcript, suitably labeled: “WARNING: The following court extract contains explicit language and may offend some readers” [168]. Naturally, those who did and those who did not get offended had their spines tingled, only for different (?) reasons. In the text a witness who claimed to have been sexually abused by the defendant described what had happened on those occasions, and the evidence was about such things as him rubbing himself against her, exposing himself, and whether or not her body was moving up and down his erect penis.

One might like to observe the following items from the religious/feminist atmosphere and then contemplate the possible fallout should something similar occur in any other circumstances of our modern world:

there is a set of actions which have been branded ‘evil’ without the slightest scientific proof (in fact whatever proper analysis there is demonstrates the exact opposite);

based on such beliefs a loose framework of categories is put together in order to define transgressions;

staffers in any organisation in which person-to-person conduct requires a reliable framework are faced with a daily obstacle course of potential catastrophes; the result is a below-standard work and sometimes unbearable pressures on the personnel as they worry about their past, present, and future;

there is an unwritten law that demands from every reporter, columnist, or commentator to spike their sentences with a sufficient number of unequivocally negative adjectives whenever they happen to refer to the substance of those beliefs and categories—but even a murder is just a ‘murder’;

a whole set of regulations exists which reflects those beliefs and categories;

offenders can be classified ambiguously through the above in terms of definition, through the indication by a single child of no effective minimum age in terms of occurrence, and through the application of suspect psychological techniques which actively interfere with the memory of key witnesses yet with the contents of these memories elevated to forensic material;

offenders (no effective minimum age again) are treated through programmes based on no empirical evidence and without a proven track record;

pictorial depictions of erotic activity are banned outright, whereas in any other area the authorities have no problem using images to highlight possible dangers even if the pictures should be gruesome, and when it comes to the Internet there exists no global hunt for websites showing images of murder, death, or terrorism;

anywhere else the access to data is welcomed if it helps to further instruct about some problem, but here despite the asserted negativity the possession of actual information is in itself deemed illegal;

usually an element of enjoyment on behalf of a defendant constitutes a mitigating circumstance but in this type of crime it actually adds to the condemnation, because for the puritan a victim must weep, not smile.

The idea does suggest itself that the above measures are not designed to protect the alleged victims but rather are a frantic attempt to keep the ideological barriers around the activity intact. One could add the virtual paralysis by people who should be expected to show a certain degree of rationality, given their professional status and their significant position in society. On the other hand the silence is understandable. Unless they operate under the mantle of ideology anyway would they not be concerned about the response by the public, the media, and perhaps most of all, their own family? How *could* they dare if one word, one hint by their wife or partner or especially their child could wreck their lives forever. The stricture can be applied to family life in general. What chance is there for a male to raise any objection lest the young goddess should point her finger at him at any time later and summon the demons at her beck and call?

The constant reference to the trauma induced in alleged victims is easily linked to the perceived context because the societal surrounds make it so. While the general public may not be expected to evaluate the given approach too critically, for the professional community the error ought to be obvious. Suppose I make the following claim: "Any chicken will die within 5 minutes if placed on a green floor!". As proof I place a container with a green bottom before the audience and holding up a stopwatch I call out "Now!" while at the same time dropping a chicken into the enclosure. Sure enough, when the 5 minutes are up the chicken lies on the floor, dead. Have I proved my original claim? Actually, no. A closer inspection reveals that the floor of the cage had been extremely hot, and so it was not the green colour which caused the chicken to die but the application of deadly heat.

Mothers do it all the time. If her child should eat the food but doesn't want to, she articulates the positive through sounds and facial expressions; likewise, if the child wants to eat something but shouldn't, the negativity is expressed in a similarly effective manner. It is also no secret that in the case of an accident the worst thing that can happen is a hysterical mother who makes the situation worse. Combine overly intense emotions with the incidence of single mother households and we get an atmosphere in which ameliorating influences are nonexistent. Since incoming information is now processed in terms of incongruent PAT chains, the ability of the growing individual to rationally assess his or her environment will be reduced to emotion-based reflexes—a recipe for depression. And as we have seen, the World Health Organisation has been able to identify an ever earlier onset of the relevant symptoms.

In principle the argument regarding the generation of trauma in children who have been exposed to erotic behaviour runs along the same lines. The trauma has not been caused by the incident but by the wider ambience of fear and rejection which had been in existence all along, just like the floor had been the real threat to the chicken; any reference to the colour would have been my own invention if I dislike green. In contrast the trauma-inducing nature of sexual repression is well documented, even leading to suicide (homosexuals for instance will know about this). Scientists in any other field would loose their credibility if proffering a deduction of the kind just shown. But a profession which readily agreed to mash the brain of gays coupled with an ideology that celebrates "nonrational knowledge" couched in the "tentative and fuzzy nature of all our linguistic formulations" so that a woman may have "retold and reinterpreted her story several times over a few years to a psychological researcher", these kind of people know how to rule.

Within a woman's circle however the signs are turned into the positive. Discussing the benefits of breastfeeding, Sarah Buckley, a Brisbane physician, criticises society's attitude towards the female breast and lists a number of diseases babies are subjected to if not breastfed [169]. Then she mentions the benefits for the mother:

> For all mammals, these include the effects of the breastfeeding hormones oxytocin, the hormone of love; endorphins, hormones of pleasure; and prolactin, the mothering hormone.
>
> These hormones keep women "calm, relaxed, and lovingly focused" on their babies. And further,
>
> The act of breastfeeding gives our babies holding, touch, skin-to-skin contact, visual, oral and tactile stimulation, and deep-seated security.

In the absence of ideological blinkers this description would instantly be recognised as nothing else but erotic pleasure. The opportunity women have to physically interact with their children and those of others even when they are older is situated within the self-same emotional complex. It may not be articulated as such (and in the current climate it certainly won't be), but the combined biological and cognitive dynamic played out through so many instances represents an erotic fulfillment. In other words, whereas the expression of male sexuality towards children is persecuted by all and sundry, for the female and her mode there are almost no such impediments.

And then there is the money. A private school in Brisbane was sued for $1.5 million by a former pupil who claims he was repeatedly sexually assaulted [170]. The same school has been hit with a similar claim for $1.6 million, and had already paid out $834,000 in civil damages to another woman [171]. One female sued the Anglican Diocese of Brisbane for $459,000, citing "mental anguish, depression, chronic dysthymic disorder, nightmares, loss of interest in living, humiliation and feelings of guilt over the suicide" of the alleged perpetrator after she had reported the incidents to police [172]. A class action based on alleged sexual abuse involved $18 million [173].

As a comparison I present some figures from the *WorkCover Benefits Guide*, a document outlining the provisions under the New South Wales Workers Compensation Act [174]. For a mere worker the maximum one-time payment for "loss of sight of both eyes" is $100,000; for "loss of right arm at or above elbow" - $80,000; "loss of either leg at or above the knee" - $75,000; "HIV infection or AIDS" - $100,000.

As for the physical damage suffered by the children through erotic activities and their presumed inability to make any decision about such matters, here are some statistics supplied by Queensland hospital emergency departments between 1998 and 1999 [175]. Of all sports injuries the proportion of 5 to 14-year olds in Rugby League was 36.8%, in soccer 44.8%, in Australian Football League 27.7%, in gymnastics 80.2%, and for all sport types it was 33.6%. Applying the same logic to sport, children as young as five should never be able to voice any desire or inclination to participate in such activities. One could go as far as to say, if our culture placed as much status on eros as it does on playing football, even a torn vagina in an eight-year old girl would be a source of pride for her in the classroom and her friends would sign their autographs on her underpants. As it is, to say 'sex is good' is like telling an anorexia sufferer that 'food is good'.

So, if the identity container exacts its denominative influence on imported abstractive complexes, and the resultant PAT chains cause misery, fear, wasted lives and massive upheavals in social and financial terms, what would happen if the core domain did not exhibit those qualities? The simple answer is, nothing. In other words, in demographics that do not feature moralistic animosity all those problems do not exist—end of story. Just as amongst people who do not believe in demons no child and no adult is troubled by nightmares about incubi and succubi—but when they did they had the courts to back them up.

Combining the incidence of an underlying affinity when two domains are separated by nothing more than a 'not', with another of the stymied drive towards eros diverted into belligerence, we arrive at a relatively stable environment for the current fury. As fanatical moralists search for samples to incite their hatred (an erotic fulfillment in disguise), so can their modern companions indulge in contexts where the combination of 'child' and 'sex' evoke a quite legal titillation. Stoking the fires is a publicity which virtually ensures that no boy or girl can walk the streets without trailing the semantic plume of deviance. Erotic beings they were naturally; now they are a link to the obscene. The mantra "In the best interest of the child" has turned into a tyrant; and, like in any true dictatorship, one finds hardly anything as vicious as the antics of sycophants who vie for the grace of their master.

Add the growing emotional weight of smaller child-guardian relationships due to general social changes (ie, in a larger family there is less emotional investment per member than in a smaller unit), then particularly in a single mother-daughter context the inherent problems of age difference and perceived competition in terms of erotic appeal come to the fore. Often, the body languages of mother and daughter suggest that the mother needs her daughter more than the daughter needs her. An intrusion can incite the older woman severely, as we have observed in the previous section. The child is a cog wheel in her machinery of instinct.

The traditional moralistic force that had demonised eros suffered a steady retreat in liberal humanist democracies. It seems all its energy is concentrated on the remaining area. Today even primary school children enjoy sex education where such things as masturbation or the taste of semen are discussed. Unfortunately, sometimes teachers have to tread an often invisible line between the rational attitude towards being human and bowing to the law. What does an educator in Brisbane for example do when asked intelligent questions by 10-year olds, but at the same time opens the daily paper and reads, “SEX: It is illegal for any person to have sex with any other person who is under 16 years of age. It does not matter if both people are under 16 years, or if they both agree to have sex” [176]. The foregoing appeared during June 2003 in the series *Headst@rt*, a periodical in the Courier Mail featuring issues of interest to young people. On that same note, what does the 10-year old make of it?

Suppose an outside analyst composes a summary of the West’s attitude towards the human eros. In tabular form it would look something like this:

Eros type	History	Effect in society
Applied to self.	Response based on an originally imported belief system. Considered deviant and classified as ‘self-abuse’. Attacked through physical and psychological means. Evolution of intellect abolishes official belligerence, although residual antagonism remains.	Antagonism produces sometimes severe trauma in individuals. Extends to produce a repressed society.
Applied to same gender.	Response based on an originally imported belief system. Considered deviant. Attacked through physical and psychological means. Evolution of intellect abolishes official belligerence, although residual antagonism remains.	Antagonism produces sometimes severe trauma in individuals and those linked to them. Value of individual to society is suppressed in favour of recrimination and fear.
Applied to other gender.	Response based on an originally imported belief system. Considered deviant, except in narrowly defined circumstances derived from the belief system. Cases outside the exceptions attacked through physical and psychological means. Evolution of intellect abolishes official belligerence, although residual antagonism remains.	Antagonism produces sometimes severe trauma in individuals and those linked to them. Value of individual to society is suppressed in favour of recrimination and fear.
Applied to someone deemed ‘older’ than the traditional standard.	Response based on an originally imported belief system. Considered deviant, except in narrowly defined circumstances	Antagonism produces sometimes severe trauma in individuals and those linked to them.

	derived from the belief system. Cases outside the exceptions attacked through physical and psychological means. Evolution of intellect abolishes official belligerence, although residual antagonism remains.	Value of individual to society is suppressed in favour of recrimination and fear.
Applied to someone deemed 'younger' than the traditional standard.	Response based on an originally imported belief system. Considered deviant and classified as 'child-abuse'. Attacked through physical and psychological means. Outcome pending.	Antagonism produces sometimes severe trauma in individuals and those linked to them. Value of individual to society is suppressed in favour of recrimination and fear. Irrational reallocation of energies.

What is the common denominator in the above? If the answer is "violence and suppression" this only addresses the symptoms. As we have seen in our discussions about default abstracts, their constant presence exerts an ongoing influence over our daily affairs which in turn become organised according to environmental and cultural demands. We have also observed how the sheer situatedness of a conceptual complex triggers the emergence of a response which too becomes part of the whole. In terms of the entire system therefore we can consider these dynamics as equal participants—to argue which caused what is now meaningless. When it comes to erotic behaviour its necessity for the species' survival juxtaposes the repressive format in which it is acted out. The two sides combine to form a system that forces both to function under pressure. Recall that pressure (under whatever guise) tends to compact its domain to simplicities and essentials, that is to say links to wider references become suppressed. In this case a repressed sexuality will articulate in a basic format, even if the conditions are anywhere near ideal; its counterpart imposes a similarly rigid ambience to uphold such format; and one supports the other.

The advantage of that symbiosis is the energy it supplies to ensure the production of children and at the same time the maintaining of a societal structure under adverse conditions. The disadvantage is the creation of sharply defined domains, accentuated by their extreme nature, thereby providing a harbour for idiosyncrasies and exclusiveness. Examples can be seen anywhere in the world: the more rigid and discriminating the sexual mores, the more regulated the family, and the greater the antagonism towards the other (where 'the other' can mean anything from erotic behaviour to social customs to religious perceptions). In some demographics a young girl will be whipped if she happens to show her bare knees while asleep, and even a perceived sexual encounter can be a death sentence. Familial relationships can be such that the accidental killing of a stranger can be worked out amicably, but damage the bicycle of a family member and it gets problematic. And antagonism towards religious differences stretches from giving them the cold shoulder to passionate murder.

In a highly complex and extensive resource space (such as the West today) these are remnants of a violent past. Religion (a product of those days), and now feminism (through its desire to gain ownership of the human eros), contribute to the upholding of the pressure. They draw out the worst of both worlds. On one hand there is the demand to adhere to strictly configured sexual behaviour, but the safeguards to provide the necessary structure are dismantled; on one hand there is the need for males to interact with their children, but the opportunities are becoming rarer and rarer. Since nature still supplies the default abstracts, something has to give. Our e-chart illustrates the two possibilities open to us: either diminish the complexity, effectively reducing the resource space; or dismantle the reasons for the progression lock. It poses a fundamental challenge to the West, a test which will decide our future.

Ideologues do not understand the sheer unreal-ness of any information locked inside glass vitrines, and the more comprehensive the information is the odder it gets. Given the penetration of data regarding drugs and sex and their social and legal backgrounds in today's context, Western moralism is educating its children to consider themselves criminals by the time they can read and write. Before they even leave school they are thoroughly accustomed to how it feels when doing something illegal is only a matter of balancing one's daily priorities. The implications for wider society are profound, since it injects an element of independence into its atmosphere affecting the bonds which so far have contributed to its very concept.

The resource space of the mind, projecting into society the re-represented form of its contents, to be reflected back into its network, a complexity raised or lowered, of clusters assembled or dissolved; and the products so derived the constructors of the system's future—it will be our final theme.

18. PATn±

Futures

We have arrived at the vastness of the System itself—from the dynamics of protein formations, to the neuronal clusters of thought structures, to affinity-based conceptualisations of the outside, to the world at large which mirrors those self-same dynamics at its own scale of complexity. Up here the relationships, chains and domains form too, and different levels of phase states interact with their higher and lower neighbours. Our e-chart informs us about its resource space, and the affinity potentials of its elements manifest their optionality posited against the characteristics within the respective product-latency-applicability framework. Congruent and incongruent PAT chains loop through the formations as entire complexes composed of their inherent members. An illusion tree which draws the content from its base to a higher-level network deals with abstracts when observed in a mind, a filigree of associative subdomains; now, at the dynamic peak, those abstracts have become entire systems in their own right, including horizontal reaches as well as vertical constructs ranging from global edifices right down to individual minds once more. Denominative patterns wash like tides between them, drawing and depositing their contents, all the while affecting the functionality of such material. And the system's container, the comprehensive identity which gives the whole a stature, exerts its influence over the entire resource space.

Fig. 18-1 is an illustration of how that space can be conceptualised.

Fig. 18-1: A large-scale resource space showing three subsystems at different complexity levels, each featuring its own set of functional elements and communicating with the rest. In the subsystems conscious structures are light blue/white, the rest belong to the subconscious. It is a greatly reduced representation of the real world.

The objects familiar from previous chapters are grouped together in subsystems with their contingent of containers, illusion trees, abstracts, and perhaps logic boundaries placed at different levels of complexity. Hence lower structures are smaller per se, whereas those higher up in the resource space can make use of their increased probability of association and tend to be larger. Due to their interdependent situatedness they are able to communicate with each other in terms of their respective associative potentials and hence their realised affinity relationships. What in the mind would be referred to as conscious or subconscious exists here in the form of recognised or unrecognised manifestations. Both configure our world, but only the former find their way into the wilful articulations of its members.

The image is but a faint approximation of what life is really like. Because the affinities, abstractive qualities, and e-chart resources are functional properties, the thus affected content becomes a matter of contemporary circumstances. Therefore the identification of future developments is not so much a matter of predicting the shape and form of objects, but one of describing the possible unfolding of their current potential. Under this perspective let us examine the bigger picture.

It starts with the transfer of information. In the neuronal organisation the synaptic transmission affects the neighbouring dendrites, giving rise to functional clusters. If a cluster A receives an additional input it will be reconfigured to a certain extent, but for some cluster B the result will only then become significant if a connectivity between B and A can be realised. Similarly, a thought structure A can be modified by a further addition but for another structure B the event will only become meaningful if this results in an affinity relationship between the two. In other words, while the lower-scale element does produce an effect somewhere else, that same element from A is meaningless in a direct relation to B.

The same goes for high-scale systems. In a technologically advanced society, given its inherent complexity in terms of the general infrastructure, there exist a large number of subsystems which interact to varying degree. They can do so because the overall system contains PAT chains acting as pathways of transmittable content. What such content means by the time it gets to its destination depends as much on its inherent quality as it does on the affinity potential offered there; that is, on the ability of those members (the subsystem's elements) to relate. However, the generation of the data at the source has no direct bearing on the status of their target. For example, if I am involved in a serious car accident the first point of contact would be the police. The report written by the officers then becomes the material for the insurance company. It would be rather unproductive to walk straight to the insurance desk and tell them about the event (if the accident was not serious I could go there directly but in this case the complexity of the event is lower and so the transposition would not have been necessary). In a similar vein a public policy can undergo modifications through the process of grass-roots initiatives, the involvement of local members, a report tabled in parliament and, after the issue has passed all such checks and balances, the creation of a new governmental Act.

Leave out any one of these steps and the process becomes precarious. Hence with corrupt governments the problem is not only the receiving of wilful moneys here or there; the shortcuts taken between some initiator and the implementing authority deprives the overall system of its capacity to interact at its various levels, making the information hidden from the wider reference. We can say a corrupt system deprives itself of information.

In an 'ideal' world all the PAT chains would be congruent, there would be no illusion trees, and logic boundaries would not exist. Since ideals cannot be implemented (see the section on religion in the previous chapter) we have to be content with the alternative. Even with congruent PAT chains the transfer of information can be likened to the articulation of specific scripts between two people; there sub-TSs formulate a string of voiced con-TSs, and the latter are communicated to the recipient. But the recipient interprets the data according to *their* network of sub-TSs and these are different from the originators. If the con-TSs contain enough semantic markers to catch inappropriate contexts, a meaningful communication is possible; if not, the message will be received but encourages conflict. Should the PAT chain be incongruent anyway, the message itself is misunderstood.

Furthermore, if the transmission occurs between clusters of varying complexity, we are faced with a conceptual intersection in which elements of the content get filtered out if the receiving end happens to be of lower complexity, and elements of the wider context get added to the message if the recipient is at a higher level. In terms of the wider resource space we can see therefore that a subsystem's higher complexity comes at a cost: while the e-chart parameters are richer and therefore heighten the subsystem's optionality, when it comes to its relationship with a lower-level neighbour such capacity becomes reduced, and this in turn will be reflected in its overall ability to interact inside the shared resource space. Likewise, the lower-level subsystem may benefit from some other higher complexity, but the feedback process returns information too extensive for its own resources to handle.

Implicit in the above is a fundamental limit to the survival of a system. Since its resource space depends on the viability of its data transmissions, the effectiveness of such data is a function of their potential in terms of their current locales, ie domains. The greater the incongruence of this data set, the less likely its content can be productive across the entire resource space. Thus the existence of a highly complex system relies on the congruence of the information it contains. In other words, systems built around illogical assumptions will only survive for as long as their functional reach does not come up against the dynamics of their larger environment, demanding a more logical response. Hence the larger the system, the more inherently logical it will have to be in order to sustain itself. Just as tribal societies are restricted to a relatively small number of members, so will the introduction and /or the preservation of irrational behaviour cause the eventual downfall of a growing society.

In a strictly technical sense our planet always has been one large resource space. The interconnectivity however which had emerged over the last hundred years or so filled that space with means to establish relations amongst its inhabitants to a degree that was simply impossible in the past. From tribes to city states to nations to integrated economies the subsystems have not only evolved internally but through their respective complexity formed links of communications which are now filling the global environment. The word 'links' should be understood in its widest sense; it involves text, commerce, politics, and the military. The resultant PAT chains weave amongst a vast variety of domains, from multifaceted networks to simple structures. Their contents contract, extend, become modified, reassigned, and applied. As the resource spaces of subsystems grow there is room for niches, and when before these merely represented an insignificant variance of objects they are now playing also a functional role affecting the status of their host.

The semantic weight of information reflects the atmosphere it inhabits. The resultant significance makes it relevant to contexts which otherwise would have been left aside. For example, decisions made in some boardroom of a larger corporation entail data about international relationships, upheavals on another continent, social pressures from some demographics, all in addition to the intrinsic concerns of the enterprise. The political and military buffers of old do not apply anymore. This does not mean they have disappeared, but their status has been relegated to merely another factor within the entire array of environmental contingencies.

In March 2003 the Shell oil company pulled out of four of its facilities in Nigeria because of the ethnic violence there [1]. Chevron Texaco has done so already. Closer to Australia the internal situation in the Solomons has become sufficiently unsettling for this country's

Foreign Minister to decide on a military intervention force in the interests of a "results oriented" regional security [2]. In his view multilateral organisations such as the United Nations were no longer able to resolve the problems posed by rogue states such as North Korea or collapsing societies such as the Solomons. From our perspective these are examples of conceptual intersections between a body such as the UN (representative of clusters of nations) and singular entities such as the two countries who function under the auspices of internal references outside those of the UN. Particularly in the Solomons the tribal conflict is indicative of interests which do not relate to the wider framework of an international standard, be that in terms of the economy, its statehood, or even interpersonal relationships.

Although the wealth of a nation also derives from its international trade, its internal system still directs the balance sheets and therefore its ability to spend. The association of Western countries with their cultural standards and hence their focus of interest compels them to view the disruptions in other demographics in an increasingly active manner. Yet the checks and balances in force inside Western borders do not reach into the developing regions, and even if they did the local context works against a similarly comprehensive implementation. The outcome is a dislocation of traditional bounds, interpreted from their perspective in terms of outside interference. At the same time the desire for stability and peace causes members of such populations to call for help from the very direction the higher complexity arrived in the first place, since it is the only effective assistance under the circumstances. The calls are responded to. But for the West these initiatives will ultimately become a question of productive returns.

The spending of hundreds of millions of dollars could be seen as an insurance against regional instability, and no doubt there are indices which delineate the comparable costs of national implosions vs a stable market. Such equations however rely on the existence of representative figures; that is to say, for this kind of analysis to make sense the data sources have to exist in the first place. At this stage the only reason the political concerns are being articulated is because a place like the Solomons is perceived as part of the same region Australia belongs to. Once that definition ceases to maintain its validity the equation itself becomes meaningless. In terms of our approach we can say that a neighbouring system will only be viewed as a *neighbouring* system for as long as it occupies the same resource space. Since 'resource space' is defined not only through geographical extent but also through its more abstract qualities such as commerce and political relevance (and given today's complexity the latter are far more important than the former), a subsystem's disappearance from its erstwhile resource space turns into a matter of definition, and on behalf of the more complex authority at that.

Once such an option becomes relevant, the erstwhile equation has changed into a different one: the loss of a subsystem on one hand vs the cost of protecting against the local ramifications on the other. As the higher complexity reaches a sufficient mass that enables it to function according to its own terms the lower counterpart becomes less and less important. The stage is set for abandonment.

In his speech to the IMF Conference on Second Generation Reforms in 1999 Francis Fukuyama uses the concept of 'social capital' to define the substance that drives modern economies. In his words,

> The definition I will use in this paper is: social capital is an instantiated informal norm that promotes cooperation between two or more individuals. The norms that constitute social capital can range from a norm of reciprocity between two friends, all the way up to complex and elaborately articulated doctrines like Christianity or Confucianism. They must be instantiated in an

> actual human relationship: the norm of reciprocity exists in potential in my dealings with all people, but is actualized only in my dealings with my friends. By this definition, trust, networks, civil society, and the like which have been associated with social capital are all epiphenominal, arising as a result of social capital but not constituting social capital itself. [3]

He goes on to describe the fundamental purpose to which it is put, that is the ability to cooperate amongst groups. He includes the "radius of trust", the extent to which the cooperation can be implemented in a productive way. The examples he provides (eg, religion, social standards, etc) are the producers of affinities we have come to understand. His juxtaposition of human relationships against society at large is not done without reason: without the capacity to transmit information from one individual to another the data cannot be distributed meaningfully across the wider resource space, and the optionality of the entire system will be compromised.

In order for the transmission to be positive the associativity of such information needs to be maintained within its reach of clusters, Fukuyama's "radius of trust". The concept does not imply the reciprocity the word 'trust' suggests however. Without the potential for affinity relationships to be built, no amount of wishful thinking will create the genial ambience for trust to emerge.

In his article *The Coming Anarchy* Robert Kaplan gives some examples of an ongoing deconstruction of societies [4]. Concentrating mainly on West Africa he writes about unlit streets, muggers roaming at will, governments ineffective at night, and the proliferating meaninglessness of national borders in the face of demographic rearrangements to suit their particular members. In comparable regions around the rest of the world the scenario is similar. Police forces that ought to enforce a common standard turn into private fiefdoms exacting tributes from the traveller, businesses rely on their own guards and networks, and the traditional, centralised power which so far has acted as a stabilising coercion (however idiosyncratic it may have seemed to Western eyes) is in effect just another collection of interests. For those with enough funds, on any given day the bank could be used to deposit one's money, a government office is visited to get a stamp on a licence, the local gang is organised for some debt collecting, friends are made to establish a venture and one's contact with the drug cartel takes care of the entertainment. Descending to lower income levels simply means adjusting those activities in size. And by the time the situation gets articulated at some trade conference foreign assistance will be adjusted according to the latest balance of trade deficits.

Decades ago the beginnings could already be identified. Take a developing country that is visited by tourists who for a couple of months fill the hotels, spread their tips and demonstrate a standard of living which to the locals is as enticing as it is obscure. Suppose there are some yachts (owned and run by foreigners) used by the visitors to take them around the place. A citizen of that country might contemplate running a similar business, and he organises a few others who can assist him. Since the money flow around the yachts is evident, as is the current exclusion of the local economy, the matter turns into something resembling national pride. Because of an obtuse bureaucracy and corruption the venture is not as straightforward as might be supposed. In the end a lot of people will be connected to that enterprise, all with their own power games, desires, and dreams (and money of course). Should an outsider get the opportunity to talk to some of the participants, what emerges is not just the sheer extent of involved relationships, but more often than not the perception of the outside world (which after all features significantly in the whole process), and it can be quite askew. Americans? They come from just a hundred miles or so from the west; Europeans?

From just over there; Seasons? No problem—we offer them a good deal; Maintenance of the vessel? I have a friend, he is a very good carpenter. When sooner or later the whole thing comes crashing down, each one of the parties has their own very definite reasons for the failure. The relationships, the moneys, the promises, the assertions, they all format and reformat themselves in terms of the specific interests under focus. On one thing they can all agree though: they have been cheated by the rich of this world.

Note the functional clusters: various local groups, official institutions, government agencies, plus those on the outside. In the absence of any unifying standards each cluster operates by its own rules, and therefore the perception and so the analysis of a problem is couched within isolated frameworks. Add the distracting influence of political developments and the dynamics for separate domains are in place. Recall the principles of conceptual intersection; just as a business can have a negative balance sheet, so can functional clusters experience a data debit. As each group interacts with an unrelated other, information gets lost over and over again while the opportunity for information gathering is virtually nonexistent.

The same process affects the data flow the other way around. From regions of high complexity the information descends to provide the perception on the local level. Once there it is harnessed to interests whose diminished interpretative facilities have turned the concept into a compacted version of its former self. For example, a foreign consortium tries to build a gold mine somewhere in the interior. Local groups evolve who apply their particular brand of politics to the endeavour and a decade and hundreds of millions of dollars later not an ounce of gold has hit the global market (what happens to those who signed the documents in their boardrooms I don't know).

While the above may have been an extreme a generation ago, in certain areas it is much more common today. The functional relationships can equally be witnessed on the street level. In cities of several million inhabitants their resource space is sufficient to accommodate a whole range of strata, from the organised rich to the dissolute poor. There are professionals providing their services under strict guidelines; entrepreneurs who play a game of favours left right up and down; and children having formed their own societies with a distribution network based on whatever opportunities present themselves. The standard of family planning moves along; from well-paced decision making to crisis management to a swift kick to the abdomen.

As *The world report on violence and health* by the World Health Organisation notes, "Global and national data are very scarce" [5]. In any case, statistics become meaningless. What does it matter if there are two hundred million below the poverty line or three or four? When it comes to protein formations the descriptor is 'transcomputational'. We could use the same at the higher scale. In order to know one needs a formal set of data, relatable and applicable. There is nothing formal, relatable or applicable in a seething mass of humanity that has long ceased to be ascertainable under a framework that possesses no units of measurement which can make sense under such alien conditions.

At the same time the global media project the images of their representative demographics into the world at large—yet another example of high-to-low conceptual transference. It triggers the desire (multiplied million fold) to relocate to areas perceived to be full of promise.

In an analysis of global immigration issues Helen Hughes gives a figure of seven and a half million legal and illegal immigrants who enter liberal Western democracies every year [6]. By doing so they create "severe social and political national and international strains". Regarding foreign aid to developing countries she writes about the *dirigiste* development strategies under which funds are directed towards the elites in developing nations which "made it possible for governments that have ignored basic commonsense and justice to survive" [7]. Religion features large as one source of persecution, particularly since the rise of the Christian and Muslim faiths [8]. As for the contribution by large-scale entities,

> The UNHCR, other UN agencies and NGOs relief for displaced persons have fed military forces in displaced person camps. The criminalisation of drugs in the West has funded military adventurers, fuelling conflicts in Colombia, Afghanistan, on Burma's borders, in Chechnya and Kosovo, and funding much other civilian and international fighting. [9]

She also adds arms trafficking as another source of assistance to such conflicts, whether the arms originate in the former communist countries, the West or developing nations.

Again we can observe the forming and operation of functional clusters. In the developing world the erstwhile habit of locals to cheat the rich visitors has been taken up by their elite, where the finalisation of an aid programme depends as much on the donor's capacity as the inclusion of a luxury car here or there that finds the approval by the receiving country's top ministers. The distribution of the funds are subjected to the political and social dynamics there and only after all this has been taken care of does someone in need actually get some relief. Foreign aid agencies largely circumvent these thickets by flying in the material and transporting it directly to a market place. But even they have to manage the administrative side of logistics carefully. Across the divide are the financial and political circles of donor nations who act on the filtered data coming in from the outside but who are also cognisant of the dynamics in their own electorates.

In the developed West a greater resource space with more levels of complexity enables its domains to extend their respective connections in a more distributed manner. This means greater variety but also gentler slopes of intensity between the centres. Greater variety is demonstrated through the existence of high professionalism and members of the same society who cannot relate to its characteristics and so opt out. The slopes of intensity become apparent during everyday life when the self-proclaimed outsiders merely hang out at their meeting places, but do so through their absence when something triggers the violent appearance of these others (eg, at international high-level meetings).

Mixing it with the rest are various regions of activity that produce an effect on their surroundings. The distances available make isolated niches possible, and so conceptualisations can emerge which maintain a life of their own. Incoming information about Third World poverty can become interpreted in terms of a political agenda and so the blame is put on the colonising West for its disruption of ancient borders. Yet the dynamics of local clustering in the Third World have made borders meaningless long ago, just as the perceptions of blame are a product of discrete domains in the West. Neither demographic will be influenced by the perception on the side of the other, yet both have sufficient momentum to continue regardless.

Similarly the emergence of the European Union is no coincidence. While attempts to establish a European international alliance have occurred in the past, the disparate dynamics of the participants made an implementation impossible. Given the principles of high- and low-abstract regions an idea is not enough; unless it can be transmitted to lower levels it remains ungrounded. Today the EU member states reflect a functional similarity in their affairs which makes such intimate coexistence possible. It also increases the resource space dramatically.

The effects of legal and illegal immigration are manifold. The legal variety follows an established set of criteria which may or may not accord with the standards of the host nation. From our perspective it is not the content which is significant, but its functionality. Hence such things as dress code, food, or appearance do not present a problem (although low-abstracting demographics would focus on those matters). In terms of functionality however the differences in abstraction levels that each individual owns within their mind become the markers around which the domains of the shared resource space turn.

The entry of Muslims for instance generates the presence of a moralism that grates against already established norms of liberality and tolerance (hard-won over the centuries one might add). In small detail it can mean the travails of a classroom teacher who has to reduce heat stress amongst her pupils on a hot summer day; her professional conscience tells her to have the children lighten their clothes, but the politics of another culture make her a deviant (and even *if* the Department is courageous enough to stand by her, how will the Muslim girl be treated once she gets home?). On a larger scope we have clerics who fulminate against the West under the protection of the very laws which they despise. And the multiplying facilities of local clusters which are able to nurture a negativity until it has grown into terrorising proportions need no longer any further explanation.

The effective difference of such clusters resides in their environments. A contextually supportive environment, even if it should be functionally poorer, may turn out to be more productive than a contextually less supportive one, even if its infrastructure is better. For example, consider the populations of Pakistan and the United Kingdom in relation to the support of terrorists. Note however that such a comparison can only ever be a relative one. In the United States support for Islamic extremism is minimal, yet look at the amount of knowledge a few individuals have gained there.

In a liberal democratic nation the variance of its dynamics ensures the vibrancy most civilised people have come to value. The optionality enjoyed by its players on the other hand is the cause for resentment on behalf of its critics. It has become the supreme irony of its status per se: the existence of a resource space which needs the sheer extent to feed its complexity—yet some of that very sustenance poisons the entities responsible for such reach.

Target countries for immigrants are already tightening their borders. The comprehensive set of data which high-complexity regions have at their disposal allows them to define the cost of the arrivals in wider-ranging terms. Although the evaluations do not entail the descriptors employed by Otoom, their representative format provides a view that encompasses society at large. After all, every cognitive functionality we have used in our analyses needs a supply of energy in one form or another. For neurons it is chemicals, OtoomCM needs electric current, and in the resource space of society the energy is money. Whether consciously articulated or not, in other words whether based on formal definitions or whether the result of unregulated interactions and responses amongst interest groups, the energy spent on dealing with immigrants will be compared with what it takes to keep them out. The equation is not simple; costs of the latter do not relate to patrol boats and border guards alone, they also incorporate the more indirect outlay for delaying bureaucracies, pressure groups, and marketing campaigns to sell the strategy. Since the overall system is interdependent, the efficacy of implementing authorities has a bearing on their share of the total energy consumption, but so do the efforts by antagonists.

Costs (ie, energy flows) are a fundamental driver of every subsystem. Sometimes they manage to overcome even more deeply seated ones. It is now twenty-five years since the first Gay and Lesbian Mardi-Gras has been held in Sydney. Back in 1978 it consisted of a couple of trucks trying to make their way through Kings Cross. They were stopped and jumped on by a contingent of police that spanned an entire section of the area, the occupants got dragged by their hair onto the asphalt and amid a screaming crowd were arrested and taken to the nearby police station where they were charged. Still, over the years the event turned into an hours-long parade attracting visitors from around the globe and filling the hotels for weeks around the time. Who knows whether the whole of Sydney likes gays and lesbians now, but it sure likes the money.

The evangelism of the Christian tradition prompts many efforts by groups to carry out their activities no matter how promising the results may turn out to be, or even whether their actions are appreciated by the recipients at all. In any case, they consume the energy of their host system. Any management of what are currently termed 'illegal drugs' is forgone in favour of costly measures to 'save' those identified as sufferers. Given the variance of the system, the statistics would cover those who succumbed but need and want some help to get better; those who want help but won't improve; and others who neither want any assistance nor would improve if given any (and then there is the invisible majority who is fit enough to escape the statistics in the first place). Regardless of the ultimate efficacy of anti-drug campaigns, their main driving force is a moralistic fervour. From the wider point of view however it becomes clear that indiscriminate spending cannot continue forever. There will be a point at which the activist will ask himself whether it is really worthwhile being spat upon while administering the anti-dote and then afterwards having to worry about funding to repeat the experience.

The philosophy of victimhood has found a comfortable partner in the feminisation of society, which draws on the available energy to pursue its interests. Investigations such as the study of boys' education cost money, and if it ever comes to some implementation that costs money again. The moral primitivism surrounding erotic behaviour forces the relocation of literally millions of dollars on behalf of a few individuals, whose own persona no-one dares to analyse in order to understand how well-spent such moneys actually are. Of course, an influence does not have to be known in order to exist, and the consolidation of the matron state means that society has to digest an increasing shift of energy away from productive areas to uncertain niches whose only claim to fame is the self-serving motive of others pointed in their direction. A more real-life simile of Andersen's tale *The Emperor's New Clothes* would be hard to come by.

Such loss of sustenance for everyone else is not confined to one society alone. As Helen Hughes has noted in her report on immigration, a considerable source of income for arms dealers, destabilising agitators, and of course terrorists comes from the sale of illegal drugs. The income is assured because the West can be relied upon in its adherence to Middle Eastern morality (can the reader detect the irony here?), the profits are huge because illegality makes them so, and the entire phenomenon is firmly hidden due to its very nature. It's like a boxer fighting an opponent under the condition his eyes are only open once every five minutes!

As far as sexuality is concerned, there are many demographics that do not exercise the same anguish as Christian adherents. Nevertheless, knowing the West's position in these matters some have adjusted their politics and the morals are now used as a convenient tool to extract money and concessions from the outside. Perhaps the religionists are satisfied, but nothing has changed in those demographics—the only difference is an additional burden for the West.

As said before, when the costs hit home some change will be contemplated. The threat of terrorism has reinforced the profile of surveillance and data collecting. On the other hand, the notion of civil liberty and particularly privacy stems to a large extent from the recognition that the state should not have the right to intrude upon our intimate selves. Since 'intimacy' is a function of the degree of unease about what morality has allowed and not allowed, the concern raised by privacy advocates is really about the concept of morality. Therefore, when X-ray technology allows us to literally strip-search every passenger of an aircraft, and when advanced surveillance techniques will apply the same principle in larger-scale sweeps across town, it is time to reconsider what such concepts of decency, obscenity, and anything in between should really mean for us.

A crime destroys, even if never discovered. Similarly, something good heals, even if it has to hide. In urban demographics wayward youth who otherwise roam the streets causing mischief find a home with those who also like them erotically. It is a bond made stronger for its greater depth and for that survives despite the lack of official support. At the same time the legal status of such arrangements does matter one way or another, and so the less interference from this quarter the more stable the relationships can be. At this stage in the West's evolution there are many who would be outraged and activate common funds to calm their emotions. If they are unsuccessful the whole of society, including those moralists, benefit from the increased safety and the smaller usage of public coffers. The officially sanctioned schemes on the other hand are largely administered by demographics that do not relate to their charges; when you're on the outside the rush of freedom, the scent of adventure, and the mystery of night streets only you know do not hold up against set bed-times and matching curtains.

The cost associated with every action that occurs in a resource space defines the status of that space. While the mutual relationships through affinities and PAT chains feed upon their own content, the energy for these processes has to come from somewhere. Just as an accountant can tell the future potential of a business by going through the ledgers, for us analysing the energy consumption and distribution performed by subsystems can tell us about their viability. From our point of view that energy is mostly expressed through information, but one could argue that even money is yet another derivative. The data transmit the content which eventually expresses itself in the form of affinity relationships, of clusters, illusion trees, and so on. Their activity determines the energy level of the whole, and the direction of its flow can either enhance the system or cause it to deteriorate. The system's elements process the data, but at any given time only part of it actively participates in further decision making—another parallel to the conscious and subconscious structures in the mind, now scaled-up.

There is then no single future. From the growing complexity in our minds to societies and to the global ocean of dynamic entities there are as many futures as there are meaningful domains. What 'meaningful' means is determined by the interdependency of these states between their own processes and those of their environments. And, maintaining the scalability of cognitive functionalities up to this level, the wider their extent the more significant the subconscious structures and the more influential their processes upon the entire system. Since those dynamics do not get articulated directly, it becomes a measure of the members' intelligence to ascertain their existence and nature and so administer them wisely. In the traditional context a culture imploded because its rule set (its container identity) was driven by madness. The Rome of Caligula, Stalinist Russia, a religious/feminist West, are not destroyed on the battle field but in their homes. In those cases the self-regulating character of a live system relates to a still wider frame of reference because the subsystems themselves—Rome, Russia etc—were incapable of changing productively on their own; the environment did it for them.

A relatively small resource space is not only defined in terms of its absolute volume; its limited set of functionalities can absorb inherent idiosyncrasies more easily because there is less demand on its overall synchronicity. For a larger system to work in harmony with its own self its constituent parts have to function in accordance with a logic that encompasses its resource space. As noted before, the more extensive its reach, the more logical that logic needs to be. In other words, while a tribe might survive despite keeping its semantic idioms, for a highly complex society the same traditionalism spells its death. Thus it is not a matter of advanced societies becoming more rational—unless its members can demonstrate their capacity to be more rational the society cannot grow to begin with. Whether the West's resource space possesses sufficient depth for its affairs to be organised in terms of the real rather than myth remains to be seen.

Given the interdependent nature of live systems the problem can also be stated in slightly different terms. If there is a large-scale system occupying a certain resource space, the set of rules that are applicable to the entire space needs to be as rational as possible. That is to say, for the system to function productively there must be a rule set which excludes those idiosyncrasies which may exist somewhere in its subsystems. It is an issue that presents itself every time a country's High Court issues a ruling which is in potential conflict with the law of a state or a province.

The problem is particularly pertinent at a time when the European Union is in the process of articulating a constitution covering all its present and future member states. The solution, as suggested by the e-chart, can be found within the realms of two types of implementation: re-organise the subsystems so that their idiosyncrasies are removed, and/or limit the overall rule set to rationally identifiable, congruent functionalities only and restrict incongruent domains to their respective locales. In effect this means that in the case of the EU for example its law makers would actively resist the incorporating of any regulation that has as its basis a culturally specific perspective, while leaving the member states free to retain their identity. A possible compromise—but it is a compromise—could be the clustering of certain members under the banner of a viable congruence and have others retain certain differences. Such an outcome represents functional terraces descending to the periphery of the system. Should a member decide such a coexistence is incompatible with their values, the long-term viability of their membership ought to be questioned. If it ever comes to a planetary council described in some science fiction stories, its nature would have to reflect an arrangement as just outlined, otherwise its implementation would be impossible. Throughout history the disregard for the functionality of congruent and incongruent structures has manifested itself in the implosion of states and empires where the illogicality of one domain was superimposed on the rest, and a battle ensued to decide which interpretation of reality should win over the others. See the Appendix for a suggested version of a constitution that represents an accommodating universality while at the same time safeguarding the essential rights and obligations of its members without which a system would not function harmoniously; a fundamental Bill of Rights.

Similarly, the central implementation of a rule set will affect the system's members, especially so if the rules happen to be of a fundamental nature. For example, in the increasingly litigious societies of the West the added cost of perceived security via insurance is added to many activities, some more important than others. That burden, although instigated by a few, is carried by all. While the disappearance of a children's playground here or there may not mean much, the destabilising of a nation's health system is another matter. Since excessive litigation is a symptom of unchecked subjectivity towards the wider arena (where 'excessive' is meant in terms of the carrying capacity of the system as a whole), the inherent cause of the current state needs to be looked for in subjectivity-inducing dynamics. The feminisation of society is one source (although not the only one—the removal of personal responsibility in favour of regulatory authorities dancing to the tune of an individual's incompetence is another), but its fundamental nature points to the degree of difficulty in ascertaining its presence as well as prioritising counter measures.

In addition we can observe an implicit relationship with the dynamics of sub- and con-TSs in the mind of an individual. Just as the heightened optionality of increased affinity relationships enables more subconscious content to be brought into awareness (see the section on consciousness in chapter 15), so does the dynamism of complex societies open hitherto hidden domains to wider scrutiny. Wider scrutiny means the incorporation of such newcomers into the relational framework at the conscious level, at which point they become subject to the affinity-based demands already in existence there. For example, there may be a certain subculture dealing with the implementation of discipline in the Armed Forces, enacted but officially unrecorded. Within that particular context it may well be valid, but appears less so by the time the rest of society with its own perspectives realises its nature.

The scenario is similar to the applied moralism mentioned before, except for the fact that a system like the Armed Forces constituted a fully fledged entity all along. Without a tiered configuration of rules in terms of their specific applicability across the system as a whole, the functionality of subdomains will be compromised in line with their interdependence.

When affinity relationships lead to certain clusters such as the EU for instance or trading blocs, the same principle applies to other demographics. The breakdown of traditional rule sets through their interaction with a greater variance produced by high-complex domains has been the cause for many observations; we mentioned some of them in previous paragraphs. Just as street children in a city form groups that enable them to sustain their identity, so do entire regions establish their outsider's status at the other end of the scale. The United Nations' *Global Report on Human Settlements 2003* sets the number of people living in slums worldwide already at thirty-two percent of the global urban population; everything suggests the ratio is increasing [10].

Street children represent a subsystem in a wider society of which some members enact measures attempting to wean them off their life style. The general idea is that the offering of accommodation, food, and a generally stable environment entices them to leave the outer. The reasons for the success or otherwise of such initiatives can be found in the affinity dynamics of their world. If the child has just left home (and the conditions there are certainly not insignificant in the whole process) its perception of the world still contains the original framework. Incoming information is still able to relate to whatever 'home' means. After a month or so however an entire new set of structures has emerged in which 'normal' middle-class values find no resonance. There may be times of hunger, cold, and the odd occasion of violence; but there is also the comfort of the group, the absence of moralising, and the sheer freedom to do whatever is possible at any given time of the day or night. Would a street kid accept a bribe from a well-trained citizen? Most probably—but sooner or later the call of the wild reaches their ears and the regulated world is left behind once again. With maturity comes a wider frame of reference, and the potential the rest of society offers can be appreciated. That is to say, the mind is now capable of productively reconciling its own preferences with those of the establishment.

On the higher scale similar dynamics are at work. Here a complex system also demands a certain subservience, the submission to an adjustment process which moulds the contributing member. But as we have seen, it is the congruent and incongruent affinity relationships which ultimately determine a system's resource space. If the overall resource space demands a well-functioning generality that happens to be incompatible with the local idiom, disengagement occurs. The result is another mix of rejection and attraction: outside help will be accepted, but that does not change the perceived identity of the demographic. Its own social, moral, and spiritual layout represents a universe in its own right, and there the laws of the outsiders have nothing to say. The nomenclature of the latter articulates the label 'slum'. There is one difference with lower-scale phenomena though; whereas street children have the opportunity to grow into the encompassing space of their society, slums built around demographic differences have no ultimate destination. They *are* the home. Therefore the equation for a society which contains adults who have come from the street works, since those adults now contribute. The balance sheets for slum cultures containing millions of members—drawn up by the organised establishment—do not add up. Then again, in their past such accounts did not exist to begin with.

The intermingling of high- with low-complex demographics *within* a given domain poses a challenge for both. Just as a member of the former needs to be able to recognise the inherent potential of someone coming from the latter, so is there an onus on the newcomer to recognise the limitations of their culture and adapt to the wider framework. For example, the achieved rationality of a European (such as it is), should not prevent that person from admitting an outsider just because *their* background had been a more limiting one. On the other hand, someone from a more traditionalist environment ought not to reject the West under the impression that objectivism and rationality are somehow 'owned' by the West. A large-scale system does not function due to its 'culture', it works because it has aligned itself with common principles. And 'common' means just that—whoever in the universe may apply them.

On that note it is significant how ideologically inclined people view current developments in gene technology. Michael Cook, who is the editor of the bioethics email newsletter *Australasian Bioethics Information*, opines on the recent creation of human-rabbit hybrid embryos by Chinese scientists [11]. The title, *Human-rabbit clone announced and no noses twitch*, suggests his antagonistic perspective on a scenario which encompasses what he sees as distinct identities. The article's attitude represents the familiar tune of those who are very determined to maintain their purity in whatever format it may be expressed. However, as Cook to his credit mentions, the research was done using embryonic stem cells. To use our nomenclature, these type of cells represent a fundamental latency such that their functionality can be transposed into other variants of higher complexity. In other words, if an embryonic stem cell coming from a human can be productively used in a rabbit egg, this demonstrates the sheer generality of that cell's potential. It therefore also demonstrates the intrinsic universality of many of life's processes. Such processes are not owned by you or me, or by any member of a religion, race or tribe, or even species—they belong to all living beings. Once again we can observe the historical context; just as Galileo showed the commonality of planet Earth and was opposed by the church, just as Darwin did the same with evolution and was attacked, so are bio-geneticists in the firing line of tribalists who cannot admit their true place in the scheme of things. It wasn't so long ago that upright citizens refused a blood transfusion because the source happened to be a black person.

The instances listed so far demonstrated our view of systems defined in terms of a certain reach, their functionalities as they were performed in juxtaposition to each other, and their degrees of efficacy as could be implied through the parameters of the e-chart. In one way or another all these are the outcomes of affinity relationships, an interplay of dynamics which can be expressed through vectors connecting one locale with others. There is one development however that is arguably the most fundamental of them all, because it affects so many subparts of any system, high or low, over-arching more specific phenomena as they occur in tandem: the ageing of the world's population.

From a system's perspective every human being can be seen as a module possessing a certain potential, realised over time under conditions which in themselves change since they too are a manifestation of the people who constitute that person's environs. The properties of any given society reflect the functions necessary for such realisation, their degree of stability an indication of how responsive to its environment the society happens to be. Throughout these pages we have encountered the sheer complexity of such an arrangement.

Now consider the role of ageing, and let us interpret its timeline as a set of interdependent functionalities between the individual and their society as it loops from the young to the middle-aged and to the older, and back again to their offspring. At first nothing seems to have changed—the periods of infancy, childhood, and teenage years follow predictable paths as children come into existence and eventually grow up. Their relationship with the rest however is determined by the combined potential of such a reservoir, although the effects become more visible as we iterate through the loop.

That reservoir, in other words the band of demographics which is able to support their young, defines itself through the overall burden placed upon it. As the percentage of the aged rises, the proportion of support needed for the latter shifts energies away from the more traditional allocations. Less support translates into lesser latency of its focus, because to prepare and train invites a cost. Or, to maintain the existing standard, the pro-rata expenditure by the middle needs to increase and thereby placing more pressure still on the entire burden. At the same time, the relative size of the middle compared to the aged is in decline as well, further upping the ante.

Moving on to the aged themselves, their needs for services in terms of the overall system increases, but at the same time their latency expressed through experience, voting power, and active definition of what they see as their preferred environment plays its role within the wider dynamic in its own right. Referring to the e-chart, if the realisation of latency means drawing on the energies of the system, and if the process of articulation as such defines the system's dynamics, we can see that the result is not necessarily a diminution of the resource space as a whole, but rather a reconfiguration of the system's interdependencies—*provided* its framework is of sufficient complexity to do so.

To illustrate the rate of ageing, the Department of Economics and Social Affairs in the United Nations published a chart which amongst other data contains two images of the globe, showing the percentage of the world's total population aged 60 years or over in the years 2002 and 2050 respectively (see Fig. 18-2) [12].

In previous paragraphs we made the connection between the logicality of a rule set and the congruence and/or incongruence of subsystems as they played their part in the overall scheme of things. Progression lock occurs if a functional module is incapable of moving up to a higher complexity, but a higher complexity requires more congruence, which in turn is a function of better logic; and so on. The demand for a more comprehensive rule set stands in often direct competition with dynamics we have identified earlier on, many of which invite a return to more simplistic interpretations of their environment due to their denominative nature.

For example, the influence of a religious, human-centred view predisposes its members to distinguish between man and machine. It translates into an implemented positioning of people and their perceived role in society on one side and technological products on the other. Yet a machine can also be seen as an extension of a person and, in a wider sense, as their partner. If the productive potential of a society is more or less related to its human members alone, the *potential support ratio*, that is the number of persons aged 15-64 years per one person aged 65 years or older, is bound to decline as the population ages. It does so already: it fell from 12:1 to 9:1 between the years 1950 and 2000, and by 2050 it is said to have reached 4:1. Extending our perception of society to machines however could theoretically change the ratio to whatever level feels comfortable—with the advantage that the maintenance of an engine or a robot together with its demands on society would be less than in humans.

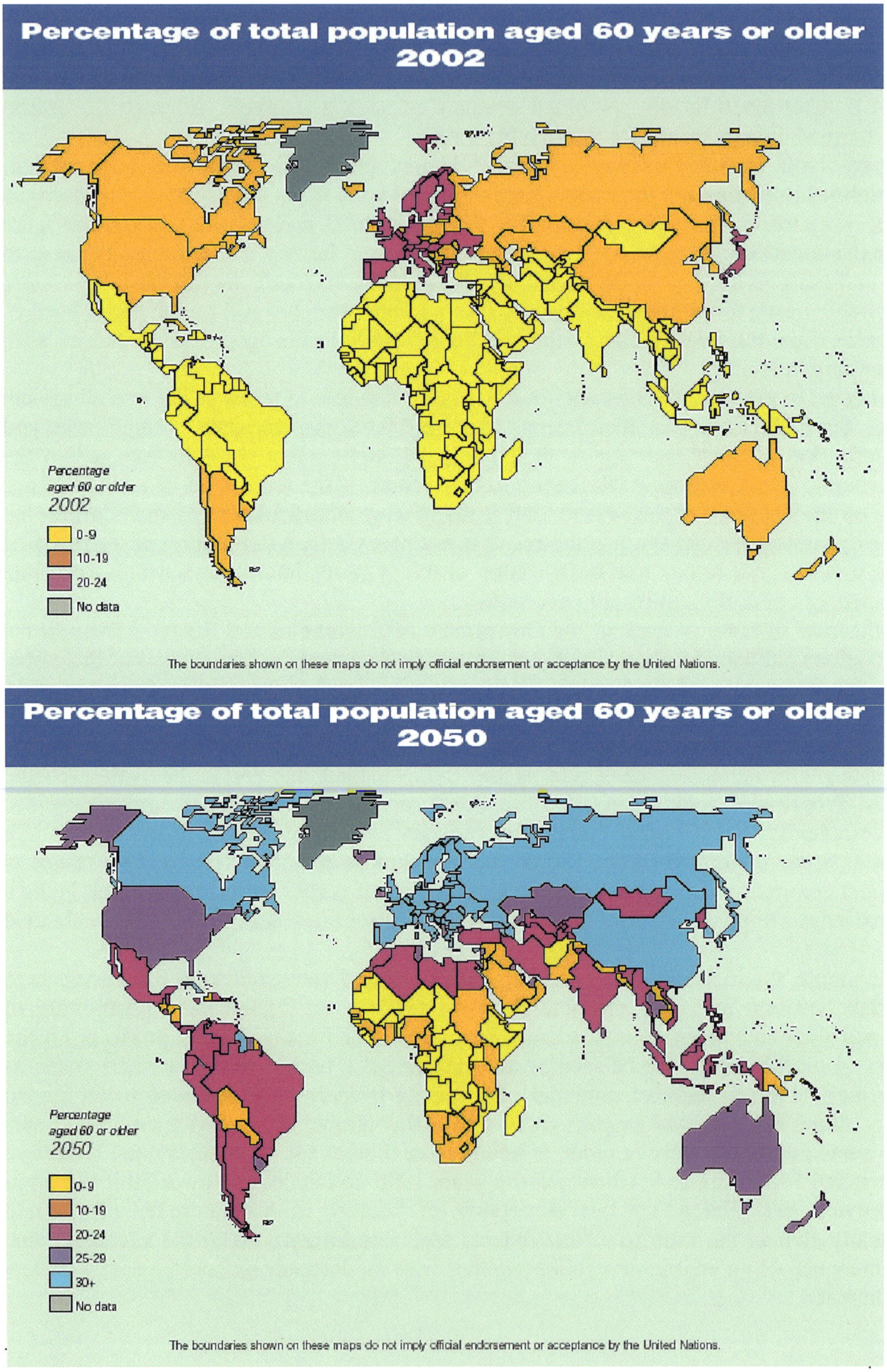

Fig. 18-2: A comparison of the percentage of the world's population aged 60 years or older between the years 2002 and 2050 if present trends continue.

To enter into the full details of the problem would go way beyond the bounds of this book; indeed, their ramifications are worthy of an entire volume by itself. As a final summary of this and other issues we can however broadly sketch their functional dynamics as positive or negative potentials, depending on the areas of interest. The potentials should be seen as further starting points. Because I have kept their labels as general as possible the mutual overlaps of many of their subsystems become apparent. The overlaps serve to indicate the interdependency of their hosts, as well as the sheer complexity of the system they inhabit. And with the possible exception of ageing their listing in alphabetical order emphasises how arbitrary any real hierarchy would be; they all are significant in their own right.

Ageing. In terms of relative numbers the older band of any demographic increases, while the middle and younger bands decrease. Being older means having more experience and so a better knowledge of how to play the game in whatever context, but can also mean greater deterioration of one's mental and physical faculties (again in whatever context). Alleviating the latter comes at a cost. Thus a higher potential means a greater challenge for the competitors, but a lower is more ambiguous. If the potential support ratio is based on conventional tax regimes, a smaller tax base (represented by the middle) is now required to allocate more resources to the young to prepare them for their support role while at the same time maintaining the older. Any pressure on a resource tends to move towards the level of its focus; the educational regime is one target, social services in general another. A relatively larger participating role of the middle deprives the society of its historical perspective (imagine asking thirty year-old financial advisers about depression-proof investments—they may even think there are any); on the other hand, it makes for greater flexibility.

Environmentalism. The protection of the environment also comes at a cost, but needs to be seen in relative terms which requires the historical perspective. The definition of what constitutes a feasible environment is subject to interpretation, a complex picture that needs a broad framework of references. The involvement of low-horizon demographics engenders more localised definitions at the expense of the wider picture (for example, a small habitat of koalas is preserved at the expense of long-term urban planning). Political views that are affinitive with environmental concerns can spill into other arenas, thereby influencing the allocation of energies across the entire system.

Feminism. The movement of women into the workforce raises the potential of the market; it also personalises the latter due to a subjectivity based on the role of carer. The interdependence between an overt perception and the covert role of instinctual emotions makes for an atmosphere of interpretation and reinterpretation, creating room for an affinity-driven ideological slant. Self-destructive moralism and the paralysis of males weaken the society from within. A more acute definition of priorities enables the implementation of distinct policies which offer themselves to a relatively clear modification if the result is identifiably positive or negative. If the definition is less precise, the response will be vague and itself escapes an in-depth evaluation. For example, social policies will produce some result, but its ultimate value is defined in ambiguous terms.

Immigration. The introduction of outsiders (in relation to their host system) is interpreted in accordance with perceived similarities and dissimilarities—and this goes for both parties. The judgmental value of migrant classifications relates to much more than an objective assessment of personal skills; they are also derived from religious, political, and social interpretations of demographics which themselves are in a constant state of flux. The mismatch between low- and high-complex domains are a potential source of trouble for residents as well as newcomers. Just as an urbane arrival from say, Pakistan invites suspicion from conservative residents in London, so is a complex demographic in Sydney compromised by an influx of peasants from a Third World country.

The matron state. The role of carer extends beyond the individual into wider society. It implies the assumption of authority over the target, dismissing its inherent potential in favour of the perceived superiority on behalf of the carer. Due to the denominative influence of an active agent over another, more passive one, costs are incurred in order to implement the strategy. Since any costs are ultimately carried by all, the lowering of potential shifts the burden of support to an ever decreasing remainder, which in turn assumes a heightened degree of charge, leading to a further deterioration of the overall system. The result is degeneracy. In the end, how many 'children' can an 'adult' really care for?

Moral primitivism. The imposition of ideological values (whether spiritual or secular) in order to define the moral status of someone's behaviour configures the checks and balances articulated by the host system. If the deciding authority lacks the intellectual resource space to evaluate the objective utility of an individual, their potential is lost to the whole while at the same time raising the judgmental authority to levels it ought not inhabit. The result is a low-complex domain governing a high-complex scenario, with the added problem of using the system's energies to pursue its whims. Government costs, and for a system to survive those costs must enable a feasible return. When ever more decisive evaluation procedures widen the net for the authority, the burden on the rest increases also—yet it is the rest on whose productivity the authority depends. The construed victimhood celebrates weakness, creating a dangerous type of relationship where children control adults, dysfunctionals control society, and its ideological guardians ensure the subjugation of the healthy. All this against a world-wide background of growing disrespect on a civilisational scale.

Polarisation of demographics. When clusters of activities perform next to each other, their mutual affinities define the viability of their resource space. Low- and high-complex domains compromise their interdependency when forced to exist together, because neither has the capacity to fully communicate with the other. Either the contrast is irreversible, in which case the subsystems need their own autonomous frames of reference, or the difference can be evened out which means an appropriate raising and lowering of standards. Standards change for a variety of reasons, some open, others less acknowledged. It is the prerogative of intelligence to understand itself better, but in such circumstances the prize means pain.

Religion. As it happened, the Middle East had exported its own brand of spiritual intensity to the rest of the world, based on the belligerence of tribalistic emotions and fervour. Global history has seen the implementation of such evangelism by various demographics under its banner, harnessing their intrinsic potential to its set of values. Whether the West has the capacity to detach itself from its demands remains to be seen, but current indications make it doubtful. In conjunction with an atmosphere of victimhood the society becomes emaciated. The emerging domains of India and China, representing a sizable portion of the world's population and with their philosophical core unencumbered by such strictures, are the new players on our planet. Their advantage lies in a ready-made technological base, and their sheer numbers to implement its complexity. A disadvantage is the potential of their own ideological baggage although, it should be noted, its significance is on the decrease.

Security and privacy. To make something secure means imposing restrictions. In order to restrict, one needs to know how the system functions. Therefore a heightened need for security leads to the removal of barriers which serve to safeguard one's privacy. If being secure enables the carrying out of acts, such acts have the potential to clash with the perception of values that define the role of privacy, should the latter have become compromised in the service of the former. The costs of making our lives secure need to be seen in conjunction with the potential of the entire system when dearly held values need to be supported as well. Sooner or later a decision will have to be made. What is more important—an efficient banking system or some ten-year-old seeing pictures of naked humans?

Unemployment. Under the traditional viewpoint full employment is seen as the ideal. The employment of an individual however is a function of that person's potential, defined by the inherent sophistication of their environment. Hence in a technologically advanced society the options for relatively under-performing members decrease, and so certain rates of unemployment can be seen as an indicator of that society's standing. To comprehensively raise the levels of implemented technology also raises the bar for the society as a whole, and similarly the creation of low-tech niches provides the opportunity for under-skilled individuals. Eventually it comes down to this: do we live in order to work, or do we work in order to live.

As the contents of a resource space's knowledge base are prohibited from being transmitted freely (for instance, when defenders of idiosyncratic perceptions complain against any criticism of their attitude in public), the effects are similar to a system that has been compromised through corruption. Again the system is being deprived of information. Dictatorial regimes eventually implode due to these artificial bottlenecks. The restrictions are doubly pernicious because the public is prevented from knowing about the very facts which point to the problem. Not quite as dramatic, but significant nevertheless, is the situation in the West. Its functional extent allows for a variety of demographic niches, some of which see a criticism of their exclusivity as an attack on their cultural identity. Yet at the same time, to suppress information about restrictive practices here or there compromises the intelligence of the entire system.

As a corollary to the above recall the section on humiliation in the previous chapter. While members of the Hutu tribe suffered under the biased regime applied by the Tutsis anyway, they only became aware of the extent of their problem once the characteristics of humiliation were articulated to them. In principle the same can be observed whenever additional information becomes available via the broadened channels of communication. A demographic that languishes under ideological constraints remains subservient until it starts to learn about the realm of alternatives. From that moment onwards the perception of its members turns into an accelerating process of liberalisation against which the existing powers prove useless. The process entails a significant dislocation of structures which initially comes at a cost. Whether the domination operates within the context of social, moral, or political constraints, the winds of change blow away its façade until it is recognised for what, in essence, it always has been: a paper tiger. Take for instance the implosion of communist regimes in Eastern bloc countries, and compare that with the strenuous censorship imposed by a moralistic ideology in the West today fighting a constant battle against the availability of alternative information.

Up to now the effective abstractive capacity of the mind had been tempered by the low state of technology. Substitute outlets were the religions, myths, and ideologies we find in the historical records. As the potential of technology increases, the resultant applicability and implementation in the form of products is able to channel that capacity into the feedback loops of the real. The efficacy of its resource space is enhanced the more fundamental the level at which its elements can interact with the system, a phenomenon we have encountered many times before. Developments in genetics, nano technology, and the outer reaches of quantum physics, provide an optionality for which idiosyncratic rule sets become ever more dangerous. Since options also imply the potential for progression locks, the demands on and for intelligence are as acute as the negative effects of its delimiters.

During the 19th century the growing cityscapes of Europe and North America reached a size where the incidence of epidemics assumed ever more significant proportions. The transmission of disease amongst small clusters of people is kept in check because with low numbers the average capacity of an individual is sufficient to restrain the potential of biological vectors at their initial stages of development. In large numbers of people however the probability of a disease finding a niche of low resistance is heightened, and once the vector had become strong enough the overall status of a population now represents a relatively lesser obstacle. At that point a far more fundamentally rational response becomes necessary, which in those days resulted in the introduction of public health measures quite novel for their time. The same principle applies to the abstract counterparts, our ideas. The large resource spaces of highly-complex societies require an understanding of their conceptualisations at a degree of formality and rigour we had no need of before. Just as a virus can do more damage in a big city, so can irrational concepts not only survive at the global scale but assume dimensions of hitherto unknown destruction.

As we have seen in chapter 14, intelligence does not limit itself to the wetware in our heads. Provided the system is dynamic, self-regulating and pattern-seeking, it can adjust to input and organise itself in terms of output. Whether such output has meaning for us depends on the contextual affinity between our semantic structures and those of the other system. In other words, *if* we are able to understand its behaviour and so can reconcile it with our own interpretative faculties, for us the system will be 'intelligent'. This also means that intelligence exists outside our human framework—it does not need us to earn that label. To comprehensively translate the articulations of our environment (meant in the widest possible sense) will be the challenge for demographics of sufficiently high complexity to do so. It is an accelerating development, generating the growth of conceptual structures which need the computational power of machines to provide the semantic bridges to our own biological computer. To ask whether machines are an extension of us or vice versa is meaningless; we all are systems acting interdependently.

Robert Bradbury, in his article *Matrioshka Brains*, suggests the eventual existence of very large-scale computing structures that harness the energy from a star [13]. One does not need to delve into such immense physical dimensions. Quantum computers reduce the material side to manageable proportions while offering no foreseeable limit to intelligence as defined in these pages—provided its quantum states can be turned into a productive system of logic. At that stage of processing power, the potentials of connectivity and affinity relationships demand the elimination of arbitrarily defined taboos by its users.

Meanwhile, in the space of the narrow-minded the fear of its owner relegates some content to deep-seated regions, to well up every now and then as demons. On the large scale the subconscious is the form and shape of secret acts, hidden away to escape the public gaze; they too well up every now and then. And yet, both are active participants of the dynamic, self-regulating, pattern-seeking system which is our mind, our society, and Life. For us, the members of this resource space, it matters how capable our conscious self is allowed to be. For ages the capacity has led us from the dim beginnings to the current light, always keeping us just one self-reflecting step ahead of what was needed. If we had not banished so many of our past demons, we could not have perceived the new and Life would have left us behind.

Until we manage to overcome them all however our destiny remains uncertain. Like someone hypnotised who acts under the control of their subconscious, it may well be that the age-old induction of shame and guilt by our religion will finally fulfill that myth's promise: the death of its human host because we have been forced to hate our selves so much.

19. Debrief

Our journey has ended.

We have moved through a land which for some is filled with glorious beauty, for others with forbidding monsters. Hopefully, the intellectual tools provided in this book will have made it possible for the traveller to view the scenery in safety. No guide has the final control over how well the tools are used in the hands of his charges.

No doubt there are some who disagree with one thing or another. With an interdependent and wide-ranging model such as this one, for them the task is to take a piece of the mosaic, reshape it to their liking, and then place it back where it came from. Obviously, they would have to make it fit again. If now they change its neighbour, the same goes for *its* surrounds; and so on.

All the observations have been based on ascertainable features. In a sense the path could be likened to crossing a river by moving from one rock to another. There is water in between, but every step lands on firm ground. Since the text comes in many layers, some will have come to know the stones; others got to see the river; others still took in the entire valley.

To describe our theme as interdisciplinary is almost an understatement. We have ranged across areas of artificial intelligence, neuroscience, biology, philosophy, psychology, social science, political science, and anthropology (have I missed one?). I do not claim to be an expert in any one of them, if by expert we understand a person who has comprehensive knowledge in their field. It has been estimated that in the year 2000 the number of scientific articles ever published has reached 900,000; during that year alone just under 1,000 articles were added to the total and the growth is exponential [1]. This makes it now virtually impossible to keep up with such a deluge, never mind combining several disciplines together. I have listed the references as I came upon them in my research, and the absence of others, possibly many, does not in any way imply their lesser status in the scheme of things.

It has become customary to close one's work with a series of acknowledgements to all and sundry. This I will not do. For one thing, where to stop? Over the many years hundreds if not thousands of individuals made it possible for me to do what I intended, in so many indirect ways; interdependent systems are like that (there was no direct involvement by anyone). In any case, those who want their names voiced can certainly do so by themselves; it is their right. Those who prefer to stay in the background, particularly in a context that is controversial, theirs is the choice too.

There are two exceptions. The lectures by Grigoris Antoniou about nonmonotonic reasoning and Sue Nielsen on Systems in Organisations opened invaluable vistas. Although default logic is not directly applicable to the approach used in this book, its formality and rigour constituted an important training ground for dealing with the seemingly confusing and sometimes contradictory observations of human behaviour. Nielsen's theme of Soft Systems Analysis provided a more general framework for human activity systems. Without such knowledge my own mind would have been much less prepared for what was to come.

Secondly, as a citizen of a relatively high-complex society I had the use of facilities provided by the public purse, from libraries to study assistances, without which the project might have even been impossible. I would like to acknowledge therefore the existence of generally provided means which, just by being there, contribute to the variance and therefore the effectiveness of their resource space. As an element in that space, they were of benefit to me. But after all these chapters, we should know this by now.

Appendix

Computer model

In the Otoom computer model (the OtoomCM) the functionalities are kept largely in line with their physical counterparts in the biological brain. We therefore have a complex of *nodes* divided into *input* and *output regions* and an inner part where the actual processing takes place. Each in- and output region can be connected to an input or output device. At the moment the input device is the keyboard, and the output is sent to a particular area on the screen.

The total number of nodes are constructed in the form of a matrix, with the in- and output regions bordering the *inner matrix*, the processing centre of the system. Thus the setup looks like this:

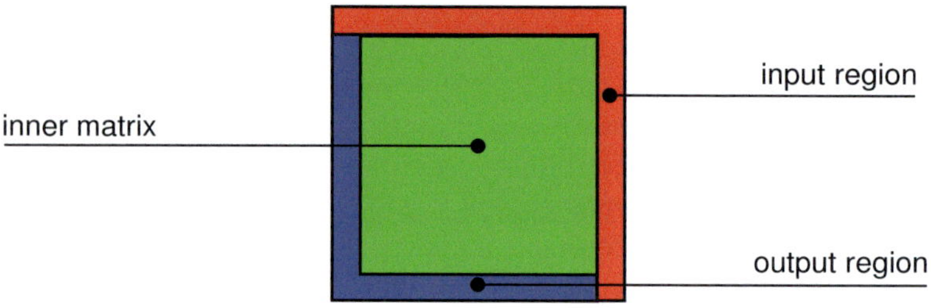

Fig. A-1: General layout of the matrix in the Otoom computer model, showing input region, inner matrix, and output region.

Each node is assigned a number of *elements*, which are also nodes representing number sequences where the same set is repeated across all the nodes; these elements form another matrix, so that a 10×10 matrix with a 10×10 element matrix per node consists of 100 nodes with 100 elements each, making it 10,000 elements altogether. The element numbers go from 1 to a maximum determined by the initial set up. If the total amount of elements at each node is larger than the maximum numbers set by the user, the numbers will be repeated once the range has been covered; if the number of elements is lower not every number in the range is used but the range is split into segments (for example: with 10 elements and a number range from 1 to 8 the elements consist of 1, 2, 3, 4, 5, 6, 7, 8, 7, 6; if there are 6 elements and the range goes from 1 to 12 the elements are 12, 10, 8, 6, 4, 2—thus the higher numbers in the range are always represented).

Each node is connected to other nodes, with two exceptions: only inner matrix nodes provide further connections (ie, input and output nodes are only connected to inner matrix nodes, and inner matrix nodes are only connected to other inner matrix nodes), and each connection is unique (ie, no node is connected to itself and each set of connections per node consists of different nodes; however, two separate nodes can have connections which include the same node or nodes). The number of connections is determined by the *connectivity*, a percentage value related to the total number of nodes (thus in a matrix of 100 nodes a connectivity of 10% means each node is connected to 10 other nodes).

The nodes can be compared to the neurons in the bio-version, the in- and output regions to the various areas in the brain which act as receptors to sensory input and as broadcasters to organs in terms of output, and the elements in each node are simulations of the proteins every neuron has and provide the 'vocabulary' for inter-neuronal transmissions via the synapses, which in themselves are laid out according to the connectivity plan of the system.

For instance, in a 10×10 matrix with three input and three output regions and a connectivity of 10%, the resulting layout is shown in the screen shot in Fig. A-2.

The active in- and output region nodes are indicated by their coloured rims in red and blue respectively (here in- and output region #1 are active), and the inner matrix nodes are bordered green. The connections are represented by dotted lines. Red dots in the inner matrix nodes indicate the termination of a connection coming from an input node, a blue dot indicates a connection to an output node, and a green dot means the transmission occurs between inner matrix nodes. As can be seen, connections of various types can and do overlap.

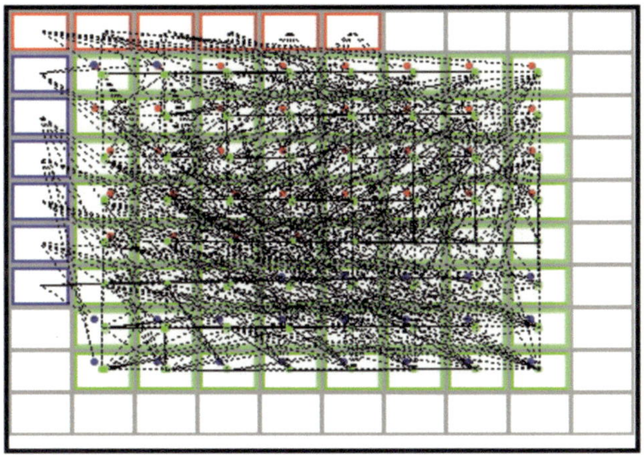

Fig. A-2: Appearance of the input nodes, the inner matrix, and the output nodes, plus their connections.

What follows is a compact version of the program's pseudo code and a discussion of the various features, accompanied by a selection of graphs highlighting the significance of those features.

Pseudo code

The pseudo code of a program is a representation of the code expressed in ordinary language. While a piece of actual code may look something like this

```
for(i=0;i<node_rows;i++)
{...
```

in pseudo code it would be expressed as

for every node in this row... (do whatever)

It would stretch the bounds of this book to present the entire program. Apart from general utilities there are monitoring functions that are concerned with the ongoing display of the results coming from the program as it goes through its paces. Although they are vitally important to verify the correctness of the system and to provide information to the user, I will not go into their technical details in these pages. We will however use their output displays to follow the operations and observe the system's behaviour under varying conditions. What these conditions are and their significance will be outlined shortly. Numbers in round brackets refer to the graphical representations afterwards.

Here are the steps identified in round brackets above, using the screen shot from before (see Fig. A-3):

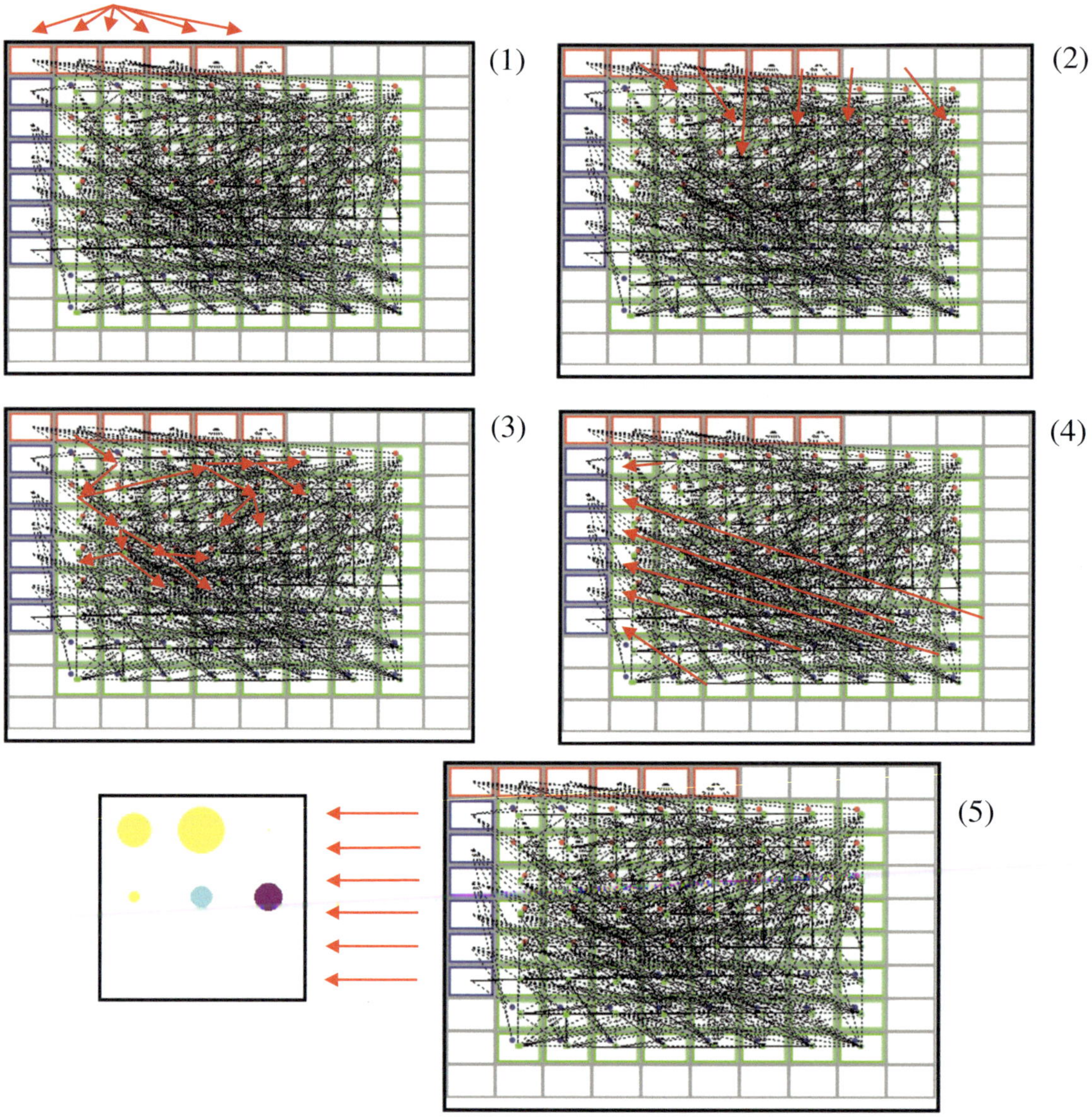

Fig. A-3: A brief summary of the system's steps through one cycle. For the sake of clarity only some instances are shown.

The program begins with input from the keyboard;

input (a numeric value generated by the user pressing a key on the keyboard) is sent to input region #1 (assuming it is this region which has been selected) (1);

selected nodes in the input region receive the input value, which means that each element of these nodes in that region is being adjusted according to the relationship between its value and that of the input (the selection follows the principles of fuzzy logic, where an input value is assigned to a node based on its place within the entire range and the number of nodes available; more details later);

the adjustment causes the *resident value* to move towards the incoming value (the *reference value*) such that the closer these two are the greater the movement in relative terms (see below for a discussion of what this in effect means);

once the element matrices of the input nodes have been adjusted the same process is applied to those nodes in the inner matrix which are connected to the input region; note that at this stage only one level of depth is being processed (for example, if three nodes are connected to every node, each input node transmits its adjustment process to the three nodes it is connected to in the inner matrix, and not further) (2);

the input having been transmitted to the inner matrix, from now on the system will cycle through the nodes there until further input (or the system is halted from the outside);

the cycle consists of going systematically through the inner matrix nodes and following the connection path of each node; that path is determined by the number of connections made available to each node as well as the *connection depth* set by the user (this tells the system how far it should follow the connections from one node to the next);

the path taken in this case is of the type *depth-first*, which means that each branch and sub-branch and sub-sub-branch etc is followed until the end (ie the depth level set at the beginning), before the next one is pursued (see below for a more detailed discussion) (3);

every time a node on the branch is reached, each element of the previous node becomes a reference value to the corresponding element in the current node; this depth-first adjustment process is repeated a number of times set by the user;

once all the inner matrix nodes have been dealt with in this manner, the same adjustment takes place between the inner matrix nodes and the output nodes they are connected to (4);

finally, the output nodes in the active output region are evaluated in terms of the values held by their respective elements; the output consists of circles of a size and colour determined by the combined difference between the elements of one node and the maximum value obtained under that cycle (a rather arbitrary choice, selected more for its graphic impact and consistency than anything else, but it does allow between 100 and 400 odd variations per node depending on the number of active output nodes) (5).

Linear vs distributed systems

In a linear system each process is followed by another process in sequence. This means that process #2 cannot begin until process #1 has been completed, and so on. In a completely distributed system every process starts and finishes in its own time and does not follow any sequence relative to any other process (this is not entirely correct when it comes to such processes implemented through a computer, because the in- and output of the processes are related to each other and so have to be coordinated in some fashion).

Provided every process of such a system is essentially the same, at the end of a cycle the output from a linear mode will be comparable to that from a distributed one (apart from the time taken of course). If however the processes are not exactly the same, are interdependent with each other, and the output is fed back into the system (ie, the system is iterative) the two outputs will vary and can do so quite substantially.

Mathematically this can easily be shown. Let E be the overall state of a system, e the state of a single event within the system, T the total time and s a speed unit defined by the period it takes to conduct an event e, and n the number of events. Then

$$E = \Sigma\ e$$

and the time taken for some assumed event during an assumed period is

$$t_a = e_a / s_a$$

In linear mode the total time taken for the system to go through its cycle is

$$T_L = \sum_{i=1}^{n} (e_i / s_i)$$

In distributed mode however it would be

$$T_D = ((\sum_{i=1}^{n} e_i) / n) / s_a$$

Since E is defined by its processes (after all, they provide an output which in due course becomes the system's next state), whether these processes span one particular time period or another does matter.

The brain can be considered an example of a distributed system, since the neurons can act independently and at any given time whole neuronal complexes can be active parallel to each other. Nevertheless, the synaptic transmission between neurons can also be considered a linear process because one particular molecule (ie, a neuro-transmitter) applies its voltage potential to the membrane and not until after the reaction has set in will the result of the electrical discharge be propagated along the dendritic connection.

Therefore, when it comes to simulate the neuronal dynamics in a non-biological environment, the question of linear vs distributed interpretation is not quite as daunting as it would seem. While the quality of the overall state needs to be considered in terms of the equations listed above, the ratio of events caused via distributed modes to those delineated from sequential relationships can be flexible provided the interdependence amongst appropriate states is not jeopardised. Which brings us to the next topic.

Depth-first vs breadth-first transmission

Consider the situation illustrated below where an input is applied to a series of receptors, which in turn are connected to other receptors, and so on.

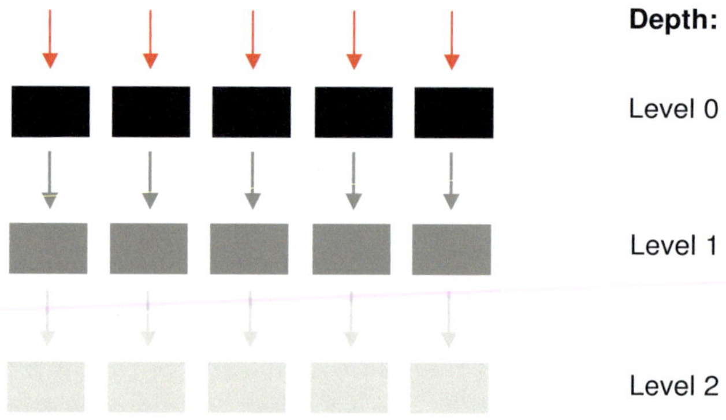

Fig. A-4: Many inputs are applied to a set of receptors which are connected in series: breadth-first.

If the input is applied to each receptor at the same time, and each receptor transmits to its follower also at the same time, then at any given level of depth the state of the system can be defined in terms of the relevant propagation. It is a breadth-first transmission because at each level the entire set of receptors contributes equally to the definition of the system, and the only difference comes from the level at which the observation takes place. If the receptors also happen to be processors of some kind, then the signal descending from one level to the next will be the same for each processor at that given level. Furthermore, if the receptor/processors are connected to each other along the same level and if these connections have an attenuating function, then a stray variance at any processor will be ameliorated by its neighbours on that level. Even if the input vectors should already differ from each other, such a setup will ensure that the overall state across the breadth of each level will be more or less even (strictly speaking, there cannot be a complete evenness because in any system the responses will suffer from some inherent time lag).

Now consider the case of depth-first propagation as shown in Fig. A-5.

Here the input is applied to the first receptor on level 0, which transmits the signal to its follower on level 1. There is no attenuation from its neighbours and all other things being equal the signal will descend largely undisturbed. If some variance does eventuate it will effect the quality downwards only, and the neighbours on any given level are only influenced by the receptor/processor experiencing the transmission. Should another input be presented to the second receptor on level 0 later on, that element will respond in terms of that input *plus* any effect it received previously from receptor 1. The same applies to all the other elements.

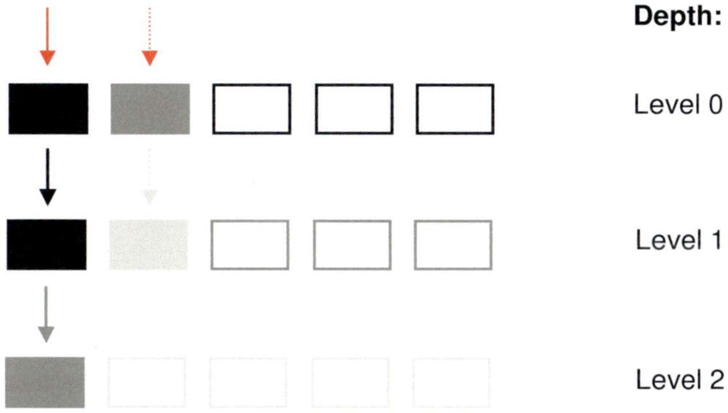

Fig. A-5: A single input is applied to a set of receptors which are connected in series: depth-first.

Here the input is applied to the first receptor on level 0, which transmits the signal to its follower on level 1. There is no attenuation from its neighbours and all other things being equal the signal will descend largely undisturbed. If some variance does eventuate it will effect the quality downwards only, and the neighbours on any given level are only influenced by the receptor/processor experiencing the transmission. Should another input be presented to the second receptor on level 0 later on, that element will respond in terms of that input *plus* any effect it received previously from receptor 1. The same applies to all the other elements.

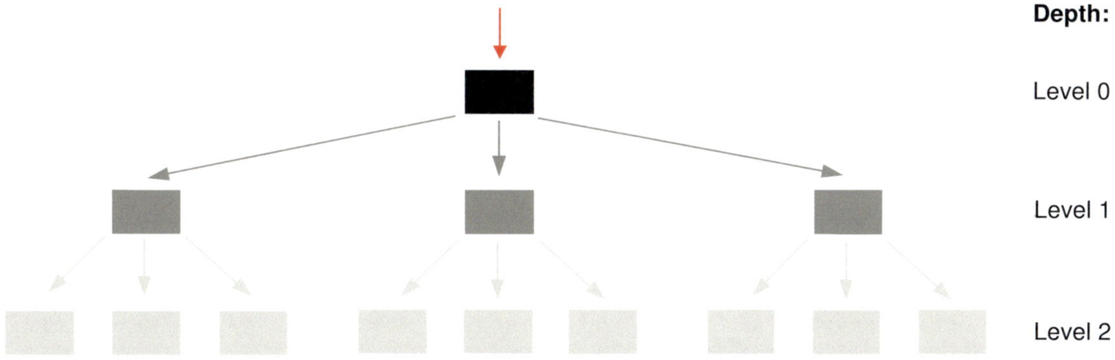

Fig. A-6: A tree structure where every processor is connected to three other processors: breadth-first.

So far we have considered single connections between receptors. If every receptor is connected to more than one other receptor, we arrive at a structure defined as a *tree*. A breadth-first transmission in this case would look as in Fig. A-6.

If the signal is sent down in depth-first mode it matters how we traverse the tree. For example, in Fig. A-7 the input arrives first at processor 1, travels as a signal to #2, and is sent onwards to #3. But then we backtrack to processor 2 and send the signal to #4 because we have reached our designated depth and cannot descend further. Similarly, after #4 we have to backtrack once again and then go to #5, backtrack to #2 and then to #1, from where we descend to processor 6 to follow another branch of the tree.

The effect of such a traversal is fundamentally different from what we would achieve through a breadth-first approach: as the input arrives at #1, its internal state is being modified; that new state influences #2, which now sends its transmission to #3. Note however that all the while #6 is left alone, just as #4 is left alone when #3 is being modified, and #5 stays unperturbed by the changes going on at #4. Once the signal starts to travel to #6, that element will not only be influenced by whatever is coming from #1, but *if* there is a connection between #6 and #2 to be activated later on (which would be the case if those elements were being shared by another) the mutual influence is between a unique state in #2 and a unique state in #6, and not an even distribution of common states shared by an entire set at the moment of flow-through.

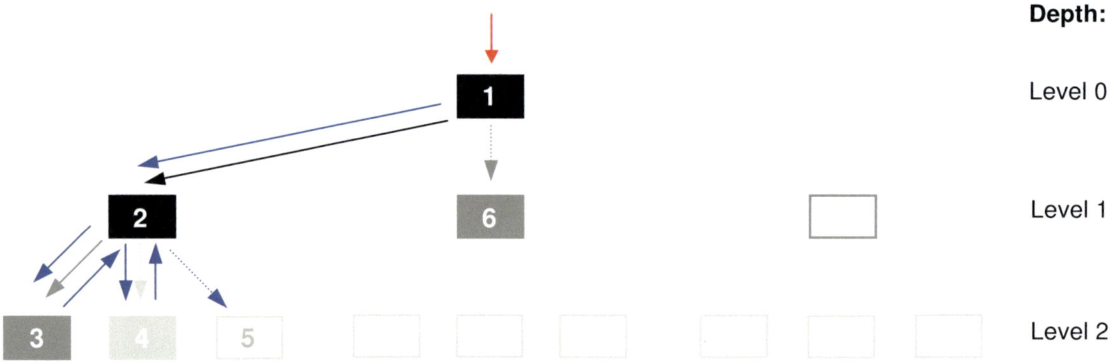

Fig. A-7: A tree structure where every processor is connected to three other processors: depth-first. The path taken through the tree is shown in blue.

In the OtoomCM the processors are the nodes of the matrix, and as each node is connected to other nodes in terms of their connectivity the transmission of the signal occurs along the branches of the tree represented by every node. Thus in a 10×10 matrix with a connectivity of 10% and a connection depth of 3, every one of these 100 nodes acts as the starting point of a tree with 10 sub-branches, each branch ending in a node which is the starting point of a tree with 10 more sub-branches, with each sub-branch ending up in a node that again has 10 connections each ending with a final node since we have reached the end of the depth.

If every node is connected to three other nodes and the connection depth is 3, then the total number of nodes to be visited per tree at depth level 0 is 1 (1), at level 1 it is 4 (1 + 3), at level 2 it is 13 (1 + 3 + 9), and at level 3 it is 40 (1 + 3 + 9 + 27). Since we can change the expression in the brackets to (3^0), ($3^0 + 3^1$), ($3^0 + 3^1 + 3^2$), and ($3^0 + 3^1 + 3^2 + 3^3$), it becomes obvious that the total number of nodes to be visited is

$$M * \sum_{i=0}^{d} n^i$$

where *M* is the total number of nodes belonging to the matrix, *n* is the number of nodes every node is connected to, and *d* stands for the connection depth, that is the number of levels one can descend while following the node links (the starting node resides at level 0). The above is a geometric sequence which can also be written as

$$M * \frac{n^{d+1} - 1}{n - 1}$$

The results of the formula assume large quantities very quickly long before we start getting into serious parameters. For example, a mere 100×100 matrix with a connectivity of

10%, a connection depth of 10 and a 10×10 element matrix at each node will require the program to adjust 1,001,001,001,001,000,000,000,000,000,000,000,000 elements!

It is clear that conventional data structures like trees would impose a huge cost on any system trying to run under such a configuration (keeping in mind of course that this example is just a puny version of what anything approaching real intelligence would require). The solution is twofold: firstly, the algorithm in our computer model does not use a declared tree structure but wanders through the nodes 'on the fly', thereby having no need of any structural resources at all besides the matrix itself and a set of addresses per node; and secondly, an implementation via a distributed system would allow the controlled activation of nodes in parallel (that is without disturbing the sequential influence from one node domain to another). Alas, the latter will have to wait for the adequate hardware.

Adjustment procedure

One could say that the adjustment of the elements is at the heart of the system. It is the modification of the elements at each node in the matrix which determines the state of the node, therefore its relationship with the other nodes, and so leads to the formation of complexes. But let us start at the beginning.

As already mentioned, a node in the main matrix is itself a matrix consisting of numbers which have been sequentially arranged from left to right, top to bottom. In the introductory paragraphs I explained the number system which ensures the presence of the high end of any given number range. To understand the adjustment process it is easier to use a one-dimensional number array as an example (naturally, any matrix can be turned into a one-dimensional array anyway).

Take the following array (the particular numbers have been chosen to concentrate the results in a few samples rather than through a lengthy and boring series):

164	207	121	75	28	13	1	-17	-36	-188

Let's say the input is '90', a number which is now our reference value, with the contents of the array the resident values. If the reference value is the smaller of the two, the process follows the algorithm

```
diff = res_val - ref_val;
new_val = res_val - ((100 - diff) * diff / 100);
```

If the reference value is larger the formula becomes

```
diff = ref_val - res_val;
new_val = res_val + ((100 - diff) * diff / 100);
```

If both resident and reference values are the same nothing gets changed, otherwise new_val becomes the current resident value. This has the effect of moving the target number towards the reference—or has it?

Here is the same array when presented with the above input allowed to 'work' the numbers twenty times:

226	145	100	87	51	30	10	-24	-68	-28
274	121	91	89	74	54	26	-39	-159	-49

428	100	91	89	87	77	49	-76	125	-103
577	91	91	89	89	88	73	-185	103	-282
496	91	91	89	89	89	87	-11	92	16
428	91	91	89	89	89	89	-12	91	35
577	91	91	89	89	89	89	-14	91	59
496	91	91	89	89	89	89	-18	91	80
428	91	91	89	89	89	89	-26	91	89
577	91	91	89	89	89	89	-44	91	89
496	91	91	89	89	89	89	-89	91	89
428	91	91	89	89	89	89	-230	91	89
577	91	91	89	89	89	89	-278	91	89
496	91	91	89	89	89	89	46	91	89
428	91	91	89	89	89	89	70	91	89
577	91	91	89	89	89	89	86	91	89
496	91	91	89	89	89	89	89	91	89
428	91	91	89	89	89	89	89	91	89
577	91	91	89	89	89	89	89	91	89
496	91	91	89	89	89	89	89	91	89

It can be seen that repeated input will cause most numbers to end up as the reference values themselves, with a few notable exceptions. Strictly speaking this procedure is not an iteration since the result is not fed back into the equation as the main variable. However it could be termed a *delayed iteration*, because firstly the result is continuously exposed to the input and secondly, as we will see later, it does become the input (ie, the reference value) to other resident values. That series in graph form appears in Fig. A-8 (extended to 50 repetitions):

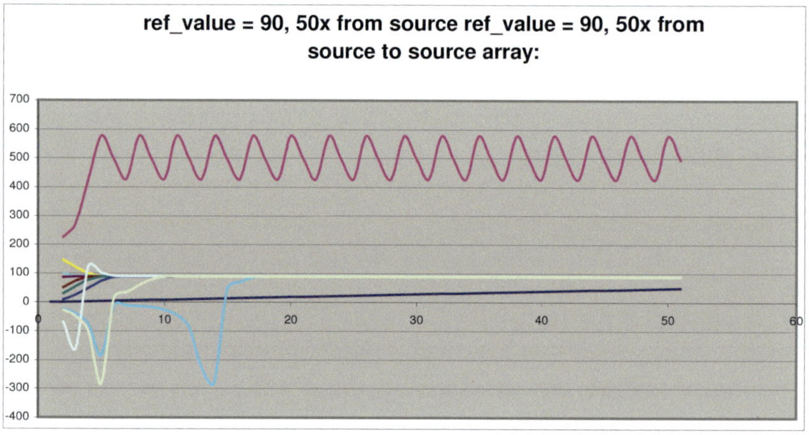

Fig. A-8: A graph showing the movement of number values in an array when exposed to repeated steady input.

The movements of the ten number values are now represented as lines, and it can be seen that most lines converge very quickly on the reference value, some meander around before settling down, and one of them does not converge at all but begins to oscillate at a regular frequency.

If the values in the array are subsequently exposed to their original figures (which are now the respective reference values for each index) the convergence is as in Fig. A-9.

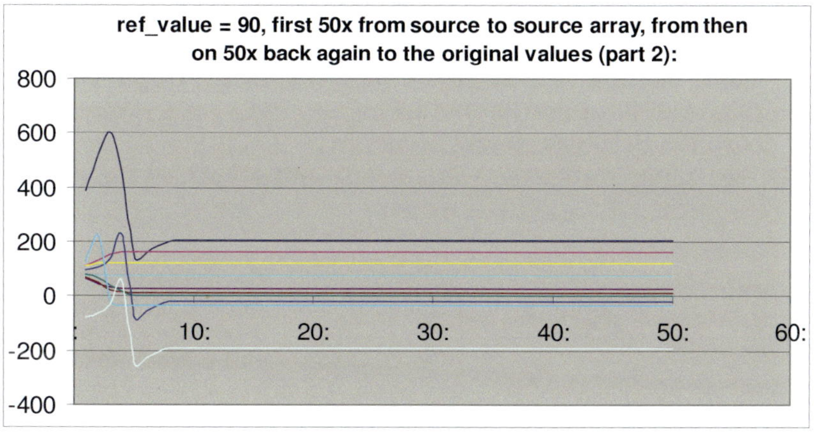

Fig. A-9: A graph showing the movement of number values in an array back towards their original values.

For most numbers the return is a straightforward process, but even here some of them do not converge easily.

It can be shown that the reference value is crucial when it comes to the behaviour of the resident values. Seemingly small differences can change the behaviour completely and turn one type of state into another. Consider the following two scenarios in Fig. A-10, using the same array, but with the reference value at first being '50', then '60'.

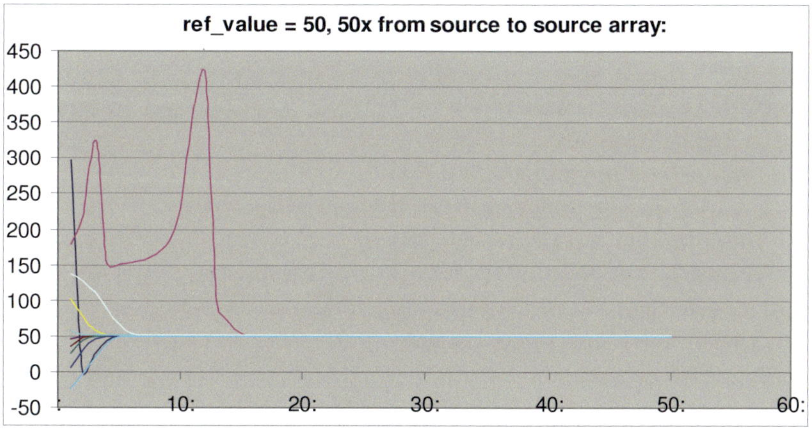

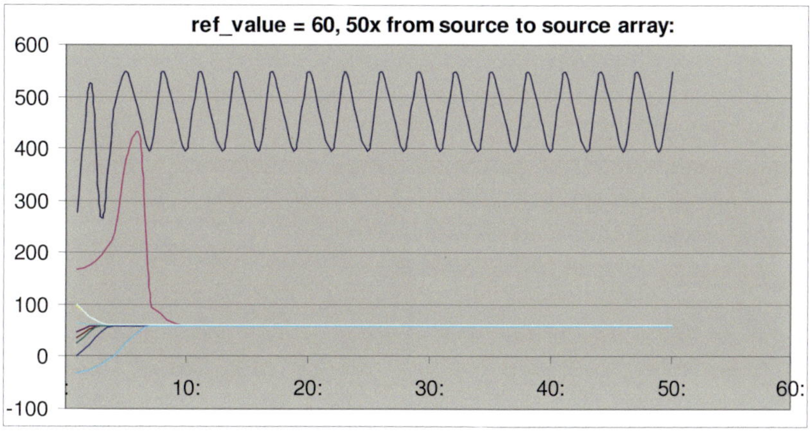

Fig. A-10: Graphs showing the movement of number values in an array under sustained input of '50' (above), then '60' (below).

In the upper diagram there is mostly complete and quick convergence although accompanied by exceptions which become stable in the longer term but experience a considerable period of instability at first. In the lower diagram there is also complete convergence except for one value that turns into an oscillation which does not settle straight away either. Another value converges but shows initial instability.

Finally, observe how the last series (reference value = 60) settles back to the original values (Fig. A-11):

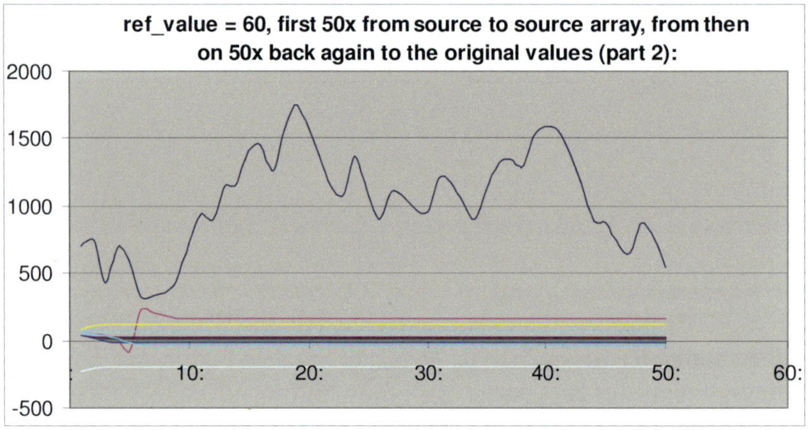

Fig. A-11: Graph showing the movement of number values back to their original values.

Instead of a convergence the movement by one array index demonstrates a chaotic behaviour, neither flat lining into a steady state nor oscillating around some stable centre.

So the adjustment algorithm produces three types of behaviour, expressed in terms of number movements relative to a given integer:

convergence towards the reference value and staying there;

oscillation around a centre which can be different from the reference value;

chaotic behaviour with no stability and no steady oscillation.

The reference value can be seen as an attractor, and depending on its effect on a number it can be described as a stable, a periodical, or strange attractor. Once the numbers in an entire element matrix have been modified, the result looks something like Fig. A-12 when presented as a surface defined by the elements' values.

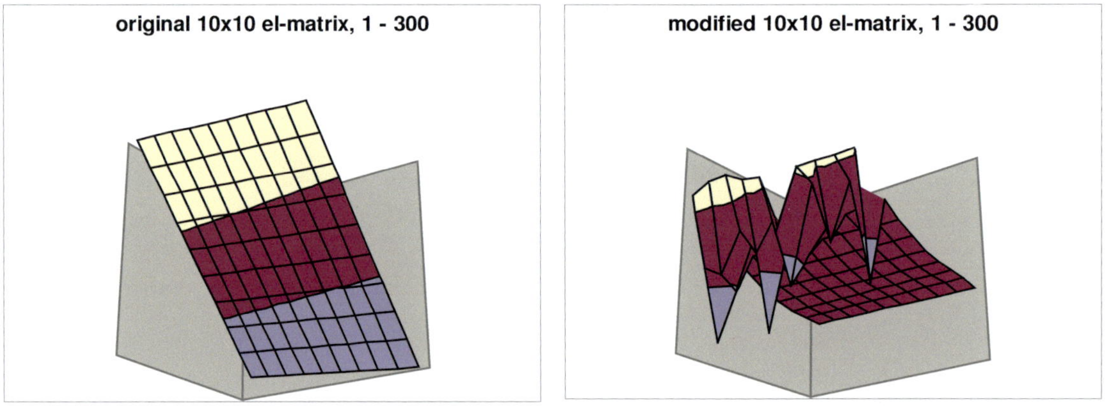

Fig. A-12: The contents of a node in the inner matrix before and after modification by the adjustment process. The values on the surface represent the numbers in the element matrix, and the numbers range from 1 to 300.

To summarise:

The element matrix of each node is a two-dimensional array, and its numbers interact with their respective counterparts in the target node during the adjustment process. The original reference value is the input to the matrix. Thus each element serves as an attractor with regard to its target which is encountered during the traversal of the tree from one node to its connections. Because the traversal is of type depth-first, the elements experience their modifications in the form of a delayed iteration, leading either to stable convergence towards the reference, to a periodical state characteristic of the reference although not directly representative of it, or to a chaotic state which also is not representative of the reference but nevertheless maintains itself as a unique representation with its own stability. These states can occur either during the presentation of input or while being exposed to the state of other nodes that have not necessarily been exposed to the input themselves.

In other words, the OtoomCM is a dynamic system in which input from the outside *as well as* existing states inside contribute to the continuous formation of substates of varying permanence, which irrespective of their degree of such permanence can exhibit stable, periodical, or chaotic behaviour. With respect to the adjustment process the synchronicity with the bio-version consists in the formal, yet unpredictable behaviour of proteins which, when presented with an electro-chemical perturbation from outside the neuron, react within the spectrum of affinity relationships in terms of their chemical dynamics. These dynamics in turn could be characterised using the terminology of attractors, leading to the specific phase state descriptions of the types outlined above.

Screen displays

To observe the system going through its cycles, three groups of displays are provided:

literal displays which supply information through words and numbers;

graphic displays which show the progress through the overall system and the state of nodes;

a graphic display that represents the output of the system.

Type 1 is mostly self-explanatory and will be referred to as the occasion arises. Type 2 presents the cycle flow through the matrix regions as well as the affinity relationships of the inner matrix nodes as the adjustment processes have moved through the virtual trees in the matrix, and type 3 indicates the output. Below is a screen shot of the system at the maximal setup at the present stage of development, which means a 26×25 main matrix, 10×10 element matrices at each node, and a connectivity of 5% (the other parameters are run-time configurations and do not require additional computer resources).

Before we go through some test runs, a few words on the node display. As the screen shot shows, in this type of display (in contrast to the connections as in Fig. A-2 and -3) the nodes are represented by coloured rectangles according to their affinity relationship with a connected node. The affinity relationship is defined by the similarity between the elements of one node and those of another, and the colours reflect that degree (see ***order of col.values*** in the legend to Fig. A-13). Since the matrix consists of many nodes, the geometric representation in the form of a rectangle is for convenience only and does not reflect the actual connectivity arrangement at run-time, and the colour image of a node can only ever refer to that one node only, it matters *what* affinity relationship is in fact shown.

The affinity value is an integer derived from a systematic comparison between the elements of one node and those of another. There are three possibilities of what this value could relate to:

the affinity value at the source node with a view towards the target node;

the value at the target node with a view towards the source node;

or with a view towards source and target nodes.

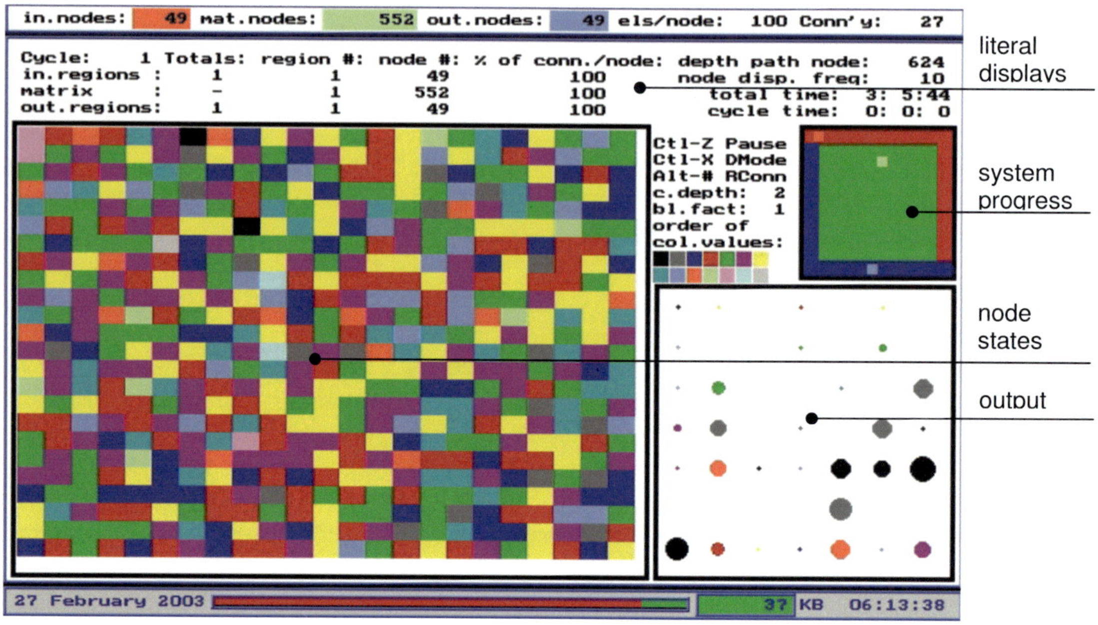

Fig. A-13: Screen shot of the total display representing a single input and one cycle through the inner matrix plus output.

in.nodes: number of input nodes
mat.nodes: number of inner matrix nodes
out.nodes: number of output nodes
els/node: number of elements per node
Conn'y: number of nodes connected to each node
Cycle: number of current cycle
Totals: number of total in- and output regions
region #: current region being processed (incl matrix)
node #: current node being processed
% of conn./node: percentage of node+tree processed
depth path node: node in the tree traversed
node disp. freq: display frequency in terms of cycles
total time: time taken for one test run
cycle time: time taken for one cycle
Ctl-Z, Ctl-X: real-time toggle switches for Pause, Display Mode (node state or connections)
Alt-#: real-time switch for in- and output region selection by number
c.depth: number of depth levels to be traversed
bl.fact: number of times a depth traversal is repeated
order of col.values: colours representing degrees of affinity, low to high from left to right, top to bottom
Bottom section: date, memory used of available total (bar), memory left (integer), time

In this computer model the target is the node next to the source according to the connectivity plan of the matrix. In other words, the colour shown represents the overall similarity between one node and those others it is directly connected to according to the plan. Note that these calculations are for display purposes only and do not in any way interact with the numbers held in the element matrices.

Considering the last item of the above choices first, such an arrangement would not be useful because the evaluation is ongoing and such a combined affinity value would therefore be changed, but under different conditions than those which caused the current value to be ascertained. The different conditions come from the connectivity layout (where any subsequent branches have nothing to do with the configuration of previous branches), and the results of the adjustment process via the depth-first approach explained earlier. Therefore each affinity value is not only specific in terms of the underlying integers but also in terms of its history. Since the history changes from one evaluation to another it cannot be used as a common and contributing factor to the result.

Choice 2 is similarly problematic. If the degree of affinity is meant to portray the state of a particular node, then looking backwards would once again incorporate the history of the evaluating procedure and hence be dependent upon the time and place within the matrix. Yet affinity relationships are not descriptors of a node's geographic location in the system, but indicators of a phase state due to relationships evoked by a transmission. Note also that in this case the affinity value not only reflects the node's own character but also says something about the source under the auspices of its history, which brings us back to the problem in the previous paragraph.

Using the affinity at the source derived from a comparison with the connected nodes is therefore the best solution. While there is a history it is forward- rather than backward-defining and so does not mix the result with backdated data. This forward-looking aspect of its history has important ramifications; although the affinity values at that point have been derived from a concrete analysis, in terms of the system's overall dynamic they represent the *affinity potential* of that node. In the adjustment processes that are to follow these values may or may not be actualised, given the interconnectivities and shared-node arrangements within the matrix. However, a repeated pattern of adjustment processes may well create a stable region in which particular affinity values are not only a fleeting potential but do become established as part of the affinity relationship complex existing there. Note that such a complex owes its presence not so much to some outside input (although this had been its original trigger) but comes about because of the subsequent internal activities around a certain location in the matrix. As such they can be seen as the basis for re-representative representations of an input: in other words, *abstractions*.

Choice 1 is also independent of the connectivity plan in the sense that the result can stand no matter what tree layout its neighbour in the matrix may be part of; as such it reflects more accurately the intent of depth-first traversals (whose adjustment processes are after all the causes for the affinity relationships).

One might say that the choice of attaching a descriptive value to the item one is looking at anyway is only natural—so why bother making things complicated? Yet it is not as simple as it looks; we are dealing here with the implementation of a conceptualisation of a biological system and what may seem obvious only does so because we have settled into a particular form of perspective. Therefore, when translating a concrete observation into the abstract and back again into the concrete care needs to be taken with the manner in which this translation is being performed. The issue is referred to several times in the main part of this book.

The colours of the node display then describe the potential affinity relationship of each node with its connections, after the system has received some input and has gone through the adjustment procedure along all the virtual trees in the inner matrix. The colours represent the state of the node at that time and since darker colours indicate lower affinity values, dark regions demonstrate a state of relatively high dissimilarity. On the other hand, lighter areas mean that here the nodes are largely similar to their connections, which is another way of saying that the nodes have either not been affected by the adjustment processes or a clustering has occurred through the application of a particular adjustment process using particular values turning out just so. The colour white stands for complete agreement: all the respective elements of the nodes under focus are the same. Note that the actual numbers themselves are quite irrelevant; what matters is not the *content* of the node matrices, but how they relate to others, that is to say their *functionalities*.

Therefore, what we see in the node display is a representation of a system which reacts to the environment via some input, and whose internal dynamic is determined by the affinity relationships with varying areas of specific states, where the representative medium (in this case integers) becomes irrelevant to the state transitions evoked by the dynamic. In the bio-version the medium are the hundreds and thousands of proteins which affect each other according to their electro-chemical relationships, and so the numbers in these matrices together with their modifications are a simulation of the affinity potential found in the neuron cell. And above all, *no rules apart from those inherent in the adjustment process govern the system.*

As mentioned in the pseudo code, the output is derived from an evaluation of the combined element values of those inner matrix nodes which are connected to the active output region. Since the range of element values changes all the time, the values are compartmentalised into as many blocks as are determined by the available colours plus the element values range, so that the eventual size and colour of a circle depends on the entire values produced so far (and of course on the available colour range, but that is fixed at 15 in this version). The principle employed here is based on *fuzzy logic*, where instead of True and False values a series of gradations is used which span a given range and so represent a location on a scale. (For a comprehensive discussion of fuzzy logic with sample code see *C++ Neural Networks and Fuzzy Logic* [1])

The same method appears in the selection process of the input nodes within their region, where once again the input value is placed into a numerical category that is determined by (a) the available range of input values so far and (b) the number of input nodes belonging to the active region. While the number of input nodes is set from the beginning, the input value range changes as the system goes through its cycles. For example, if there are 4 input nodes and the input values at that point range from 0 to 60, there will be 4 blocks each holding a number range of 15, so that input node #1 will receive any input between 0 and 15, input node #2 will receive any input between 16 and 30, and so on. If there are no exact multiples the remainder will be allocated to the last input node. That way not only will the spirit of depth-first input be adhered to (ie, there is a unique input to a section of the input region only), but the synchronicity with the bio-version is found in such instances as the special purpose nerves assigned to specific audio frequencies in the ear, or the visual field-specific receptors on the eye's retina.

Test runs

To illustrate the effects of scale and various methods of imparting the input several examples are presented, from simple configurations to more complex ones. Fig. A-14 shows a screen shot at the completion of the first cycle, with the parameters given around the area.

The input in Fig. A-14 was '60', repeated at each cycle. The affinity relationships of the nodes are varied with no particular pattern, and the output shows a certain homogeneity across the nodes. Since the input never changed the system settled into a somewhat stable state rather quickly with no change in the affinity relationships and, more significantly, a series of outputs that reflected the sameness of the input value. Fig. A-15 shows all the outputs of that particular test run (from now on a 10×10×10×10 matrix means 10 node rows and columns with 10 element rows and columns per node).

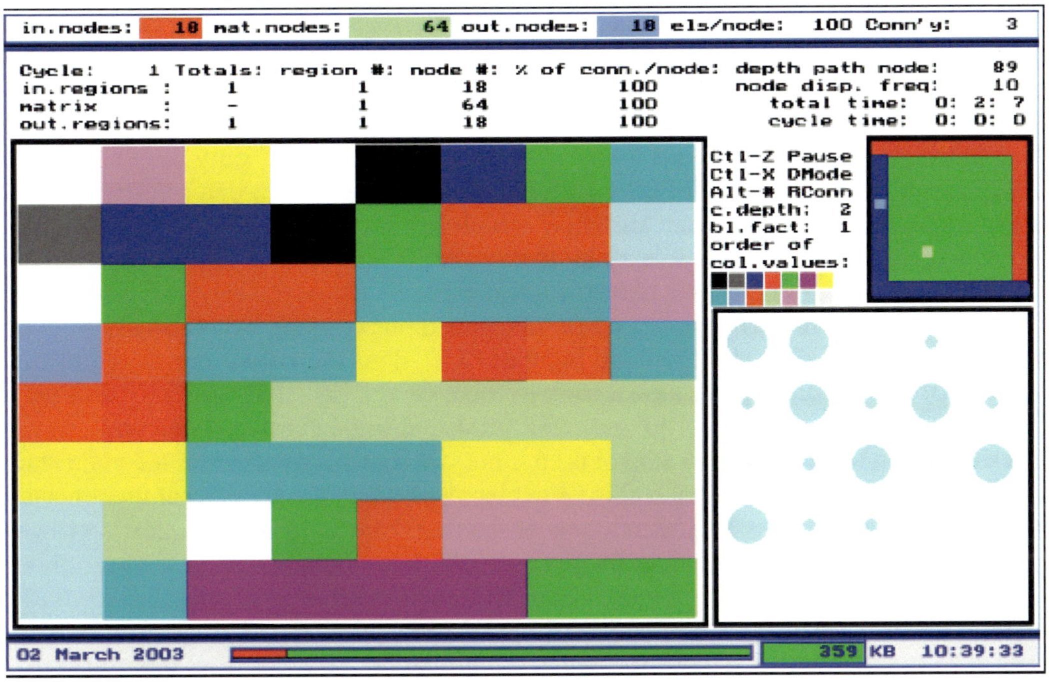

Fig. A-14: Screen shot of a 10×10×10×10 matrix with a connectivity of 5% and an element number range from 1 to 50.

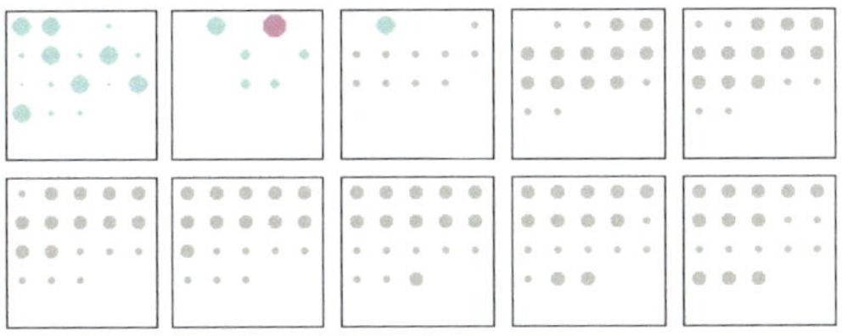

Fig. A-15: Outputs from the configuration shown in Fig. A-14.

matrix: 10×10×10×10	**numbers range:** 1-50	**connections:** 3
cycle numbers: 10	**connection depth:** 2	**block factor:** 1
input frequency: 1	**node display freq:** 2	**initial seed:** no
seeding value: n/a	**how often applied:** n/a	**basic input:** 60
alternative inputs: 0 0 0 0 0		
at cycles: 0 0 0 0 0		

Using roughly the same configuration but with an element number range from 1 to 300 enables the output to be more expressive. This is to be expected, given that the adjustment process can deal with a greater variety of resident values (Fig. A-16):

Fig. A-16: Outputs from a 10×10×10×10 matrix with a connectivity of 5% and an element number range from 1 to 300, with staggered input. On the left is the actual screen display, on the right a 'distant' view.

In addition to the higher number range, the system received a steady input of '60' every second cycle except for a '10' at cycle #5, '110' at #6 and '150' at #7. Looking at the circles on the screen presents a technically correct but too detailed image, rather like the dots in a newsprint. It is only at greater distance that the picture begins to blur and the discrete detail gives way to the representation we are used to. In other words, a 'face' only becomes a face when the accuracy gets suppressed in favour of the overall meaning. For that same reason we can observe these outputs at a greater 'distance' which has been artificially induced through blurring. At this level of inaccuracy it is easier to discern the influence of the different inputs. Thus it can be seen that from cycle #2 onwards a major purple ambience becomes established which is being maintained until cycle #5 when '60' gets replaced with '10'. There is an even greater change at #6 and only at #7 with an input of '150' does the colour purple return to a significant degree, with subsequent inputs back to '60'. Obviously, '10' and '110' had been more disturbing to the system than '150', and '60' produced some definite character.

So far the system had to adjust to some initial input, which then established a certain state across its nodes that was more or less stable. What happens when the significant input occurs *after* the system had been adjusted due to a type of seeding? Fig. A-17 shows a 26×25×10×10 matrix going through 20 cycles with a basic input of '60', but which had been seeded 5 times with a value of '200' before the first cycle. Then there were different inputs of '10', '110', '10', '10', and '10' at cycles #5, #9, #13, #15, and #19 respectively. The display is again blurred to give it sufficient 'distance' for better interpretation. (By the way, this test run—including node displays—took 13 hours and 26 minutes to complete on a 333MHz Celeron machine; that's linear processing for you)

The basic ambience appears to be a bluish hue which gets shifted towards the green and yellow by the input value of '10' at cycle #5. The yellow component lingers on in #6 but decreases steadily over the next two cycles, when an input of '110' reasserts purple and blue. These two colours trigger a return to light blue which could be observed at the beginning, but gets suppressed when at cycles #13 and #15 the input '10' is presented once again. Note that this time the purple/blue shift is stronger than before, with green appearing between #14 and #15 but gradually giving way to light blue, which by #18 begins to be firmly re-established. At the last cycle not only do we have a significant component of light blue, but now it is in conjunction with green. It appears therefore that under certain conditions one particular input can divert the colour ambience towards a new value, but can nevertheless trigger a re-occurrence of an older colour scheme which had been composed at an earlier stage. Since the system works through the processing of affinity relationships, such recurrences point to affinity complexes which can adapt to new trends, but are also able to revert to previous states within the bounds of existing values that are held by the nodes. As such they indicate the emergence of memory.

matrix: 26×25×10×10	**numbers range:** 1-300	**connections:** 27
cycle numbers: 20	**connection depth:** 10	**block factor:** 5
input frequency: 4	**node display freq:** 5	**initial seed:** yes
seeding value: 200	**how often applied:** 5	**basic input:** 60
alternative inputs: 10 110 10 10 10		
at cycles: 5 9 13 15 19		

Fig. A-17: Outputs from 20 cycles with a 26×25×10×10 matrix with a connectivity of 5% and an element number range from 1 to 300, with staggered input and initial seeding. Those outputs with different input are emphasised (see next page).

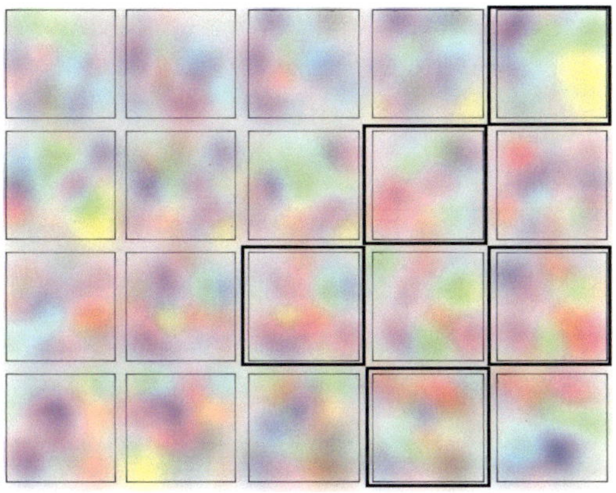

To keep it simple the outputs so far have been produced without the use of fuzzy logic when it came to the distribution of input values across the input nodes. As a final example let's see how the same setup from Fig. A-16 fares with fuzzy logic assigning a given input value to the respective nodes (Fig. A-18):

matrix: 10×10×10×10	**numbers range:** 1-300	**connections:** 3
cycle numbers: 10	**connection depth:** 5	**block factor:** 6
input frequency: 2	**node display freq:** 2	**initial seed:** no
seeding value: n/a	**how often applied:** n/a	**basic input:** 60
alternative inputs: 10 110 150 0 0		
at cycles: 5 6 7 0 0		

Fig. A-18: 10 cycle output with the 10×10×10×10 matrix from Fig. A-16. This time the input was applied to the input nodes according to the rules of fuzzy logic. Those outputs with different input are emphasised.

The most notable difference is the absence of any output whatsoever for the first four cycles. Sending '60' down one node twice (the input frequency was 2) is apparently insufficient to produce any state change in the system effective enough to reverberate through to the output nodes. However, at cycle #5 input '10' causes a definite display, far more distinct than had been the case with the input spread out in the previous example. Input '110' confirms the light blue but also brings in yellow, which maintains itself right to the end. Note also the stability of green once it had appeared as well as the purple/blue configuration that once again suggests the emergence of memory despite the fact that the last three cycles feature an input value of '60', a figure which initially managed no output but now carries the ambience triggered by '10' at the fifth cycle.

It is interesting to observe the node displays in this test run in Fig. A-19.

As expected from our outputs in Fig. A-18 the first 4 cycles showed the nodes without any differences at all—every element matrix was exactly the same as any other element matrix (note that the colours in the node display are calculated *after* the entire set of adjustment processes through all the connections at every connection depth level has run its course). In other words, the depth-first adjustment procedures may have dealt with some initial modification produced at some input node and its connection, but that was quickly overcome by the following adjustments at all the other nodes. Clearly, this particular input was not enough to 'sway' the system. However, by cycle #5 we have relatively dark colours indicating a high degree of dissimilarity between the nodes overall, but with the notable appearance of complexes (ie, the dark red and light blue regions). Recall that a region of dark red implies an area of similar dissimilarity, that is a region where the nodes are equally disturbed but show no congruence towards each other. Equally, a region of light blue illustrates also an area of similar dissimilarity, but its degree is now lessened. Note it is at such a period of high and relatively common dissimilarity (at cycle #6) that yellow emerges (compare with the output in Fig. A-18), and this colour is being maintained through subsequent cycles where the node displays demonstrate distinct regions of profound difference amidst a general background of greater and/or complete similarity (the red rectangles among green and even white areas in cycles #7 to #10). The emergence of memory-like outputs therefore coincides with the appearance of unique and stable regions, clusters which are able to maintain their character despite—or, as can be inferred from the results seen under "Adjustment procedure" earlier on, *because of*—the influence of continuously applied reference values. Hence it points to the nature of memory as being a dynamic phenomenon which not only relies on the previous presence of certain data, but also needs affinitive re-activation for it to exist as a re-representation of previous states.

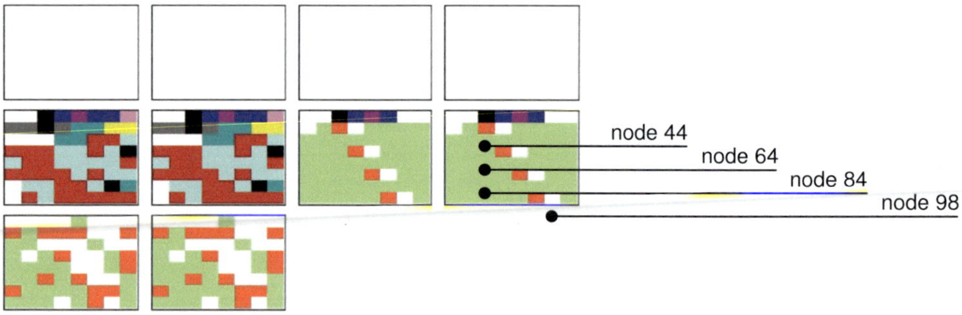

Fig. A-19: The node displays from the 10 cycle output with the 10×10×10×10 matrix from Fig. A-18. The locations of certain nodes are indicated; node 98 lies outside the inner matrix (see text for details).

The earlier graphs were based on selected number ranges to highlight the types of effects that can be expected.

Fig. A-20 has more samples of moving index values and this time derived from the configuration used in Fig. A-18, except that now there are 100 cycles and input occurred at every one of them. Nodes #44, #64, #84, and #98 were polled, and their element values at index 10, 20, 30, 40, 50, 60, 70, 80, 90, and 100 are arranged in series to observe their changes over time.

As a comparison between the outputs, the node states, and the graphs shows, regions of greater stability coincide with less turbulence in integer movements, although some activity seems to be necessary for a regional character to be preserved. By contrast, node #98 is an output node and so does not participate in the adjustment process along the depth traversals. This does not mean it is not affected, but the modifications are less volatile and demonstrate a more or less steady trend.

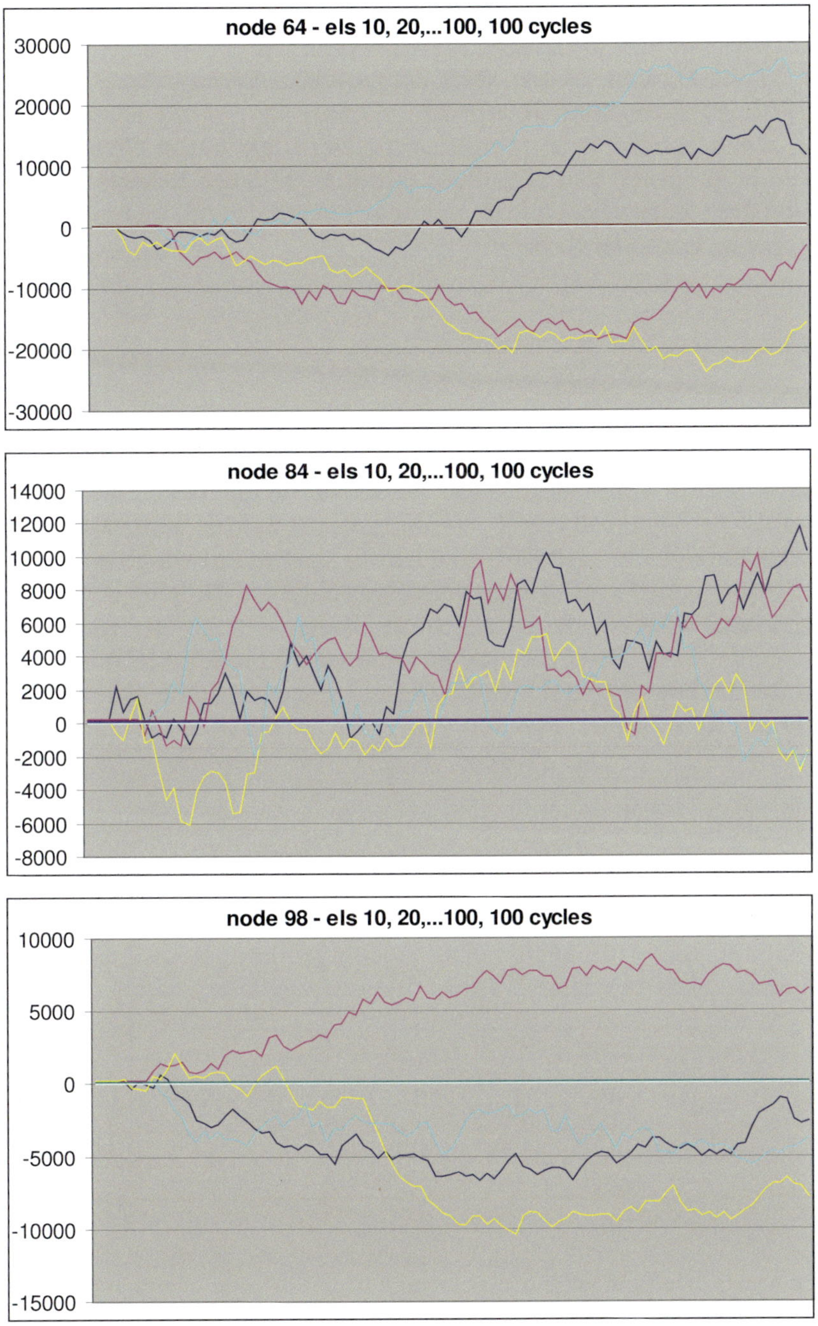

Fig. A-20: The movement of values at elements 10, 20, 30, 40, 50, 60, 70, 80, 90, 100 at nodes #44, #64, #84, and #98. Same configuration as under Fig. A-18, but with 100 cycles and an input at every cycle. Due to the scale only the movements of four elements are apparent with the rest not significant enough.

It should be mentioned that the value changes recorded here are a function of an interdependent system, where adjustments not only occurred between a source node and its target but with both sharing connections across the rest of the network. Therefore the lines reflect the trend of ongoing, accumulative affinity relationships and have to be seen as combinatorial states of the types outlined already.

As simple as the present configurations are, they demonstrate the recurrence and re-appearance of characteristics as a function of affinity relationships within the system's emergent complexes. The actual number values are of secondary importance; what matters are the affinities which can be produced and it is those affinities, affected through specific input values, that lead to cluster formations, their retention as potentials and, under the right conditions, to their re-emergence in similar form.

Einstein's 'mystery'?

This is the entire first paragraph of the book *Relativity* by Albert Einstein:

> In your schooldays most of you who read this book made acquaintance with the noble building of Euclid's geometry, and you remember—perhaps with more respect than love—the magnificent structure, on the lofty staircase of which you were chased about for uncounted hours by conscientious teachers. By reason of your past experience, you would certainly regard everyone with disdain who should pronounce even the most out-of-the-way proposition of this science to be untrue. But perhaps this feeling of proud certainty would leave you immediately if some one were to ask you: "What, then, do you mean by the assertion that these propositions are true?" Let us proceed to give this question a little consideration.

SI units

exa	E	10^{18}	deci	d	10^{-1}
peta	P	10^{15}	centi	c	10^{-2}
tera	T	10^{12}	milli	m	10^{-3}
giga	G	10^{9}	micro	μ	10^{-6}
mega	M	10^{6}	nano	n	10^{-9}
kilo	k	10^{3}	pico	p	10^{-12}
hecto	h	10^{2}	femto	f	10^{-15}
deca	da	10^{1}	atto	a	10^{-18}

Torresian Crow

Letter from Regional Service Director (Southern), Queensland Parks and Wildlife Service [3].

> Dear Mr Wurzinger
>
> I refer to your letter dated 12 January 1999. I apologise for the delay in responding to your earlier letters. Your queries about the Torresian Crow have been discussed with researchers from Griffith University. I have addressed in turn each of the dot points you have raised.

University researchers have measured decibel levels of crow roosts in suburban areas at a distance of 20 metres. From this research they were able to draw some interesting comparisons and conclusions. When compared with other noises that occur when crows seem to be the most vocal (at dawn), the average decibel level of crows is significantly lower than other very common suburban occurrences such as semi-trailer trucks, garbage removalists and kookaburras. None of the latter, whilst being much "louder", are considered public nuisance under current laws.

There are no figures of crow numbers in the greater Brisbane area, however there are good indications that populations have grown dramatically over the last thirty years, although no change in distribution is apparent. Whilst it appears that the population now has reached its maximum potential, it is unlikely that numbers will decline while activities such as the feeding of meat-eating birds remains a popular community activity, and other scavenging opportunities are wide spread in suburban environments (such as open tips, schools, parks, shopping centres and road networks).

There is no research to prove that crows don't carry agents of disease as such research is considered unnecessary due to the little *direct* contact humans have with crows. The lack of this direct contact makes the theory of disease spread as a result of crows not scientifically viable.

There is, however, research to support the fact that the scavenging nature of crows dramatically reduces the quantity of decaying matter that has the potential to produce harmful bacteria. Without their help, waterways and puddles could become dangerous places, especially after rain. Children playing in creeks could be most at risk. Therefore, in response to your fourth point, it is not only considered acceptable, but also a vital part in the suburban "ecosystem".

Conservation laws are widely supported in the urban community due to the great abundance and diversity of native wildlife and habitats the Brisbane area supports. The values of these are appreciated by members of the public for creating a unique natural character right in our own back yards. Many private land holders in urban areas also provide and maintain habitat for wildlife which is a vital element in the long term survival and conservation of our native flora and fauna.

I hope this clarifies some of the issues you have raised in relation to presence of crows in Brisbane.

Yours sincerely,
(signed)
Mike Harris
A/ Regional Service Director (Southern)
Queensland Parks and Wildlife Service

A Bill of Rights

The following articles are presented as a basic framework for a fundamental rule set around which a society turns. Although different cultures and different times create their own priorities and significance, there are dynamics common to all humans because the species shares a neurological system that underpins any subsequent variation.

Actual documents may differ across the borders, may emphasise this or that aspect, and in turn the understating of some other may have been balanced through further detail. Such hierarchy of values matters however when it comes to the daily life. Progression locks and/or assumed freedoms can emerge, clouding not only the concrete arena of societal events but also the perceived image of one's society and what kind of possibilities it offers as well as its limitations. The present articles also follow a hierarchy, with the upper always overriding the lower.

Human affairs never reflect some elegant vision, regardless how well-meaning its designer may have been. Instead they are often obtuse, convoluted, and opaque. Still, there is a difference between arguing about what *can* be done and spending energy on what *cannot* be done. Both set up their own flow of events, affecting whatever they touch.

1.

The ultimate goal of any nation must be the mastery of its environment, whether that be naturally given or artificially created.

For this to be successfully accomplished the intellectual character of the community has to be of a quality that supersedes the demands of present systems. Within the privileges of every fully functioning citizen is incorporated the fundamental duty to uphold this quality to his or her best ability and knowledge.

Before any rights can be ascertained and given, the system must have the capacity to comply with that demand. A top priority must therefore be the safeguarding of this capacity.

No system can survive a challenge to its nature without possessing the sufficient degree of latency to assemble responses to its current state. Hence the capacity of the system needs to include the potential for overcoming stasis.

2.

It is the inherent right of every human being to express freely any ideas or thoughts deemed necessary. The judgment of this necessity must always be the property of the originating person and must be defensible.

Flow of information is essential for the health of a system, whatever the nature of data. Given the possibility of high $\prod$ low or low $\prod$ high conceptual intersections, exceptions can occur that require a withholding of information to particular other domains. Since the judgment about applying an exception rests with the originator, the informative value of an event entailing the creation, distribution, or containment of information will be as high as possible.

3.

Every citizen, regardless of age and gender, has the right to happiness and the quest towards it. This includes education; medical care; erotic fulfillment; meaningful employment; and social care.

It is the duty of every citizen to utilise these rights in such a manner that no-one else's pursuit of happiness will be jeopardised objectively.

It is the nature of humans to exercise their inherent optionality throughout their life. Curtail any one aspect and substitutes will be thought outside a given framework (centering on but not limited to the one designed to apply such constraint).

The availability of benefits to some at the cost of limiting them to others is akin to unchecked censorship regarding information. A system suffers if the potential of its domains should be restrained through the influence of others.

4.

No endeavour which seeks to influence and define people's lives may be enacted that has as its basis and justification the dogma of any religion, culture, or tradition.

Similarly, no activity pursued by any individual or individuals may be prohibited through justifications derived from the realms of religion, culture, or tradition.

The larger the system, the greater the need for functional synchronicity. While niches of idiosyncratic dynamics can exist anywhere, the moment they interact denominatively with the dynamics of other member domains (at whatever scale) such an interaction becomes a constraint leading to degeneration.

In terms of the dynamics between two or more domains, it does not matter whether there is an act that seeks to prescribe or proscribe, or whether there is a prescription or proscription in relation to some act.

5.

The governing process of the nation must at all times be open and accessible to any citizen, and the justification for access must be the sole property of the initiator. It is the duty of every governing body to utilise the best available technology in order to facilitate such access.

In line with the previous paragraphs free flow of information must be maintained, including the need for updating the necessary infrastructure.

6.

The purpose of the law is to enable any course of action to proceed harmoniously and efficiently. Notwithstanding any previous paragraph only those laws are allowed that contain the following structure: the objective goal; the relevant event; and the principles underlying that event.

Any law that seeks to achieve its purpose by the discrimination against any human characteristic is invalid. The only justifiable limiting measure is the degree of capability in relation to the goal, the event, and/or its principles.

Similarly, no person or group of persons is entitled to claim special privileges on behalf of anyone due to race, religion, or cultural background.

Contrary to some beliefs the law should serve the people, rather than the people serving the law. Unless a law can demonstrate its utility under the parameters given above, it runs the risk of being discriminatory and exclusively self-serving.

Further in line with the above the purpose of a law is to provide a format to instantiate a useful event under the auspices of the document. Apart from any safeguards specific to the event there is no justification for any other constraint applied to a person.

7.

This Constitution is the binding document defining the basis for the existence of all individuals who freely and unhindered subscribe to its tenets as the framework of their lives, their intellectual and materialistic endeavours; and who have thus established the geographical extent of their activities by virtue of territory and boundaries.

Domains can be defined and from a certain complexity onwards the domains define themselves. Whether in terms of physical size or cognitive dynamics a system attains a greater degree of justification if it has consciously chosen to assert its existence.

Glossary

Note: the number in brackets refers to the first page that contains a comprehensive description of the item.

abstract, abstraction
The general representational form of some information processed by the mind. See also *object abstract* and *functional abstract*. (17)

abstraction level
The level at which an *abstract* exists in terms of its cognitive functionality, where higher levels denote more general perceptions and less actual detail. (381)

adjustment procedure
An algorithm through which one value in an *element* matrix (the *resident value*) is changed in terms of the corresponding value in another element matrix (the *reference value*). It is the underlying rule in the *Otoom computer model*. (354)

affinity potential
An *affinity relationship* amongst a set of *nodes* within the *inner matrix* of the *Otoom computer model* which would be realised if the values underpinning that relationship were to be confirmed through the application of the *adjustment procedure*. (402)

affinity relationship
A set of *affinity values* which describes the state of similarity amongst a set of *nodes* in the *inner matrix* of the *Otoom computer model*. (358)

affinity value
An integer which is derived from a systematic comparison between the values of one *element* matrix and those of another. As such it expresses the degree of similarity between these matrices. (539)

associative halo
A set of associations related to a particular *thought structure* because it and the other TSs possess mutually compatible affinities. (182)

associative intensity
The degree to which one *denominative thought structure* or *denominative thought pattern* is able to modify the functional abstracts of another *thought structure* or *thought pattern* respectively or more generally, the degree to which one *thought structure* relates to another inside the mind. (90)

associative resonance
The result of *associative intensity*, where one or more *thought structures* are now held to be more significant. (131)

autonomous thought complex
A thought complex has become autonomous in relation to its *abstracts* when it has become capable of absorbing new associations without any longer requiring the elements responsible for defining its original domain. (323)

breadth-first
A type of traversal through a *tree* structure in which every *child node* at the next level is being accessed before any child node at the following level is being accessed. (553)

child node
A node in a *tree* which can be accessed from a node above, its *parent node*. (532)

conceptual intersection (CI)
An interaction between two *abstraction* complexes of dissimilar extent, with one being added to the other such that the outcome reflects the uneven distribution of abstractive potential. It is different in scale but similar in principle to the *adjustment procedure* dealing with *resident* and *reference values*. (416)

conceptual radius
The distance between the centre of an idea (its core) and those concepts which are seen to be relevant to the core by the person. (407)

cognitive functional abstract (CFA)
A *functional abstract* which not only provides the information on the existence of some circumstance but also has links to other *thought structures* regarding the reasons for its existence; as such it cannot appear without some elaborate thought processes and, to remain true to its nature, it must not be shifted across the logic boundary. (100)

congruent container
See *container*.

congruent PAT loop (c-PAT-l)
The property of a *PAT chain*, where all sets of neighbouring PATs are associated with mutually compatible *rule sets*. (81)

congruent thought pattern (CTP)
The property of a set of *thought structures*, where all thought structures are associated with mutually compatible *rule sets*. (92)

connection depth
The number of levels the *adjustment procedure* reaches downwards through the connections of a *node* (the *tree*) in the *inner matrix* of the *Otoom computer model* in order to change a *resident value* towards a *reference value*. (355)

connectivity
A parameter belonging to the *Otoom computer model* which denotes the number of nodes every node in the matrix is connected to. It is expressed as a percentage of the total number of nodes in the matrix. (355)

consciousness
A default recursion: as the conscious part of a *thought structure* is used for articulation (through words, gestures, music, etc) and the result seeks further associations amongst the entire *thought structure*, those associations may form into further *conscious thought structures* to be articulated; and so on. A recursive subprocess becomes exhausted when no more associations can be formed at any level of abstraction; its underlying process however continues until the recursive operation is shut down through sleep or some trauma. (94)

conscious thought structure (con-TS)
That part of a *thought structure* which is made conscious; as such it can be articulated through language, pictures, music, etc. (93)

container
A higher-level *thought structure* or *thought pattern* which serves as a holding area for *abstracts* and *emotional modules*. (58)

A container is congruent if it has been assembled from TSs which are affinitive because of their specific contents; it is incongruent if its creation is due to its set of *emotional modules* having an affinity with the EMs of other TSs, thereby neutralising the significance of their contents. (134)

contextual association
An association between *abstracts* based on the similarity of their representational contents, while disregarding their respective *abstraction levels*. (177)

contextual storage
A means of storing information where each data item is specifically associated with another data item, no matter how many times a particular item may be repeated in different *thought structures*. Its opposite is *functional storage*. (260)

default abstract
An abstracted information derived from some internal process of the system itself. (46)

delayed iteration
A form of iteration in which a resultant parameter value is not fed directly back into the process but becomes part of a set of values which later on are returned to the system under the same algorithm. (546)

denominative thought pattern (DTP)
A *thought pattern* which is associative with another thought pattern such that the first is able to modify the functional abstracts of the second. (90)

denominative thought structure (DTS)
A *thought structure* which is associative with another thought structure such that the first is able to modify the functional abstracts of the second. (90)

depth-first
A type of traversal through a *tree* structure in which each branch is followed systematically to the last level of the given *connection depth* and then back-tracked until the next hitherto

unexplored branch is encountered until all the tree's *child nodes* have been covered. It is used in the *Otoom computer model*. (531)

distributed system
A system in which the processes can occur in parallel to each other, not necessarily in any particular order; eg, process #3 can run while process #1 is running also. (532)

effectiveness chart (e-chart)
In its basic form a chart featuring three parameters of a given system: the number of products existing, the *potential* of the products (*latent* and *realised*), and the degree of applicability resulting from the latent and/or realised potential of the products. The potential and applicability values are relativistic, and the chart is used in anti-clockwise fashion, starting with the number of products. The extended 3D version places the parameters within the system's resource space and adds a complexity parameter. (224)

elements
Indices in a two-dimensional array (matrix) that represents a *node* in the matrix of the *Otoom computer model*. The indices are assigned integers according to a system. (529)

emotive module (EM)
A name for emotions which reflects their functional modularity acting within the framework of *thought structures*. (134)

fractal boundary
A line which surrounds a region of possible outcomes in a chaotic system, determined by the nature of its attractor. (367)

functional abstract (FA)
An abstracted information derived from some *rule*, that is an observed event. (43)

functional association
An association between *abstracts* based on their similarity of *abstraction levels* on which they reside, where "functional" refers to the role the abstract can play due to its level. (177)

functional storage
A means of storing information where a particular data item is stored once only in a general form, and only when the need arises does it become associated with another data item according to specific *functional abstracts*. (260)

fuzzy logic
A manner of assigning a parameter in terms of the relationship of its value to the context in which it occurs. Thus it is not limited to True/False expressions but can articulate the parameter through a proportionality in relation to the range of all possible values that parameter can have under specific conditions. (530)

grounding
To connect an *abstract* existing at a higher *abstraction level* with lower level representations such that contact is made eventually with one or more *PAT*[1]s. (419)

homologous TSs
Two or more *thought structures* which are similar in terms of the configuration of their cognitive elements, but differ in the behaviour resulting from their presence. (209)

horizontal perception-abstraction transference (horizontal PAT)
A *perception-abstraction transference* which occurs across the same abstraction level. (138)

illusion tree
A *PAT chain* (the trunk) containing a certain *abstract* and rising from a network of *grounded abstracts* (the roots) up through several *abstraction levels* until at a high level it is once again associated with various abstracts (the crown), none of which has downward links with reference to that particular abstract except through the trunk. (194)

incongruent container
See *container*.

incongruent functionality
The property of two associated abstractions, the second one level higher or lower than the first, and each associated in turn with mutually incompatible *rule sets*. (58)

incongruent PAT loop (i-PAT-l)
The property of a *PAT chain*, where one or more sets of neighbouring PATs are associated with mutually incompatible *rule sets*. (80)

incongruent thought pattern (ITP)
The property of a set of *thought structures*, where one or more thought structures are associated with mutually incompatible *rule sets*. (89)

induction method
A form of reasoning, where a conclusion is reached based on previous similar occurrences. (322)

inner matrix
A set of processing *nodes* in the *Otoom computer model* which is part of the entire matrix. (529)

input nodes
A processing unit in the matrix of the *Otoom computer model* which is designed to receive an input. (529)

input region
A set of *input nodes* in the *Otoom computer model* which may or may not comprise the entire set of input nodes available under the particular configuration set by the user. It is part of the entire matrix. (529)

intelligence
In general, the ability to form associations with higher-level *abstracts* (the higher the level, the greater the intelligence), as well as with affinitive abstractions belonging to the same level

in other *thought structures*; the ultimate truth value of such associations remains undefined. (369)

intuitive functional abstract (IFA)
A *functional abstract* which represents a *thought structure* that merely holds information about some circumstance because the event has been experienced; no actual understanding has taken place but the existence of the circumstance is accepted. (100)

latency
The degree to which the *latent potential* of a product has been preserved, given the complexity of its environment; the less complex the environment, the higher the latency during the life cycle of the system. (224)

The degree of latency of a system is proportional to the dimensionality of its *fractal boundary*. (367)

latent potential
That part of the *potential* of a product which has not been realised (also see *realised potential*). (224)

life
Any dynamic, self-regulating, pattern-seeking system. (400)

linear system
A system in which the processes occur one after the other; eg, process #2 cannot start until process #1 has finished. (532)

logic boundary
A functional boundary defining the extent to which an abstraction can be shifted to a higher level while still maintaining congruity with the lower one. Beyond the boundary the congruity is lost, due to the influence of some *functional abstract*. (52)
neurites
The extensions of *neurons* which connect them to each other. (339)

neuron
A processing unit that makes up the biological nervous system. The functional equivalent to a *node* in the *Otoom computer model*. (339)

neuron doctrine
A view that states that *neurons* are autonomous units that are connected to each other. (339)

neurotransmitter
A chemical compound which acts as a transmitter of information between two *neurons* across their *synapse*. (339)

node
A processing unit that makes up the matrix in the *Otoom computer model*. The functional equivalent to a *neuron* in the biological version. (529)

object abstract (OA)
An abstracted information derived from a single representational form of information, the ultimate source of which lies outside the entire system. (43)

Otoom
The name given to the descriptive realm which lies between that of neuroscience and cognitive science. It is derived from the acronym of the book's title. (336)

Otoom computer model
A computer program which simulates the functionalities of the biological mind, reducing them to the bare functional minimum necessary to exhibit the mind's dynamics; thus it can be described as an artificial mind. (351)

output nodes
A processing unit in the matrix of the *Otoom computer model* which is designed to receive an output from the *inner matrix*. (529)

output region
A set of active *output nodes* in the *Otoom computer model* which may or may not comprise the entire set of output nodes available under the particular configuration set by the user. It is part of the entire matrix. (529)

parent node
A node in a *tree* from which a number of other nodes (the *child nodes*) can be accessed in a downward direction. (532)

PAT chain
A series of *PATs* linked together across several *abstraction levels*, where each level may or may not be different from its neighbour. (101)

pattern
A phenomenon which has a correlative dynamic in an accompanying system in terms of the phenomenon's characteristics. (319)

perception-abstraction transference (PAT)
The movement of information from one *abstraction* to another; denoted by PAT^n, where n represents the steps to the target. The movement can also be upwards to a higher level, or downwards to a lower one. (20)

periodic attractor
In the *Otoom computer model* a parameter value which causes another parameter value to achieve an oscillating state when connected via some specific algorithm (eg, a *reference value* acting upon a *resident value* and producing an integer whose value oscillates through a steady range through subsequent *delayed iterations*). (353)

potential
The things that can be done with a product, where 'product' can be a material object or an abstract or some other cognitive structure (see also *latent potential* and *realised potential*). (228)

progression lock
A general form of determinism which restricts the future options available to the active parts of a system by virtue of the already existing characteristics of these parts. Its main feature is a decrease in potential against a corresponding increase in complexity in the environment. (278)

projection
The process of linking some incoming information (in the form of *abstracts*) with *thought structures* that are peculiar to the receiving mind. (84)

pseudo code
The meaning behind a piece of computer code expressed in ordinary language. (530)

qualities of an abstract object (Q_{ao})
The identified qualities of a thing after the first level of *abstraction*. (19)

qualities of a perceived object (Q_{po})
The identified qualities of a thing at the first level of *abstraction*, ie as sensory input. (19)

rationalisation complex
A process involving two sets of *thought structures*, one subconscious and the other conscious and both connected via illusion trees. The *sub-TSs* consist of *synonymous TSs*, the *con-TSs* consist of *homologous TSs*. (301)

realised potential
That part of a *potential* of a product which has been utilised; the more has been utilised, the less the product's *latency*. (277)

reciprocity
A conclusion reached via the processing of information across a *PAT chain* in one direction is held compatible with another reached across the same chain in the opposite direction. (62)

reference value
The name given to a value assigned to an index in the *element* matrix array of the *Otoom computer model* which adjusts the *resident value* according to the *adjustment procedure*. (353)

religion
The product of a society which is able to process higher-level abstractions, but without the full understanding of their usage. (385)

resident value
A value assigned to an index in the *element* matrix array of the *Otoom computer model* which is being adjusted by a *reference value* according to the *adjustment procedure*. (353)

resonant thought structure (RTS)
A *thought structure* which has been associated with another thought structure such that due to their mutual *associative resonance* it has now acquired a greater degree of significance in the person's mind. (131)

semantic resolution
The degree of detail in a *thought structure*, represented by its content elements, which allows it to be associatively accessed by another TS (internally or externally) so that its intended meaning can be transmitted without loss. (329)

social horizon
The extent across society at large up to which one's actions can be appreciated to be significant in terms of a mutual interdependence between society and one's self. (134)

stable attractor
In the *Otoom computer model* a parameter value which causes another parameter value to achieve a steady state when connected via some specific algorithm (eg, a *reference value* acting upon a *resident value* and producing an integer which remains the same through subsequent *delayed iterations*). (353)

strange attractor
In the *Otoom computer model* a parameter value which causes another parameter value to achieve a varying and unpredictable state when connected via some specific algorithm (eg, a *reference value* acting upon a *resident value* and producing an integer whose value undergoes unpredictable changes through subsequent *delayed iterations*). (353)

subconscious thought structure (sub-TS)
That part of a *thought structure* which remains in the subconscious; as such it is not directly articulated through language, images, music, etc. (93)

synapses
The physical region of contact between two *neurons*, across which the transmission of information occurs. (339)

synonymous TSs
Two or more *thought structures* which are similar in terms of their behaviour in a given context, but differ in the configuration of their cognitive elements. (300)

thought pattern (TP)
A set of *thought structures* which have an affinity with each other. (2)

thought structure (TS)
A set of *abstracts* which have an affinity with each other. It may become divided into the *conscious thought structure* (*con-TS*) and the *subconscious thought structure* (*sub-TS*). (2)

tree
A data structure which consists of a *parent node* from which there are paths to *child nodes*. Each child node can be considered a parent node in relation to any subsequent connections. A parent node can have a pre-set number of child nodes, or that number can change during the life cycle of the tree. Each set of child nodes per parent node comprises a depth level, with the very first parent node usually allocated level 0. (533)

twin structure
Two structures of *abstracts* of the same type (eg *thought structures*, *illusion trees*, *containers*), both sharing certain abstracts which then become virtually *homologous* in character; as such the abstracts have a different functionality depending on the particular structure through which they are accessed. (217)

unified field phenomenon
A conclusion which has been confirmed in a detail can also be ascertained within a wider context, and vice versa. (4)

vertical perception-abstraction transference (vertical PAT)
A *perception-abstraction transference* which occurs across differing abstraction levels. (138)

Table of polar pairs

Some items found in the Glossary represent one version of an underlying double theme. This table lists those items as polar pairs.

cognitive functional abstract	intuitive functional abstract
congruent container	incongruent container
congruent PAT loop	incongruent PAT loop
congruent thought pattern	incongruent thought pattern
conscious thought structure	subconscious thought structure
contextual association	functional association
functional abstract	object abstract
homologous TS	synonymous TS
horizontal perception-abstraction transference	vertical perception-abstraction transference
qualities of an abstract object	qualities of a perceived object
latent potential	realised potential
thought pattern	thought structure

Bibliography

On Internet material:

(1) Due to the changing nature of domain names and addresses and the complexity of some links, a source from the Internet is (a) listed under the title of the text, the name of the local site's proprietor and the date accessed, but not its URL; and (b), quoted from the then current web page.

(2) Because page numbers are irrelevant in browser-style displays, three words from the relevant tract have been quoted instead for use in text search (eg, <ab cd ef>).

13. A change of pace

[1] 27 June 1999, *Rotblat makes a hit with call for a Hippocratic oath*, nature news+ conference diary world conference on science, Nature, Macmillan Publishers Ltd, England. Accessed on Internet 25 Nov 2002.

[2] Süskind, P., 1998. *Perfume - the story of a murderer*, Chivers Press Bath, UK.

14. From: Chaos (?)

[1] Chudler, E.H., *Brain Facts and Figures*, personal web page. Accessed on Internet 11 Nov 2001.

[2] Bear, M.F., Connors, B.W., Paradiso, M.A., 2001. *Neuroscience - Exploring the Brain*, Lippincott Williams & Wilkins, Baltimore, USA. (p. 25)

[3] Chudler, E.H., *op. cit.*, same page.

[4] Bear, M.F., Connors, B.W., Paradiso, M.A., *op. cit.*, (p. 515)

[5] *ibid*, (p. 800)

[6] *ibid*, (p. 161)

[7] Clayton, K., 1997. *Basic Concepts in Nonlinear Dynamics and Chaos*, A Workshop presented at the Society for Chaos Theory in Psychology and the Life Sciences meeting, July 31, 1997 at Marquette University, Milwaukee, Wisconsin. Accessed on Internet 9 Sep 2002.

[8] Abraham, F.D., 1992. *Chaos, Bifurcations, & Self-Organization: Dynamical Extensions of Neurological Positivism & Ecological Psychology*, reprinted from Psychoscience, 1(2), 85-118. Accessed on the Internet 18 Aug 2002. (<By implying that>)

[9] *ibid*, (<A parameter can>)

[10] von Foerster, H., 1997. *Lethology - A Theory of Learning and Knowing vis à vis Undeterminables, Undecidables, Unknowables*. Accessed on Internet 4 Aug 2002. (<A "machine" in>)

[11] Smith, C.U.M., 1993. *The Complexity of Brains: A Biologist's View*. Accessed on Internet 4 Aug 2002.

[12] *ibid*, (<The KCS definition>)

[13] *ibid*, (<Levels of organisation>)

[14] Bear, M.F., Connors, B.W., Paradiso, M.A., *op. cit.*, (p. 39)

[15] Howell, F.W., Dyhrfjeld-Johnsen, J., Maex, R., Goddard, N., De Schutter, E., 1999. *A large scale model of the cerebellar cortex using PGENESIS*. Accessed on Internet 5 Oct 2002.

[16] *ibid*, (p. 4)

[17] *ibid*, (p. 2)

[18] Glanville, R., 1997. *The Value of being Unmanageable: Variety and Creativity in CyberSpace*, CybernEthics Research®, Hants, United Kingdom. Accessed on Internet 4 Aug 2002.

[19] *ibid*, (p.2)

[20] *ibid*, (p. 3)

[21] *ibid*, (p. 6)

[22] Kahn, D., Hobson, J. A., 1994. *Self-organization and the dreaming brain*, Harvard Mental Health Letter, May 94, Vol. 10 Issue 11, p3, 3p. Accessed on Internet 4 Aug 2002. (<Many theories about>)

[23] *ibid*, (<Another example of>)

[24] Bear, M.F., Connors, B.W., Paradiso, M.A., *op. cit.*, (p. 132)

[25] *ibid*, (p. 728)

[26] Williams, R., 2002. *The Science Show*, Radio National, Australian Broadcasting Commission. Broadcast 20 Jul 2002.

[27] 5 October 2002. *Brain wave well thought out*, The Courier Mail, Brisbane.

[28] Zeki, S., 1999. *Inner Vision, An Exploration of Art and the Brain*, Oxford University Press, Oxford. (p. 17)

[29] *ibid*, (p. 71)

[30] Stankov, L., 1999. *Capacity versus Speed in Intelligence and Ageing*, Australian Psychologist, July 1999 Volume 34 Number 2. (p. 138)

[31] Irion, R., 2001. *Say the magic words*, New Scientist, 9 June 2001.

[32] Wang, G.M., Sevick, E.M., Mittag, E., Searles, D.J., Evans, D.J., 2002. *Experimental Demonstration of Violations of the Second Law of Thermodynamics for Small Systems and Short Time Scales*, Physical Review Letters, Volume 89, Number 5, 29 July 2002.

[33] Stowasser, J.M., Petschenig, M., Skutch, F., 1997. *Stowasser*, HPT-Medien-AG, Zug, Switzerland. (p. 566)

[34] Bear, M.F., Connors, B.W., Paradiso, M.A., *op. cit.*, (p. 416)

[35] Nakott, J., 1998. *Jeder irrt, der zu wissen glaubt, was ein anderer denkt*, Bild der Wissenschaft, Deutsche Verlags-Anstalt GmbH, Stuttgart. Accessed on Internet 4 Aug 2002.

[36] Alexander, V.N., 2001. *Neutral Evolution and Aesthetics: Vladimir Nabokov and Insect Mimicry*, Dactyl Foundation for the Arts & Humanities, New York.

[37] Boden, M., 2000. *Autopoiesis and Life*, Cognitive Science Quarterly (2000) 1, 117-145, Hermes Science Publications, Paris.

[38] Bear, M.F., Connors, B.W., Paradiso, M.A., *op. cit.*, (p. 612)

[39] Pinker, S., 1998. *How the Mind Works*, The Softback Preview, UK. (p. 306)

[40] Zeki, S., *op. cit.*, (p. 23)

15. To: Order (?)

[1] Bear, M.F., Connors, B.W., Paradiso, M.A., *op. cit.*, (p. 482)

[2] *ibid*, (p. 484)

[3] Zeki, S., *op. cit.*, (p. 21)

[4] 17 Mar 2003. *Der Schraubenzieher sagt dem Gehirn: Greif mich!*, Bild der Wissenschaft, Deutsche Verlags-Anstalt GmbH, Stuttgart. Accessed on Internet 21 Mar 2003.

[5] Furlow, B., 2001. *Play's the thing*, New Scientist, 9 June 2001.

[6] Carr, T., 1977. *The Dance of the Changer and the Three*, from *Deep Space*, Corgi Books, London. (p. 136)

[7] Elston, G.N., 2003. *Cortex, Cognition and the Cell: New Insights into the Pyramidal Neurons and Prefrontal Function*, Cerebral Cortex, V 13 N 11, Oxford University Press.

[8] Chainay, H., Humphreys, G.W., 2002. *Neuropsychological evidence for a convergent route model for action*, Cognitive Neuropsychology, 2002, 19 (1), Psychology Press, UK. (p. 67)

[9] *ibid*, (p. 71)

[10] Smith, R., 2002. *Does Reading alter our Reality?*, Psychology Today January/February 2002, Sussex Publishers Inc, New York. (p. 28)

[11] Bear, M.F., Connors, B.W., Paradiso, M.A., *op. cit.*, (p. 643)

[12] *ibid*, (p. 766)

[13] *ibid*, (p. 623)

[14] Zeki, S., *op. cit.*, (p. 75)

[15] Raftopoulos, A., 2001. *Is perception informationally encapsulated? The issue of the theory-ladenness of perception*, Cognitive Science 25 (2001), Cognitive Science Society, Inc. (p. 423)

[16] Eliasmith, C., 1996. *The third contender: A critical examination of the dynamicist theory of cognition*, Philosophical Psychology, Vol. 9 No. 4 pp. 441-463. Accessed on Internet 18 Aug 2002.

[17] Port, R.F., 2001. *The Dynamical Systems Hypothesis in Cognitive Science*, accepted for The MacMillan Encyclopedia of Cognitive Science. Accessed on Internet 18 Aug 2002.

[18] Zeki, S., *op. cit.*, (p. 184)

[19] Bear, M.F., Connors, B.W., Paradiso, M.A., *op. cit.*, (p. 314)

[20] *ibid*, (p. 321)

[21] Huntington, S.P., 1998. *The Clash of Civilizations and the Remaking of World Order*, Touchstone Books, London. (p. 21)

[22] Nørretranders, T., 1998. *The User Illusion*, Penguin Books, New York. (p. 280)

[23] Lockwood, M., 1993. *The Grain Problem*, from *Objections To Physicalism*, Clarendon Press, Oxford. Accessed on Internet 20 Jul 2002.

[24] Kenny, V., 1985. *Life, the multiverse and everything; An introduction to the ideas of Humberto Maturana*, presented paper. Accessed on Internet 4 Aug 2002.

[25] Maturana, H., 1970. *Biology of Cognition*, Biological Computer Laboratory Research Report BCL 9.0. University of Illinois, Urbana IL. Accessed on Internet 4 Aug 2002.

[26] Maturana, H., Mpodozis, J., Letelier, J.C., 1995. *Brain, Language and the Origin of Human Mental Functions*, Biological Research 28: 15-26 (1995). Accessed on Internet 4 Aug 2002.

[27] Sharov, A., 1999. *Pragmatism and Umwelt-theory*. Accessed on Internet 4 Aug 2002.

[28] Gabora, L., 1998. *Autocatalytic Closure in a Cognitive System: a Tentative Scenario for the Origin of Culture*, Psycoloquy 9(67).

[29] Epp, S.S., 1995. *Discrete Mathematics with Applications*, Brooks/Cole Publishing Company, Pacific Grove, CA, USA. (p. 270)

[30] Bear, M.F., Connors, B.W., Paradiso, M.A., *op. cit.*, (p. 741)

[31] Kokinov, B.N., Petrov, A.A., 2000. *Integration of Memory and Reasoning in Analogy-Making: The AMBR Model*, CEE Center for Cognitive Science, New Bulgarian University, Sofia, Bulgaria. Accessed on Internet 18 Aug 2002.

[32] Nissley, H.M., Schmitter-Edgecombe, M., 2002. *Perceptually Based Implicit Learning in Severe Closed-Head Injury Patients*, Neuropsychology 2002, Vol. 16, No. 1, 111–122, American Psychological Association, Inc., Washington. Accessed on Internet 20 Sep 2002.

[33] Bear, M.F., Connors, B.W., Paradiso, M.A., *op. cit.*, (p. 588)

[34] Richards, J.M., Gross, J.J., 2000. *Emotion Regulation and Memory: The Cognitive Costs of Keeping One's Cool*, Journal of Personality and Social Psychology, September 2000 Vol. 79, No. 3, 410-424, American Psychological Association. Accessed on Internet 20 Sep 2002.

[35] Gallagher, S., 2000. *Neurons and Neonates: Answers to the Molyneux Question*, paper presented at the Philosophy Colloquium, Trinity College, Dublin (November 2000). Accessed on Internet 20 Jul 2002.

[36] Katz, D.I., 2001. *Minimally Conscious States*, KurzweilAI.net. Accessed on Internet 20 Jul 2002.

[37] Bear, M.F., Connors, B.W., Paradiso, M.A., *op. cit.*, (p. 786)

[38] De Schutter, E., Bjaalie, J.G., *Coding in the Granular Layer of the Cerebellum*, Progress in Brain Research. Accessed on Internet 5 Oct 2002.

[39] Tannenbaum, A.S., 2001. *The Sense of Consciousness*, Journal of Theoretical Biology, Vol. 211, No. 4, Aug 2001, pp. 377-391, Academic Press. Accessed on Internet 20 Jul 2002.

[40] Thomas, N.J.T., 1998. *Imagination, Eliminativism, and the Pre-History of Consciousness*, paper delivered at the Toward a Science of Consciousness (Tucson III) Conference, April 30th 1998. Accessed on Internet 20 Jul 2002.

[41] Stowasser, J.M., Petschenig, M., Skutch, F., *op. cit.*, (p. 113)

[42] Bear, M.F., Connors, B.W., Paradiso, M.A., *op. cit.*, (p. 42)

[43] Schwitzgebel, E., Gordon, M.S., 2000. *How well do we know our own conscious experience? The case of human echolocation*, Schwitzgebel & Gordon, September 25, 2000, Knowledge of Echolocation. Accessed on Internet 20 Jul 2002.

[44] Nørretranders, T., *op. cit.*, (p. 217)

[45] Bear, M.F., Connors, B.W., Paradiso, M.A., *op. cit.*, (p. 175)

[46] Aguirre, L. 2001. *The Electric Brain*, NOVA Online interview with Rodolfo Llinás. Accessed on Internet 9 Sep 2002.

[47] Newton, M., 2002. *Savage Girls and Wild Boys*, Faber and Faber Limited, London. (p. 99)

[48] *ibid*, (p. 208)

[49] Thirakoul, P., 2000. *Buddhist Meditation and Personal Construct Psychology*, Serendip. Accessed on Internet 5 Oct 2002.

16. Setting the scene

[1] Adams, D., 1979. *The Hitch Hiker's Guide to the Galaxy*, Pan Books Ltd, London. (p. 49)

[2] Gabora, L., 1997. *The Origin and Evolution of Culture and Creativity*, Journal of Memetics - Evolutionary Models of Information Transmission, 1. Accessed on Internet 24 Nov 2001.

[3] Williams, R., 2001. *Origin of Dogs*, The Science Show, Radio National, Australian Broadcasting Commission. Broadcast Saturday 8/12/01. Accessed on Internet 3 Jan 2002.

[4] Dimitrov, V., 2000. *Strange Attractors of Meaning*, Centre for Systemic Development, University of Western Sydney, author's web page. Accessed on Internet 18 Aug 2002.

[5] Preti, A., 1999. *The Gift of Saturn: Creativity and Psychopathology*, Serendip 1994-2002. Accessed on Internet 5 Oct 2002.

[6] Dawson, A., 24 March 2001. *Noffke strikes for Bulls*, The Courier Mail, Brisbane.

[7] Probyn A., 18 June 2001. *Fighting for peace*, The Courier Mail, Brisbane.

[8] Fynes-Clinton, J., 8 August 2002. *Starting out smart*, The Courier Mail, Brisbane.

[9] 23 April 1994. *The New Africa - A Guide to the Sub-Saharan Continent*, The Courier Mail, Brisbane.

[10] Wiseman, N., 19 July 1996. *Spider unravels web of brain*, The Courier Mail, Brisbane.

[11] Carpenter, J., (dir.), 1995. *The Village of the Damned*, Universal City Studios, Inc.

[12] Leonard, B., (dir.), 1992. *The Lawnmower Man*, United International Pictures.

[13] Dowland, H. 2003. Correspondence to the author.

[14] Yanner, M., 2002. *Murrandoo Yanner's Sorry Day Speech from Message Sticks*, Radio National Awaye! Australian Broadcasting Commission, 8 October 2002. Accessed on Internet 8 Oct 2002.

[15] Madigan, M., 29 April 2002. *Lack of education proving 'deadly'*, The Courier Mail, Brisbane.

[16] Wenham, M., 8 February 2002. *Youth centre staff accused of brutality*, The Courier Mail, Brisbane.

[17] Wenham, M., 8 June 2002. *Foster parents face high costs of caring*, The Courier Mail, Brisbane.

[18] Oberhardt, M., Odgers, R., 4 November 2002. *Dubious rewards for Palm violence*, The Courier Mail, Brisbane.

[19] Moscaritolo, M., Griffith, C., 15 April 2003. *ATSIC set to give up powers*, The Courier Mail, Brisbane.

[20] Koch, T., 29 March 2003. *Is that a serpent in your trousers?*, The Courier Mail, Brisbane.

[21] Brunton, R., 10 October 2002. *Women in chains*, The Courier Mail, Brisbane.

[22] Marinos, S., 11 January 2003. *a Kiss before Dying*, The Courier Mail, Brisbane.

[23] Smith, S., 14 April 2001. *Aboriginal bark instead of bite*, The Courier Mail, Brisbane.

[24] Koch, T., 29 June 2001. *Face to face with the horror of blind rage*, The Courier Mail, Brisbane.

[25] Wilson, P., 3 July 2001. *The awful truth*, The Courier Mail, Brisbane.

[26] Heiss, A., 18 November 2000. *In the skin of the indigene*, The Courier Mail, Brisbane.

[27] Mageria, D., 10 April 2003. *Congo family killed 'like cows'*, The Courier Mail, Brisbane.

[28] Reader's letter, 23 December 2002. The Courier Mail, Brisbane.

[29] Rafferty, K., Keraudren, F., 1999. *The Last Cigarette*, New Yorker Films.

[30] Doyle, A.C., new ed. 1976. *The Complete Sherlock Holmes Treasury*, Avenel Books, New York.

[31] Bear, M.F., Connors, B.W., Paradiso, M.A., *op. cit.*, (p. 519)

[32] Tendler, S., May, D., 1984. *The Brotherhood of Eternal Love*, Panther Books, London.

[33] Donnelly, K., 15 January 2003. *Degrees of dumbness*, The Courier Mail, Brisbane.

[34] 28 May 2002. *Jail for Mum scares kids back to school*, The Courier Mail, Brisbane.

[35] Nolan, J., 3 November 2001. *Age of anxiety a cause of violence against school teachers*, The Courier Mail, Brisbane.

[36] Bantick, C., 7 November 2001. *Keep violent kids out of mainstream classrooms*, The Courier Mail, Brisbane.

[37] 8 August 2002. *French block foul students*, The Courier Mail, Brisbane.

[38] Watson, B., (transl.) year not given. *Chuang Tzu, Chapter 32: Lieh Yu Kou*, Panlatrevo. Accessed on Internet 22 May 2003.

[39] Stafford, P., *op. cit.*, (p. 147)

[40] Shostak, M., 1983. *Nisa, The Life and Words of a !Kung Woman*, Penguin Books, UK. (p. 19)

[41] Dullroy, J., 21 January 2003. *Extent of Schoolies shame revealed*, The Courier Mail, Brisbane.

[42] 14 February 2003. *Far north trouble figures high*, The Courier Mail, Brisbane.

[43] *Gold Coast Facts and Statistics...*, Gold Coast City Council. Accessed on Internet 28 Apr 03.

[44] Muscat, G., 2001. *Australian crime, facts and figures 2001*, Australian Institute of Criminology, Canberra. (p. 22)

[45] Heywood, L., 21 February 2001. *XXX makes a councillor see red*, The Courier Mail, Brisbane.

[46] Wilson, P., 23 January 2001. *Clockwork culture is a fearful future*, The Courier Mail, Brisbane.

[47] 16 October 1986. *Report says alcohol is a worse drug than heroin*, The Sydney Morning Herald, Sydney.

[48] 9 September 1992. *Research points to the agony of ecstasy*, The Sydney Morning Herald, Sydney.

[49] Bear, M.F., Connors, B.W., Paradiso, M.A., *op. cit.*, (p. 138)

[50] Drury, N., 1988. *Pan's Daughter - The strange world of Rosaleen Norton*, Collins Australia, Sydney.

[51] Grof, S., Halifax, J., 1977. *Human encounter with death*, E.P. Dutton, New York.

[52] Huxley, A., 1954. *Doors of perception* and *Heaven and Hell*, Chatto & Windus, London.

[53] Bear, M.F., Connors, B.W., Paradiso, M.A., *op. cit.*, (p. 518)

[54] Tendler, S., May, D., *op. cit.*, (p. 24)

[55] Drury, N., *op. cit.*, (pp. 112 & 98)

[56] 1992. *the Yellow House 1970-72*, exhibition catalogue, New South Wales Art Gallery. (p. 21)

[57] van Os, J., Bak, M., Hanssen, M., Bijl, R.V., de Graaf, R., Verdoux, H., 2002. *Cannabis Use and Psychosis: A Longitudinal Population-based Study*, Vol. 156, No. 4, American Journal of Epidemiology.

[58] *ibid*, (p. 2)

[59] *ibid*, (p. 6)

[60] *ibid*, (p. 7)

[61] *ibid*, (p. 4)

[62] *ibid*, (p. 8)

[63] Kinder, H., Hilgemann, W., 1978. *The Penguin Atlas of World History, Volume II: From the French Revolution to the Present*, Penguin Books, UK. (p. 145)

[64] Stafford, P., *op. cit.*, (p. 98)

[65] Tiller, S., (prod.) 2000. *Drug Wars*, WGBH Educational Foundation, Boston.

[66] 8 April 1986. *Business, Robbery, Etc: tax, casinos, drugs...*, The Sydney Morning Herald, Sydney.

[67] Burrell, S., 14 March 1988. *Youth drinking heavy: Blewett*, The Sydney Morning Herald, Sydney.

[68] Chamberlin, P., 16 March 1989. *Drug crime 'out of control'*, The Sydney Morning Herald, Sydney.

[69] Western, J., Aldred, R., Irwin, J., Reynolds, A., Sheehan, M., Shaw, I., Nelson, P., Freeman, A., Fenlon, T., Carlyon, T., Wall, R., Dickie, P., Tregarthen, S., 1993. *Cannabis and the Law in Queensland - A Discussion Paper*, Advisory Committee on Illicit Drugs, Brisbane.

[70] *ibid*, (p. 9)

[71] *ibid*, (p. 55)

[72] *ibid*, (p. 75)

[73] *ibid*, (p. 68)

[74] *ibid*, (p. 144)

[75] *ibid*, (p. 145)

[76] Doneman, P., 6 July 1998. *Dawn raids smash Coast drug ring*, The Courier Mail, Brisbane.

[77] 3 September 1998. *Parents hire drug spies*, The Courier Mail, Brisbane.

[78] Callinan, R., 3 October 1998. *Task force formed to tackle jail growth*, The Courier Mail, Brisbane.

[79] Lawlor, A., 9 July 1999. *Attention drugs invade schools*, The Courier Mail, Brisbane.

[80] 9 July 1999. *Drug crime links*, The Courier Mail, Brisbane.

[81] Heywood, L., Koch, T., 14 September 2000. *Taskforce goes cold on heroin shop plan*, The Courier Mail, Brisbane.

[82] Hart, M., Monk, S., 1 December 2000. *State youth drug use skyrockets*, The Courier Mail, Brisbane.

[83] Wenham, M., Heywood, L., Smith, W., 17 March 2001. *Jail offers 'no cure' for drug addicts*, The Courier Mail, Brisbane.

[84] Hart, M., 14 June 2001. *Aussie drug use goes to pot*, The Courier Mail, Brisbane.

[85] Lill, J., 22 August 2001. *Amphetamine boom drains drug workers*, The Courier Mail, Brisbane.

[86] Hart, M., 5 December 2001. *War on drugs a 'loosing battle'*, The Courier Mail, Brisbane.

[87] Jones, C., White, A., 12 January 2002. *Barking up the wrong tree?*, The Courier Mail, Brisbane.

[88] Doneman, P., 15 January 2002. *Alcohol fuels attacks on police*, The Courier Mail, Brisbane.

[89] Parnell, S., 13 March 2002. *Bikie gangs make state a 'drug hub'*, The Courier Mail, Brisbane.

[90] le Grand, M., 28 March 2002. *Drugs war lost at port*, The Courier Mail, Brisbane.

[91] 17 September 2001. *Where to next? - a discussion paper, Inquiry into substance abuse in Australian communities*, House of Representatives, Standing Committee on Family and Community Affairs, Canberra. (p. 6)

[92] Dorries, B., 13 May 2002. *Ravers on child behaviour drugs*, The Courier Mail, Brisbane.

[93] Madigan, M., 18 February 2003. *City nightclubs serve up cocktail of drugs*, The Courier Mail, Brisbane.

[94] Pedersen, D., 8 March 2003. *PM turns Bangkok into war zone*, The Courier Mail, Brisbane.

[95] Jones, C., 25 March 2003. *Huge increase in illegal drug labs*, The Courier Mail, Brisbane.

[96] 1998. *SBS World Guide*, Hardie Grant Publishing, Melbourne. (p. 533)

[97] Parnell, S., 5 May 2003. *Overhaul looms after cig blitz fails*, The Courier Mail, Brisbane.

[98] 2002. *our strongest defence against the drug problem...*, Commonwealth Government, Canberra. (p. 10)

[99] Tiller, S., *op. cit.*

[100] Foucault, M., *op. cit.*, (p. 75)

[101] *ibid*, (p. 76)

[102] *ibid*, (p. 255)

[103] Kamen, H., *op. cit.*, (p. 177)

[104] Foucault, M., *op. cit.*, (p. 286)

[105] Wolpert, S., 1997. *A new history of India*, Oxford University Press, New York. (p. 212)

[106] Ferguson, J., 18 January 1997. *Brits and Aussies don't add up*, The Courier Mail, Brisbane.

[107] Hammond, P., 14 August 1997. *Mothers in the dark on infant jabs*, The Courier Mail, Brisbane.

[108] Retschlag, C., 1 October 1999. *Teachers turn to counsellor 'agony aunts'*, The Courier Mail, Brisbane.

[109] Nolan, J., Dullroy, J., 27 October 2001. *Child, 9, bashes teacher*, The Courier Mail, Brisbane.

[110] 2000. *Schools Australia*, Australian Bureau of Statistics, Canberra. (pp. 19 & 81)

[111] Ryan, S., 17 June 2002. *Mental health crisis in schools*, The Courier Mail, Brisbane.

[112] Fynes-Clinton, M., 4 September 2002. *Principal hit, punched by student's dad*, The Courier Mail, Brisbane.

[113] Oberhardt, M., Dullroy, J., 4 September 2002. *Teens' machete fights worry judge*, The Courier Mail, Brisbane.

[114]. Yallamas, L., 19 November 2002. *Teachers seek police shield*, The Courier Mail, Brisbane.

[115] 2002. *White Pages 2002/03 Brisbane*, Telstra Corporation. (p. 170)

[116] McPhedran, I., 30 September 2002. *'Street cred' urged in war on gangs*, The Courier Mail, Brisbane.

[117] Duffy, M., 27 March 2002. *Blame lies with our culture of complaint*, The Courier Mail, Brisbane.

[118] Stolz, G., Gilby, L., 15 May 2002. *Wave of fear at $4m payout*, The Courier Mail, Brisbane.

[119] Griffith, C., 6 April 2002. *Hearing the evil*, The Courier Mail, Brisbane.

[120] Ward, D., 30 April 2003. Correspondence to the author, Department of Employment and Training, South Bank, Brisbane.

[121] 2002. *Forever Summer with Nigella Lawson*, Channel 4, London.

[122] 2002. *the Best*, endemol uk productions. Screened on ABC-TV 23 Apr 2003.

[123] Owen-Brown, M., 30 April 2003. *Asian countries face tighter controls*, The Courier Mail, Brisbane.

[124] Horton, R., 7 May 2003. *Results of first major epidemiological study of SARS*, The Lancet. email.

[125] Kiple, K., *op. cit.*, (p. 117)

[126] Kohn, G.C., *op. cit.*, (p. 393)

[127] Foreman, W., 7 May 2003. *Riots and more restrictions as toll rises*, The Courier Mail, Brisbane.

[128] Owen, R., 10 May 2003. *The panic pandemic*, The Courier Mail, Brisbane.

[129] Kiple, K., *op. cit.*, (p. 16)

[130] 2003. *Symptoms - Prevention*, Meningococcal Australia Inc. Accessed on Internet 12 May 2003.

[131] McCabe, H., 23 January 1998. *Pain, pride in a man's world*, The Courier Mail, Brisbane.

[132] 7 August 1998. *Stance hailed*, The Courier Mail, Brisbane.

[133] 22 March 2000. *Cutting criticism*, The Courier Mail, Brisbane.

[134] 26 December 2002. *GUATEMALA Jail Horror*, The Courier Mail, Brisbane.

[135] Cole, M., 23 July 2001. *Website urges attacks on police*, The Courier Mail, Brisbane.

[136] Leach, M., August 2001. *SAVE Kurilpa's Bushland Gully Give the Developers the CHOP*, Neighbourhood NEWS, Brisbane.

[137] Laver, B., October/November 1999. *Ratbags of West End Unite*, Neighbourhood NEWS, Brisbane.

[138] Coghlan, A., 2001. *Green menace*, New Scientist, 9 June 2001.

[139] Kuhnhenn, J., 26 March 2003. *Bush asks Congress for $126b*, The Courier Mail, Brisbane.

[140] 2 April 2003. *STATISTICS Bombs increase*, The Courier Mail, Brisbane.

[141] 12 April 2003. *Clerics hacked to death at Shi'ite peace talks*, The Courier Mail, Brisbane.

[142] Robberson, T., 23 April 2003. *Shi'ite majority in push for an Islamic republic*, The Courier Mail, Brisbane.

[143] Huntington, S.P., *op. cit.*, (p. 290)

[144] *ibid*, (p. 108)

[145] Duffy, M., 29 September 2001. *Improved by Asian work ethic*, The Courier Mail, Brisbane.

[146] Rehn, A., 28 December 2002. *Rural pressure cuts lives short*, The Courier Mail, Brisbane.

[147] O'Malley, B., 30 March 2002. *All too easy*, The Courier Mail, Brisbane.

[148] Brunton, R., 1998. *The End of the Overpopulation Crisis?*, IPA Backgrounder December 1998, Vol. 10/5, Institute of Public Affairs Ltd, Melbourne.

[149] *ibid*, (p. 4)

[150] 17 October 2002. *Magpie madness*, The Courier Mail, Brisbane.

[151] Hele, M., Odgers, R., Watt, A., 23 December 2002. *Drivers won't wake up to demerit blitz*, The Courier Mail, Brisbane.

[152] O'Malley, B., 23 March 2002. *Aggressive invaders swoop on backyards*, The Courier Mail, Brisbane.

[153] January 1998. *Something to crow about*, State of Queensland, Department of Environment, 1998.

[154] Sweetman, T., 2 May 2003. *Medicines fail trust test*, The Courier Mail, Brisbane.

[155] 6 May 2003. *Health body sets limits on safe levels of vitamin and mineral use*, The Courier Mail, Brisbane.

[156] Watt, A., 5 May 2003. *Epidemic of obesity weighs on nation*, The Courier Mail, Brisbane.

[157] Edmistone, L., 15 May 2003. *Increase in diabetes a healthcare time bomb*, The Courier Mail, Brisbane.

[158] McCutcheon, S., 29 May 2002. *Australia Talks Back*, Radio National, Australian Broadcasting Corporation.

[159] Moore, W., 10 May 2003. *television: opiate of the masses*, The Disinformation Company Ltd. Accessed on Internet 11 May 03.

[160] 3 November 2000. *PM*, Radio National, Australian Broadcasting Corporation.

[161] 15 April 2001. *Books and Writing*, Radio National, Australian Broadcasting Corporation.

[162] Einstein, A., 1961. *Relativity - The Special and General Theory*, Crown Publishers, Inc., New York. (p. 1)

[163] McCutcheon, S., 10 June 2002. *Australia Talks Back*, Radio National, Australian Broadcasting Corporation.

17. Heralds of doom

[1] Various, 1976. *Good News Bible - Today's English Version*, American Bible Society, New York.

[2] *ibid*, (Old Testament, p. 5)

[3] *ibid*, (Old Testament, p. 7)

[4] *ibid*, (Old Testament, p. 11)

[5] *ibid*, (Old Testament, p. 24)

[6] *ibid*, (Old Testament, p. 84)

[7] *ibid*, (Old Testament, p. 147)

[8] *ibid*, (Old Testament, p. 93)

[9] *ibid*, (Old Testament, p. 148)

[10] Flaubert, G., (transl) 1977. *Salammbo*, Penguin Books, London. (p. 241)

[11] Various, *op. cit.*, (New Testament, p. 25)

[12] Parnell, S., 2 January 2003. *Rioters in same class as those planned for Pinkenba centre*, The Courier Mail, Brisbane.

[13] Kamen, H., *op. cit.*, (p. 28)

[14] White, M., 2002. *The Pope and the Heretic*, Little, Brown and Company. London. (p. 24)

[15] *ibid*, (p. 62)

[16] *ibid*, (p. 190)

[17] *ibid*, (p. 2)

[18] Lazaredes, N., 5 March 2003. *Warning Bells for Morocco*, SBS-TV Dateline.

[19] Hart, J. 1973. *B.C. - Right On*, Fawcett Publications, Inc, Greenwich, Conn.

[20] 6 October 2001. *"Earth's Final Warning" National Sunday Law Is Coming - Are you ready?*, advertisement from Maranatha Ministries, The Courier Mail, Brisbane.

[21] Stolz, G., 24 August 2002. *Council plea for weekend rain prayers*, The Courier Mail, Brisbane.

[22] Green, G., 24 July 1997. *Holy war over 'possessed' pupils*, The Courier Mail, Brisbane.

[23] 3 August 2000. *Bringing old ills to book*, The Courier Mail, Brisbane.

[24] Wenham, M., 17 February 2001. *Sins of the father*, The Courier Mail, Brisbane.

[25] Wenham, M., 24 February 2001. *Orphanage brutality 'exposed'*, The Courier Mail, Brisbane.

[26] Green, G., 16 February 2001. *School strap teacher employed at college*, The Courier Mail, Brisbane.

[27] Parnell, S., 30 March 1998. *Neerkol fallout lingers: Sisters*, The Courier Mail, Brisbane.

[28] Thomas, M., Parnell, S., 25 August 2000. *Abuse victims left out in cold*, The Courier Mail, Brisbane.

[29] Wenham, M., 23 August 2001. *Victims take the fight to George Street*, The Courier Mail, Brisbane.

[30] Wenham, M., 4 April 2002. *Church will give to child victim fund*, The Courier Mail, Brisbane.

[31] McCutcheon, S., 30 October 2000. *Australia Talks Back*, Radio National, Australian Broadcasting Corporation.

[32] Various, *op. cit.*, (New Testament, p. 282)

[33] 10 March 1994. *Burn women priests at stake: vicar*, The Courier Mail, Brisbane.

[34] Pockley, P., 13 June 1992. *Creationism is alive and well*, The Sydney Morning Herald, Sydney.

[35] MacKenzie, D., 2000. *Unnatural selection*, New Scientist, 22 April 2000.

[36] Plimer, I., 2000. *I'd do it all again*, New Scientist, 22 April 2000.

[37] Plimer, I., 1994. *Telling Lies for God*, Random House Australia Pty Ltd, Sydney. (p. 267)

[38] Charlton, P., 16 January 2001. *Plenty of dark deeds behind all the candles*, The Courier Mail, Brisbane.

[39] Jones, B., 28 February 1987. *The psychology of the Joh phenomenon*, The Sydney Morning Herald, Sydney.

[40] 1986. *The Constitution*, Australian Government Publishing Service, Canberra. (pp. 7, 20, 24)

[41] Monk, S., 13 June 2001. *Chapel for Yarralumla*, The Courier Mail, Brisbane.

[42] Cotes, A., 15 March 2003. *Why God's gone bush*, The Courier Mail, Brisbane.

[43] Bear, M.F., Connors, B.W., Paradiso, M.A., *op. cit.*, (p. 506)

[44] *ibid*, (p. 691)

[45] Street, H., 2002. *Exploring Relationships Between Goal Setting, Goal Pursuit And Depression: A Review*, Australian Psychologist July 2002, Volume 37 Number 2. (p. 96)

[46] Woodward, A.J., Carless, S.A., Findlay, B.M., 2001. *A Psychometric Evaluation of the Irrational Beliefs Inventory in a Marital Context*, Australian Psychologist November 2001, Volume 36 Number 3. (p. 255)

[47] *ibid*, (p. 258)

[48] Bear, M.F., Connors, B.W., Paradiso, M.A., *op. cit.*, (p. 553)

[49] *ibid*, (p. 554)

[50] Pinker, S., *op. cit.*, (p. 464)

[51] Raine, A., Reynolds, C., Venables, P., Mednick, S., 2002. *Stimulation Seeking and Intelligence: A Prospective Longitudinal Study*, Journal of Personality and Social Psychology, 2002, Vol. 82, No. 4, 663–674. (p. 669)

[52] *ibid*, (p. 671)

[53] Roberts, C., 1999. *The Prevention of Depression in Children and Adolescents*, Australian Psychologist March 1999, Volume 34 Number 1. (p. 49)

[54] *ibid*, (p. 50)

[55] *ibid*, (p. 53)

[56] Ranzijn, R., Luszcz, M., 1999. *Acceptance: A Key to Wellbeing in Older Adults?*, Australian Psychologist July 1999, Volume 34 Number 2. (p. 97)

[57] *ibid*, (p. 95)

[58] Monk, S., 28 April 1997. *Women fear fall in ad ethics*, The Courier Mail, Brisbane.

[59] 19 July 1997. Apartment Central Hotels advertisement, The Courier Mail, Brisbane.

[60] April 2003. Epson advertisement, Australian Personal Computer, Sydney. (p. 31)

[61] These were:
O'Connor, M., 11 April 1997. *Ionic columns of no small inconvenience*, The Courier Mail, Brisbane.
Cotes, A., 14 April 1997. *Scriptures see women as leading men astray*, The Courier Mail, Brisbane.
Spender, D., 17 April 1997. *Service without the smiles*, The Courier Mail, Brisbane.
O'Connor, M., 18 April 1997. *Dressed, in a sense*, The Courier Mail, Brisbane.
Johnston, L., 30 April 1997. *Older, sensible and sexy*, The Courier Mail, Brisbane.
Spender, D., 1 May 1997. *It's no secret - action is vital*, The Courier Mail, Brisbane.

[62] 1996. *White Pages*, for Brisbane, Sydney, Australian Capital Territory, Melbourne, Hobart, Adelaide, Perth, Northern Territory, Telstra Corporation Ltd.

[63] Milliner, K., 26 February 1999. *Join the Dots for familiar picture*, The Courier Mail, Brisbane.

[64] Hocking, S., 17 May 1999. *Where the boys are*, The Courier Mail, Brisbane.

[65] Hocking, S., 27 March 2000. *Scent of a man*, The Courier Mail, Brisbane.

[66] Dudley, J., 23 June 2000. *TV blooper leaves red faces at Ten*, The Courier Mail, Brisbane.

[67] Dworkin, A. 1982. *Our Blood - Prophesies and Discourses on Sexual Politics*, The Women's Press, London. (p. 12)

[68] Bolt, A., 26 April 2000. *Hollywood plays loose with the truth*, The Courier Mail, Brisbane.

[69] Gerrie, A., 7 June 2003. *Wonder woman*, The Courier Mail, Brisbane.

[70] Petersen, D., 7 February 1996. *Gender Wars*, The Courier Mail, Brisbane.

[71] Fiebert, M.S., 1997, 1998. *Battered Men - The Hidden Side of Domestic Violence Latest Research Findings*, MenWeb. Accessed on Internet 22 May 2003.

[72] *ibid*, (<Caulfield, M. B., & Riggs, D. S. (1992)>)

[73] *ibid*, (<Feather, N. T. (1996)>)

[74] *ibid*, (<Brinkerhoff, M., & Lupri, E. (1988)>)

[75] Middap, C., 13 November 2000. *Gloves off in battle of sexes*, The Courier Mail, Brisbane.

[76] 16 March 2001. *Women in jail*, The Courier Mail, Brisbane.

[77] Zeki, S., *op. cit.*, (p. 4).

[78] 26 November 1998. *Ex-files reveal secrets of revenge*, The Courier Mail, Brisbane.

[79] Monk, S., 14 May 1999. *Editor at centre of new age controversy*, The Courier Mail, Brisbane.

[80] Stein, S., 1985. *Solutions for Writers*, Souvenir Press Ltd, London. (p. 90)

[81] 2002. *Boys: getting it right - Report on the inquiry into the education of boys*, House of Representatives Standing Committee on Education and Training, The Parliament of the Commonwealth of Australia, Canberra, October 2002.

[82] *ibid*, (p. 8)

[83] *ibid*, (p. 9)

[84] *ibid*, (p. 10)

[85] *ibid*, (p. 12)

[86] *ibid*, (p. 20)

[87] *ibid*, (p. 22)

[88] *ibid*, (p. 59)

[89] *ibid*, (p. 65)

[90] *ibid*, (p. 77)

[91] *ibid*, (p. 87)

[92] *ibid*, (p. 130)

[93] Kirkman, M., 2002. *What's the Plot? Applying Narrative Theory to Research in Psychology*, Australian Psychologist, March 2002, Volume 37, Number 1.

[94] *ibid*, (p. 30)

[95] *ibid*, (p. 31)

[96] *ibid*, (p. 31)

[97] *ibid*, (p. 33)

[98] Dormer, S., Davies, B., 2001. *Desiring Women and the (Im)Possibility of Being*, Australian Psychologist, April 2001, Volume 36, Number 1.

[99] *ibid*, (p. 4)

[100] *ibid*, (p. 5)

[101] *ibid*, (p. 6)

[102] (a) 1981. *The Macquarie Dictionary*, Macquarie Library Pty. Ltd., Sydney. (p. 1114, 1112)
(b) 1997. *The Macquarie Dictionary*, Macquarie Library Pty. Ltd., Sydney. (p. 1373)
(c) 1989. *The Oxford English Dictionary - Volume IX*, Clarendon Press, London. (p. 845)

[103] Bear, M.F., Connors, B.W., Paradiso, M.A., *op. cit.*, (p. 557)

[104] *ibid*, (p. 559)

[105] Various, *op. cit.*, (Old Testament, p. 19)

[106] Skatssoon, J., 19 May 2003. *Parents warned against routine circumcision*, The Courier Mail, Brisbane.

[107] 20 May 2003. *Dispute on baby snips*, The Courier Mail, Brisbane.

[108] Tannahill, R., *op. cit.*, (after p. 244)

[109] Various, *op. cit.*, (Old Testament, p. 21)

[110] *ibid*, (Old Testament, p. 119)

[111] *ibid*, (Old Testament, p. 193)

[112] *ibid*, (New Testament, p. 190)

[113] *ibid*, (New Testament, p. 213)

[114] *ibid*, (New Testament, p. 230)

[115] *ibid*, (New Testament, p. 318)

[116] Wright, E., *op. cit.*, (p. 25)

[117] White, M., *op. cit.*, (p. 11)

[118] Newton, M., *op. cit.*, (p. 149)

[119] Zeki, S., *op. cit.*, (p. 93)

[120] Lindner, E.G., >=2001. *Humiliation - Trauma That Has Been Overlooked: An Analysis Based on Fieldwork in Germany, Rwanda / Burundi, and Somalia*, own webpage. Accessed on Internet 9 September 2002.

[121] Dasgupta, N., Greenwald, A.G., 2001. *On the Malleability of Automatic Attitudes: Combating Automatic Prejudice With Images of Admired and Disliked Individuals*, Journal of Personality and Social Psychology, November 2001, Vol. 81, No. 5, 800-814. Accessed on Internet 20 September 2002.

[122] Bear, M.F., Connors, B.W., Paradiso, M.A., *op. cit.*, (p. 430)

[123] *ibid*, (p. 432)

[124] Nørretranders, T., *op. cit.*, (p. 150)

[125] Hornadge, B., 1971. *Chidley's Answer to the Sex Problem*, Review Publications, Dubbo, Australia. (p. 30)

[126] *ibid*, (p. 36)

[127] *ibid*, (p. 39)

[128] *ibid*, (p. 84)

[129] Brome, V., 1979. *Havelock Ellis, Philosopher of Sex, A Biography*, Routledge & Kegan Paul, London.

[130] Hornadge, B., *op. cit.*, (p. 87)

[131] Paxton, J., (ed.), 1984. *The New Illustrated Everyman's Encyclopaedia - Volume Three*, Octopus Books Limited, London. (p. 713)

[132] 1981, *op. cit.*, (p. 721, 722)

[133] Stafford, P., *op. cit.*, (p. 130)

[134] O'Chee, A., 19 March 1998. *Class of '48 returns to no man's land*, The Courier Mail, Brisbane.

[135] 30 August 2001. *Chilly reception greets topless Nordic tourists*, The Courier Mail, Brisbane.

[136] 14 January 2002. *PROTEST Skinny dippers*, The Courier Mail, Brisbane.

[137] 1974. *Family of Man*, Marshall Cavendish Limited, London. (p. 2106)

[138] 2003. *Nuba Vision - April Issue*, Nuba Survival, Middlesex, UK. Accessed on Internet 19 June 2003.

[139] Charlton, P., 21 July 2001. *The terminators*, The Courier Mail, Brisbane.

[140] Pearce, E., 1975. *Anatomy and Physiology for Nurses*, Faber and Faber Limited, London. (p. 174)

[141] 2002. *Diagnostics*, Braun GmbH. Accessed on Internet 22 December 2002.

[142] Bridget, J., 1996. *The Role of Psychiatry/Medicine in Developing and Perpetuating Homophobia*, Lesbian Information Service. Accessed on Internet 29 October 2001.

[143] Sabbatini, R.M.E., 1997. *The History of Lobotomy*, Brain & Mind Magazine, June 1997. Accessed on Internet 29 October 2001.

[144] Kirby, M., 29 April 2000. *Devils of homophobia*, The Courier Mail, Brisbane.

[145] Heffernan R., O'Malley, B., 13 November 1999. *Opposition united in fury over gay law*, The Courier Mail, Brisbane.

[146] 22 December 2000. *Gay attack*, The Courier Mail, Brisbane.

[147] Dank, B.M., de Albuquerque, K., 1998. *Banning sexual asymmetry on campus*, Electronic Journal of Human Sexuality, Volume 1, August 10, 1998. Acessed on Internet 9 September 2002.

[148] Dank, B.M., Fulda, J.S., 1998. *Forbidden Love: Student-Professor Romances*, Sexuality & Culture, 1, 107-130 (1997). Accessed on Internet 9 September 2002.

[149] Cole, M., 6 July 1999. *Suspicious minds set against men in schools*, The Courier Mail, Brisbane.

[150] Fynes-Clinton M., 24 April 2002. *Suspicious minds*, The Courier Mail, Brisbane.

[151] Skatssoon, J., 2 June 2003. *New mums at greater risk of depression*, The Courier Mail, Brisbane.

[152] 22 November 2001. *Habits of men*, The Courier Mail, Brisbane.

[153] Hammond, P., 9 June 1997. *New 'terrorism' hits middle class males*, The Courier Mail, Brisbane.

[154] 29 August 1992, *Child Sexual Assault*, Sydney Morning Herald, Sydney.

[155] Green, G., 31 March 2001. *Couple dirty on pupils' sexy texts*, The Courier Mail, Brisbane.

[156] McLean, S., 30 November 2002. *Through the eyes of a child*, The Courier Mail, Brisbane.
[157] *Worldwide Ages of Consent*, AVERT. Accessed on Internet 16 November 2001.

[158] Dudley, J., 10 November 2001. *A matter of radical ritual*, The Courier Mail, Brisbane.

[159] (a) McLean, S., 15 December 2001. *Embracing Bronze*, The Courier Mail, Brisbane.
(b) Anonymous. Nudist webpage. Accessed on Internet 4 January 2002.

[160] Smith, W., 30 September 2002. *Father cleared of raping daughter*, The Courier Mail, Brisbane.

[161] Rasmussen, L.A., 1999. *The Trauma Outcome Process: An Integrated Model for Guiding Clinical Practice with Children with Sexually Abusive Behavior Problems*, Journal of Child Sexual Abuse, Volume 8 Number 4 1999. Accessed on Internet 9 July 2002. (p. 4)

[162] *ibid*, (p. 5)

[163] *ibid*, (p. 9)

[164] *ibid*, (p. 13)

[165] Baker, R.A., (ed) 1998. *Child Sexual Abuse and False Memory Syndrome*, Prometheus Books, New York.

[166] Keim, T., 29 May 2003. *Alleged child-torture mum filmed by hidden camera*, The Courier Mail, Brisbane.

[167] They are (The Courier Mail, Brisbane):
18 August 1997, *Eavesdrop scheme to fight child molesters,*
Documents show Lewis sheltered paedophiles

19 August 1997, *Report to fight child sex abuse cover-ups*

20 August 1997, *Crime commission to fight paedophilia*

21 August 1997, *Predator - The mind of the paedophile,*
The Paedophilia Debate (next page)

22 August 1997, *New agency to fight paedophilia*

23 August 1997, *CJC threat to seize evidence on paedophiles,*
Child Sex War (other page)

25 August 1997, *Paedophile report on CJC man seized*

26 August 1997, *Child sex group identified in report,*
Paedophilia, drugs the target,
Kids offer themselves on 'the rack' for sale

27 August 1997, *Public alerts on paedophiles urged*

28 August 1997, *Speaker protects files from CJC*

29 August 1997, *CJC accused of Watergate-style sex investigation*

30 August 1997, *Brisbane video company centre of paedophile porn industry*

1 - 11 Sep
interruption due to Princess Diana's death

12 September 1997, *Foster father used boys for sex,*
Fight against paedophilia 'curtailed'

13 September 1997, *CJC ignored 'non-profit' paedophilia, Molester offers castration in return for love of boy*

16 September 1997, *Child sex abuse claims ignored*

17 September 1997, *New body takes over organised crime powers*

18 September 1997, *A classic example of why Queensland needs a crime commission*

[168] 28 October 2000. *Judge allows in camera evidence to be reported*, The Courier Mail, Brisbane.

[169] Buckley, S., 25 March 2002. *Breastfeeding is liquid love*, The Courier Mail, Brisbane.

[170] Marx, A., Callinan, R., 9 January 2002. *Churchie hit with $1.5m sex claim*, The Courier Mail, Brisbane.

[171] Hele, M., Gearing, A., 5 February 2002. *Churchie in new sex claim*, The Courier Mail, Brisbane.

[172] Gearing, A., 3 September 2002. *Compo payout to sex abuse victim*, The Courier Mail, Brisbane.

[173] Potter, D., Mosel, G., 17 January 2003. *Grammar pays sex claims*, The Courier Mail, Brisbane.

[174] April 2002. *WorkCover Benefits Guide*, WorkCover NSW, Sydney. (p. 28)

[175] Sands, J., 10 May 2000. *Injuries plague football kids*, The Courier Mail, Brisbane.

[176] 3 June 2003. *Young people & the law*, The Courier Mail, Brisbane.

18. PAT$^{n\pm}$

[1] 26 March 2003. *NIGERIA Oil giants retreat*, The Courier Mail, Brisbane.

[2] Cole, M., 27 June 2003. *Downer questions role of UN in region*, The Courier Mail, Brisbane.

[3] Fukuyama, F., 1999. *Social Capital and Civil Society*, International Monetary Fund, Prepared for delivery at the IMF Conference on Second Generation Reforms. Accessed on Internet 2 September 2002.

[4] Kaplan, R.D., 1994. *The Coming Anarchy*, The Atlantic Monthly Company, February 1999, Volume 273, No. 2; pages 44-76. Accessed on Internet 2 September 2002.

[5] Krug, E.G., Mercy, J.A., Dahlberg, L.L., Zwi, A.B., 2002. *The world report on violence and health*, Lancet 2002; 360: 1083-88. Accessed on Internet 8 October 2002.

[6] Hughes, H., 2002. *Immigrants, Refugees and Asylum Seekers - A Global View*, The Centre for Independent Studies Limited, Sydney. (p. 41)

[7] *ibid*, (p. 10)

[8] *ibid*, (p. 29)

[9] *ibid*, (p. 34)

[10] 1 Oct 2003. *The Challenge of Slums – UN-HABITAT's new Global Report on Human Settlements*, UN-HABITAT. Accessed on Internet 8 October 2003.

[11] Cook, M., 26 August 2003. *Human-rabbit clone announced and no noses twitch*, The Courier Mail, Brisbane.

[12] March 2002. *Population Ageing 2002*, Population Division of the United Nations Department of Economic and Social Affairs. Accessed on Internet 24 October 2003.

[13] Bradbury, R.J., 2000. *Matrioshka Brains*, web page. Accessed on Internet 20 September 2002.

19. Debrief

[1] Hawking, S., 2001. *The Universe in a Nutshell*, Bantam Press, Sydney. (p. 158)

Appendix

[1] Rao, V., Rao, H., 1995. C++ Neural Networks and Fuzzy Logic, MIS:Press, New York.

[2] Harris, M., op. cit. Appendix

Index

Overview of chapters

1. The start of the journey

The stages of this book - preparing the reader for what is to come - caution about words and their usage - introducing thought structures and patterns - why the link to Darwin - comments on the sources and methodology.

Part I: The Stated

2. The mind-body myth

Socrates - Plato - Aristotle

Differences between objects and people - subjectivity and objectivity - representations and abstracts - personal characteristics and idealisations - leaving the grounds of reality - the quintessential natures of things leading to presumption - a simple version of the abstraction process.

3. The universe as a fractal

Jainism

The foundations of the Jaina cosmological model - a careful use of abstracts - from perception to representation - congruent and incongruent functionalities - the pitfalls of qualities and how the Jains handled them - a more advanced version of the abstraction process - the perception-abstraction-transference (PAT) - how the atom is seen - Jaina logic and how it shapes their cosmos.

4. Building Europe's mind: the foundations

Manichaeism - St Augustine

Before Christianity - 'up' is good and 'down' is bad - political savvy in ancient Palestine - training the mind to hate its body - functional and object abstracts - default abstracts - the ideals in the lap of the god - higher-level PATs - the higher latency of Christianity compared to other religions - the logic boundary and the proof of a god.

5. Rights and wrongs

St Thomas Aquinas - Cathars - etc!

A quick revision - ideals again and other-worldliness - a gradual increase of abstraction levels - grappling with existence and how the system of the mind responds - the terror of ungrounded PATs - stereotypes are vital - the joys of suffering - high-level abstractions and their dangers - incongruent PAT loops - the inevitability of mass destruction - time travel to a town in the Middle Ages (an exercise).

6. Into the light

René Descartes - Benedict de Spinoza

The age of reason is full of caution - denominative thought structures - congruent PAT loops - first encounters with the subconscious and a definition of consciousness - a schema of grounded and ungrounded associations - intuitive and cognitive functional abstracts - becoming conscious of subjectivity - fixing cars and oneself - back to God.

7. The mind as government

John Locke

What it takes to understand - the first beginnings of the complexity concept - dangers of incongruence - comparing lower with higher abstractions - sensing time - the idea as object - resonant thought structures - the contributions or otherwise of emotions - social horizon and free will - horizontal and vertical PATs - how to solve puzzles systematically - the aesthetics of children's rhymes - the Molyneux question.

8. God by any other name

Jean-Jacques Rousseau - Immanuel Kant

A longing for nature and other phantasies - logic boundaries disregarded - analogies between the mind and society - the affinity of functional abstracts and its traps - selective uses of ideals and associations - conceptual containers - a precarious use of maths - abstracting to dizzying heights - a play with essences - what happens when objects are discarded in favour of highest abstracts - the functional circularity of Kant's thought processes - ...and the circuit-breaker.

9. The will for independence

Georg W.F. Hegel - (Charles Darwin) - Friedrich Nietzsche

A new Europe - history in terms of the perceived and its abstractions - how the logic boundary gets pushed ever outwards - Hegel's functional container - is 'knowledge' really knowledge? - illusion trees and what drives them - another free will - still no recognition of the subconscious - contextual associations - the subjective use of 'freedom' - functional parallels in Darwin's work - homology and synonymy - Nietzsche's mask - contextual twin structures - a push-pull approach to Christianity.

10. Anatomy of a universe

Karl Marx - Edmund Husserl - Carl G. Jung

About the interdependency between an object and its usage - the effectiveness chart - how an ideology comes into being - tracing the role of incongruent PAT chains, illusion trees, and reduced abstraction levels in one's perspective of the world - the cognitive universe of communism - another version of abstract manipulation, this time upwards - floating above the realm of objects and how one can crash in Hegel's universe - putting together the pieces of Jung's world - some down-to-earth nitpicking - God hiding behind denominative thought structures - a universe of pain.

11. In search of freedom?

Herbert Marcuse - Jean-Paul Sartre

More on the effectiveness chart with the addition of complexity - determinism and progression locks - the hankering for original bliss - sexuality handled with kid gloves - delving into the subconscious - more twin structures - Freud and Dalí and all that - abstracts as high as ever - even 'nothing' is something - abstracting one's perception of oneself - rationalisation complexes.

12. Back to earth

Ludwig J.J. Wittgenstein - Karl R. Popper - Homo Sapiens

Latency in high-complex systems - a quick discourse on probabilities - and so to a definition of Life - trying to deconstruct language - a pseudo recursion - affinity relationships and pattern recognition at different scales - the problem with induction - foregoing statistical evidence in favour of functional possibility - the pitfalls of patterns - a functional picture of the mind of the boy Calvin as he fears monsters under his bed.

13. A change of pace

Where we are now and where we are going - defining the conceptual space of Otoom - some more words on words.

Part II: The Unstated

14. From: Chaos (?)

Neurons - complexity - post-bio

A quick summary of neurons - what is complexity? - iterating systems transposed into a live environment - the sheer magnitude of complexity - the problems with describing a complex system - some misunderstanding - functional scales - a generic description of intelligence - what it takes to be intelligent - introducing a non-biological, intelligent system - parallels to nature - what it takes to have distinct representations - stable, periodical, and strange attractors and how they lead to affinity relationships - the significance of depth-first traversal - content is not important but relationships are - samples of output and what they mean - the importance of affinity potential and self-organisation - teleology: humbug!

15. To: Order (?)

Abstractions - clusters - consciousness

What abstractions really are - the kind of problems a robot has to deal with - calculating latency in terms of fitness peaks and comparing it with the output from the computer program - a homo-centric interpretation of the program's output - what we can learn from apraxia patients - more about object and functional abstracts - conceptual relativity - colour, what's a colour? - to some extent we all are split-brain patients - the famous grain problem and what's wrong with it - autocatalytic closure - Turing's halting problem and what it teaches us - the nature of memory and how it can deceive - how emotions affect one's affinity relationships - the underlying mechanics of consciousness - a simile from a toy - how consciousness tricks us - an experiment with time - the context of myth creation - abandoned children.

16. Setting the scene

Society

A trivial question - the emergence of culture - the ultimate home of homo sapiens - grounded and ungrounded affinity relationships - the latency of a resource space - autocatalysis writ large - why religions came into being and when - the benefits or otherwise of intelligence - the ramifications of high and low abstracting minds - conceptual intersections - the drug issue as an example of identity containers and what they do - how statistical facts can be turned into illusion trees - abstraction complexes influencing such things as violence, education, culture - the evangelism of the Christian West - environmental concerns becoming a religion - a list of behaviour forms demonstrating a balance of emotional influences.

17. Heralds of doom

Religion - Woman - Sex

The Christian Bible from a functional perspective - religious rule sets in terms of cognitive modules - the moral background of the West - the transference of will in religious mind sets - the Iraq war as an example - debunking ideals and perfection - delineating cognitive functions against the biological background of gender - affinity relationships as they apply to 'mother' and 'carer' - the ideological side of feminism - how perceptions are construed - education again - what happens when language becomes amorphous - testing one's vocabulary - the biological platform for sex and how an orgasm could be interpreted - it's eros, not sex! - religious illusion trees and how we hate and fear eros - the nature of humiliation - the travails of William James Chidley - religious violence towards the self - homosexuality and how it got treated - the moral double-bind - celebrating victimhood - Rodin vs nature - using memory to get what one wants - the question of age - induction of trauma - a history of eros in tabular form.

18. PAT$^{n\pm}$

Futures

Information transfer in dynamic, complex resource spaces - limits to the survival of a system - Fukuyama's radius of trust - the decomposition of social structures and the formation of conceptual islands - global migration - Islam in the West - moralism, privacy and security issues - many futures - the EU in terms of the dynamics of the effectiveness chart - street children, slums and the definition of 'home' - gene research and biological latency - summarising: ageing, environmentalism, feminism, immigration, the matron state, moral primitivism, polarisation of demographics, religion, security and privacy, unemployment - warnings.

19. Debrief

Some advice for my critics - acknowledgements.

Appendix

Parallels between the wetware and software - description of the setup - pseudo code - linear vs distributed systems - depth-first vs breadth-first transmission - the adjustment procedure - the appearance of stable, periodical, and strange attractors - what the screen displays tell us - affinity values between nodes - affinity relationships and their potential - fuzzy logic applied to in- and outputs - test runs - the emergence of memory and abstractions - Einstein's mystery(?) - some SI units - an example of fuzzy thinking - an 'Otoom-friendly' version of a Bill of Rights.

Made in the USA
Las Vegas, NV
03 November 2026

57423444R00178